Sustainable Production of Pulses in Diverse Agro-ecosystems

Vol. 2. Stress management and livelihood security

Other Related Books

Title	Author
Agri Numeric Solutions	Biswa Ranjan Nayak
Agrotechnology for Dryland Farmaing 2nd Revised Ed	A.M. Dhopte
Biofuels: Potential and Challenges	A.K. Pandey
Challenges and Strategies of Dryland Agriculture	S.C. Rao
Climate Agriculture and Man Shrinking Biodiversity and Sustainability	V.K. Sehgal
Crop Management 2nd Ed	U.S. Walia
Crop Production in Salt Affected Soils	I.C. Gupta
Doubling Farmers Income Through Agroforestry	Abhay Kumar
Drought Mitigation and Management	Suresh Kumar
Drought Proofing and Water Resource Management	T.S. Chouhan
Farm Mechanization for Production	Dhirendra Khare
Farming System and Sustainable Agriculture	S.S. Walia
Forage Legumes	J.V. Singh
Glossary of Indian Crops	N.R. Das
Glossary of Terms in Crop Production	S. Ramamoorthy
Guar in India	D. Kumar
Horsegram in India	D. Kumar
Legumes in Dry Areas	D. Kumar
Measuring Agricultural Sustainability	Kaushalya Ramachandran
Moth Bean in India	D. Kumar
Nutrient Management Practices in Crops and Cropping Systems	C.P. Ghonsikar
Organic Seed: Traditional Varieties and Technologies	K. Vanangamudi
Physiological Efficiency for Crop Improvement	A. Hemantaranjan
Physiology of Nutrition and Environmental Stresses on Crop Productivity	A. Hemantaranjan
Plant Abiotic Stresses Physiological Mechanisms Tools and Regulation	A. Hemantaranjan
Quinoa (Chenopodium Quinoa Willd)	Dharm Singh
Resource Conservation Technology in Pulses	P.K. Ghosh
Resource Conserving Techniques in Crop Production	A.R. Sharma
Rice Science	B. Chandrasekaran
Sorghum Medicinal Food (Medicinal and Industrial Perspective)	U. D. Chavan
Sustainable Agriculture	R.S. Meena
Sustainable Development of Dryland Agriculture India	R.P. Singh
Sustainable Development of the Indian Arid Zone	R.P. Singh
Sustainable Dryland Farming	A.M. Dhopte
Upland Rice in India	R.K. Singh

For more details visit website

Sustainable Production of Pulses in Diverse Agro-ecosystems

Vol. 2. Stress management and livelihood security

Narendra Kumar
Ph.D., ARS.
Pr. Scientist (Agronomy) & Head
Division of Crop Production
ICAR-Indian Institute of Pulses Research
Kanpur – 208 024 (U.P.), India

C.P. Nath
Ph.D., ARS. Scientist (Agronomy)
Division of Crop Production
ICAR-Indian Institute of Pulses Research
Kanpur – 208 024 (U.P.), India

C.S. Praharaj
Director
ICAR-Directorate of Groundnut Research
Junagadh – 362 001 (Gujarat), India

Uma Sah
Ph.D., ARS.
Pr. Scientist (Agric. Extension) & Head
Division of Social Sciences
ICAR-Indian Institute of Pulses Research
Kanpur – 208 024 (U.P.), India

N.P. Singh
Vice Chancellor
Banda University of Agriculture and Technology
Banda – 210 001 (U.P.), India

SCIENTIFIC PUBLISHERS

Published by

SCIENTIFIC PUBLISHERS (INDIA)
5 A, New Pali Road, P.O. Box 91
Jodhpur 342 001 (INDIA)
E-mail: info@scientificpub.com
Website: http://www.scientificpub.com

ISBN : **978-93-90749-19-5** HB (Vol 2)
EISBN : 978-93-90749-35-5 (Vol 2)
ISBN : **978-93-90749-77-5** HB (SET)
ISBN : 978-93-90749-95-9 HB (Vol 1)

Printed in India

सत्यमेव जयते
त्रिलोचन महापात्र, पीएच.डी.
सचिव, एवं महानिदेशक
TRILOCHAN MOHAPATRA, Ph.D.
SECRETARY & DIRECTOR GENERAL

भारत सरकार
कृषि अनुसंधान और शिक्षा विभाग एवं
भारतीय कृषि अनुसंधान परिषद
कृषि एवं किसान कल्याण मंत्रालय, कृषि भवन, नई दिल्ली 110 001
GOVERNMENT OF INDIA
DEPARTMENT OF AGRICULTURAL RESEARCH & EDUCATION
AND
INDIAN COUNCIL OF AGRICULTURAL RESEARCH
MINISTRY OF AGRICULTURE AND FARMERS WELFARE
KRISHI BHAVAN, NEW DELHI 110 001
Tel.: 23382629; 23386711 Fax: 91-11-23384773
E-mail: dg.icar@nic.in

Foreword

Sugarcane is the second important industrial crop of the country and contributing significantly to rural economy. This crop is emerging as a multipurpose crop involved in the production of sugar, ethanol, electricity, paper and allied products. The area under this crop is not likely to increase and therefore, it is important to enhance the sugarcane production by growth in productivity and sugar recovery to meet increasing demand.

National level institutes, namely Sugarcane Breeding Institute, Coimbatore, Indian Institute of Sugarcane Research, Lucknow and National Sugar Institute, Kanpur besides All India Coordinated Sugarcane Improvement Project are contributing immensely by developing modern technologies of sugarcane production. With a view to provide ready access to the stakeholders, the information on technologies has been compiled in the form of a book "Modern Technologies of Sugarcane Cultivation". This publication has been developed under the able guidance of Dr. Shankar Lal, Former Director, Indian Institute Pulses Research, Kanpur.

I am sure, this document will add value to available literature and benefit the sugarcane growers, extension workers and sugar industry.

(T. Mohapatra)

Date: 22nd September, 2021
Place: New Delhi - 110001

PREFACE

Pulses are an important group of food legumes, which are grown in diverse agro-ecosystems of the world. They have been an integral part of agriculture since time immemorial. These crops are widely acknowledged as super foods with double the protein in wheat and three-fold that in rice. Besides, they are also rich in complex carbohydrates, micro-nutrients, protein and B-vitamins and minerals like folate, iron, calcium, magnesium, zinc and potassium. Pulses are important crops for food security, combating malnutrition, alleviating poverty, improving human health and enhancing agricultural sustainability. By virtue of their ability to improve soil health and being rich in protein, pulses offer one of the viable options for crop diversification. Considering the importance of pulses, Food and Agricultural Organization (FAO) has declared year 2016 as 'International Year of Pulses' and 10th February as 'World Pulse Day'. Despite these, pulses could not received due attention by Indian farmers and continue to be primarily grown on marginal rainfed land with poor management.

Pulses have far greater role than the traditionally acknowledged role in 'nitrogen fixation' and 'insect-pest break' in the modern agriculture. Their role to reverse the negative effects of continuous following of cereal-cereal cropping systems and in conservation agriculture is well established. Also, they can be used as cash crops, green /brown manuring, cover crop and to control weeds. India is the largest producer and consumer of pulses in the world contributing around 24-28% of the total global production. At present India is near to the self sufficiency in pulses with production level of about 25 Mt. However, to fulfill the demand of ever increasing population of India, there is need to produce 39 Mt of pulses by 2050 A.D. at an annual growth rate of 2.2%. To achieve the target, for every 5-year interval, productivity needs to be enhanced by an average of about 80 kg/ha to achieve a final productivity rate of 1,335 kg/ha by the end of 2050 A.D. following inclusion of about 4 million ha additional area under pulses cultivation. Therefore, concerted efforts are required to develop new & advance production technology to achieve the target.

The present book is outcome of valuable contributions made by various scientists and researchers working on pulse crops in diverse agro-ecosystems of the country. This book has comprehensive coverage of advanced production technologies related to abiotic stress management and livelihood security in the form of 18 chapters that provided the technological back-up to make the country self-sufficient in pulses production. This book would serve as an information source for all R&D personnel involved in promotion of pulses in the country. The editors are sincerely thankful to all the contributors for their valuable chapters to this book, thereby sharing their experiences with the readers.

Editors

CONTENTS

CHAPTER 1

Pulses for food and nutritional security

C.P. Nath*, Narendra Kumar, K.K. Hazra, C.S. Praharaj and Asik Dutta
ICAR-Indian Institute of Pulses Research, Kanpur 208 024
**Corresponding author email:* cpnath4@gmail.com

Abstract

Pulses have an important place in common food of the majority of house hold. The common Indian food viz., *dal-chawal* and *dal-roti* is being consumed since long. Similarly, they remained the integral part of Indian agriculture since time immemorial. Pulses though a very important component of agricultural sustainability as well as food and nutritional security, have received less attention by both researchers and policy makers and thus moved to marginal environment. Therefore, pulses production in the country remained stagnant for longer period (1970-2010) and thus demand and supply gap increased with time. Being a cheap source of protein, pulses play greater role in nutritional security of large vegetarian population of the country. They are considered as climate smart crops due to their deep root system, ability to fix nitrogen, withstand drought situations and perform well even under marginal environment. Hence, pulses are playing greater role in food and nutritional security of the country.

Keywords: Biodiversity, Climate resilient, Crop diversification, Protein, Resource conservation, Sustainability

1. IMPORTANCE OF PULSES

Food security is a current major concern. Protein malnutrition, in particular, must be overcome urgently. Simultaneously, food manufacturers are increasingly interested in proteins from plant sources. Hence, pulse crops could be exploited as sustainable, climate change resistant and high quality protein sources. Scope and approach: Pulses, such as peas, common beans, cowpea, fababeans, chickpea, lentil, or lupin, contain high levels of globulins and albumins, providing around 33% of dietary needs. Nevertheless, there are different compositional and technological problems (e.g., digestibility, allergenicity, or anti-nutritional factors: ANF) that need to be overcome to consider pulse proteins as feasible alternative sources of protein. Food security is one of the basic supports of the human condition, being associated with a homeostatic health status in addition to guarantee the dietary nutritional needs (Fig 1). However, the world's population has been growing at a worrying rate since the beginning of the 20th century. Actually, and as reported in the 2015 Revision of the United Nations Population Estimates and Projections, the world population is expected to reach 8.1 billion in 2025, further increasing to 9.6 billion by 2050 and 10.9 billion around 2100 (UN, 2016). Due to the worldwide increasing population, the imbalance in food production associated with the climate changes may result in a future food shortage. The global management of agricultural production, as well as the adoption of alternative and sustainable sources of food and water, are needed to feed people and avoid malnutrition (Day, 2013; Jones, 2016), a physiological state that results from undernourishment, or poor absorption/use of ingested nutrients due to some pathological condition.

There is no doubt that the main nutritional contribution of pulses is related with their protein content. Nonetheless, some recent studies highlight the possibility of using pulse by-products, particularly their pods, as potential sources of ingredients with food application. According to a recent review on this subject (Zhong *et al.,* 2018), pulse seed coats are far from being utilised in human food at their full potential. However, these by-products could represent a natural "nutritious dietary fibre" with favourable effects in (1) filling the "fibre intake gap", (2) providing minerals and antioxidants, and (3) achieving an optimized utilization of pulses by exploiting value-added applications of their by-products. Despite the lack of studies on the biochemical and nutritional properties of pulse seed coats, it seems evident that these tissues are mainly constituted by structural polysaccharides (non-starch polysaccharides), which are mainly cellulose, hemicellulose, pectin, mineral elements such as Ca, Mg, Mn, Cu, Zn, B, Al and Na, polyphenols (mainly phenolic acids: benzoic/cinnamic acids and their derivatives, flavonoids: particularly flavone and flavonol glycosides, and condensed tannins) (Zhong *et al*., 2018).

Pulses

Rich source of mineral content (mg/100 g) also

Pulses	Ca	P	Fe	Mg
Chickpea	114	387	6.2	168
Pigeonpea	124	304	5.8	133
Mungbean	124	326	7.3	171
Urdbean	154	385	9.1	185
Pea	75	298	5.1	124
Lentil	69	293	4.8	94

Fig. 1. Nutritional composition of pulses

2. PULSES AS A CLIMATE SMART CROP TO MEET FUTURE FOOD DEMAND

Pulses are climate resilient and can be sown in rainfed areas. They can fix nitrogen in the soil and reduce dependence on nitrogenous fertilizers. They enrich the soil in nitrogenous compounds and are beneficial for crop rotation and mixed cropping. Hence increasing area under pulses or planting pulses as an inter-season crop promotes sustainable agriculture; the requirement of nitrogenous chemical fertilizers is reduced for the succeeding crops and periodical crop disease cycles are disrupted by the decreased use of chemical pesticides and weedicides (Venkateswarlu *et al*., 2008; FAO, 2016). According to FAO and United Nations, to fulfill the sustainable development goal's three-fold objectives viz. to achieve food security, to end hunger, to improve nutrition requirements and promotion of sustainable development, pulses consumption and its production should be enhanced in line. Food and nutritional security are said to be achieved when adequate food (quality, quantity, safety, socio-economic acceptability) is available and accessible for and satisfactorily used and utilized by all individuals at all time to live a healthy and active life (UN, 2016). Food security has always been a very important public policy issue in India emanating from severe food scarcity until the mid 1960s. The impressive growth of Indian agriculture no doubt has helped the country in achieving self-sufficiency with respect to availability of food grains at national level. But at the same time climate change, population explosion, food insecurity, and associated risks directly influencing the economic systems of under developed and developing countries where agriculture contributes a considerable share to the economy. Globally more than eight hundred million people are suffering from severe hunger and malnutrition (WHO, 2018). India is house for 445 million poor i.e. 35 per cent of Indian are living on less than $1.25 a day. Half of the pregnant women are anemic in India while in the case of children under the age of five years. 74 per cent are reported to be anemic and 43 per cent underweight (World Bank, 2012).

Hence, ensuring household food and nutritional security is still a challenge for the country, particularly when a huge proportion of population (approximately 1.2 billion) is poor and malnourished. A large percentage of Indians suffer from nutrition problems including protein deficiency. Under these circumstances, a solution for nutritional security for a large mass of people is urgently required which should be cheap, easy to access, requires minimum input, moisture, care, and management practices and having a high nutritive content. Pulses are the best answer to all these requirements as it fulfills all the above mentioned criteria. World Health Organization recommends 80 g pulses/man/day (Pathak *et al.,* 2017). Given the levels of poverty and incidence of vegetarianism, ameliorating the problem of protein deficiency implies an important role for pulses.

3. PULSES PROTEIN QUALITY

Recently, pulses have been gaining interest as vegetable and alternative protein. These species vary in colour, size, shape and texture, but they are, in any case, rich sources of protein, dietary fibre, minerals, vitamins and several phytochemicals (Omosebi *et al.,* 2018). Throughout legumes seed development, the content can change, being the mature seed the stage corresponding to the largest protein level. Attending to their AA profile, pulses have high lysine (a limiting amino acid in cereals), leucine, aspartic acid and arginine content, but are usually poor (sometimes absent) in sulphur-containing AA (methionine and cysteine) and tryptophan (Boye *et al*., 2010). Therefore, different pulse species might be especially adequate to complement cereal-based diets, also because it has been validated that key bioactive components of whole grain cereals and pulses are structurally different, providing complementary health benefits, particularly anti-inflammatory properties, and positive impact on gut microbiome. In either case, the proximate composition of different pulses, as well as their protein content, may present some variation within different species or in result of environmental factors (such as geographic origin or phenological cycle). Pulses are mainly constituted by storage proteins: albumins, globulins and glutelins, being globulins and some albumins often used as nitrogen source for embryo germination (Barac *et al.,* 2015).

4. ROLE OF PULSES IN FOOD/NUTRITIONAL SECURITY

4.1 Affordable source of protein and minerals

In many countries meat, dairy and fish are expensive and thus out of the reach of many, especially the poor. These populations therefore depend on plant foods to fulfill their protein needs. Protein and energy deficiencies, in both quantity and quality, are often the culprit for widespread malnutrition, which is manifest in the form of stunting or wasting. In addition, iron deficiency is an important micronutrient deficiency worldwide, especially for people who do not have access to balanced diets. These issues are further compounded as the world's population is growing rapidly and agricultural production must intensify accordingly, if it is to meet global food needs. However, increases in agricultural production must also be sustainable. Pulses are an important source of accessible protein, especially for the smallholder farmers who consume part of their agricultural produce. In fact, the protein obtained from pulses is significantly less expensive compared to animal foods. In some countries, protein from pulses costs much less than the protein sourced from milk. Additionally, the iron absorption of pulses and the protein quality of the diet are enhanced when pulses are eaten with cereals and vitamin C rich foods.

4.2 Low food wastage footprint

Food waste is one of the main problems related to food security. It is estimated that one-third of the food produced for human consumption worldwide is lost or wasted. Through the entire agricultural supply-chain, losses and wastage occur. In developing countries, most losses occur during production or transportation while in developed countries, a large proportion of food is wasted at the consumption stage. Since pulses are shelf stable, the proportion of food waste at the consumption stage due to spoilage is very low and thus, a very good option to ensure household food security. From all the described, pulses present innumerable possibilities of assisting food security demands, and might be considered as a suitable answer to the increasing interest of food manufacturers in plant proteins to be further included in food formulations. From a nutritional point of view, pulses stand out for their protein content, which include high percentages of lysine, leucine, aspartic acid and arginine, being especially adequate to complement cereal-based foods.

Nevertheless, it is obvious that beyond their nutritional richness, pulse proteins will be useful to decrease the persistent environmental impact imposed by excessive exploration of animal species. Actually, the importance of pulses was evidently upraised when the 68th UN General Assembly declared 2016 as the International Year of Pulses. Other advantages of using pulses are associated with their minor constituents, which have been reported, as described herein, as having anti-inflammatory, antimicrobial, antioxidant, ACE inhibition, and hypoglycaemic effects, benefits against cancer, diabetes, and cardiovascular disease. Despite the highlighted advantages, exploring pulses as protein sources still needs a great deal of effort, namely in selecting the most suitable species, or implementing optimized production and processing technologies to improve their rheological and sensorial properties. Under this topic, the structural versatility of pulse proteins, which are diverse enough to form fibril like structures (to mimic meat) or soft gel particles (similar to whey proteins), is an additional advantage. Nonetheless, and in terms of consumers' acceptability, it might be necessary to reduce/eliminate undesirable flavours (e.g., the well-known "beany" flavour generally associated to legumes).

In what concerns the major limitation typically associated to pulse proteins, the information gathered in this review allowed to conclude that ANF might be greatly reduced by physical operations, heat treatment, vegetative cycle modulation, or specific enzymatic treatments. Despite the former limitations (digestibility, allergenicity, and presence of ANF), it is clear that pulse proteins will be of utmost importance in satisfying food protein demands, in that way contributing in great extent to food security.

4.2 Suitability of some pulses for marginal areas

There are many drought-resistant pulses, such as pigeonpea (*Cajanus cajan* (L.) Huth), bambara beans (*Vigna subterranea* (L.) Verdc.) and lentils (*Lens culinaris* Medik.). These pulses can be cultivated in arid climates that have limited, and often erratic, rainfall of 300-450 mm/year. These are lands where other crops can fail or produce low yields. Additionally, drought-resistant and deep rooting species such as pigeonpea are not only able to improve food security and nutrition of farmers in marginal environments, but pulses can also supply groundwater to companion crops when planted in intercropping systems. Therefore, people living in dry environments, where food security represents a huge challenge, can intensify their production systems in a sustainable manner using locally adapted pulses. Nevertheless, appropriate policies and programs to support both the marketing of pulses in local trade systems and modern consumption habits, must also encourage production systems for pulses, in order to increase the availability and consumption of drought-resistant pulses.

4.3 Long shelf life

If properly stored, pulses remain edible for several years. Farmers have learnt to store pulse seeds with low water content, in dry places, and to discard those seeds that have been attacked by insects or have spoiled. Additionally, pulses show orthodox seed storage behaviour, meaning that they are seeds that are able to germinate after being stored for a long period. In some cases, farmers can store their pulses and then plant them in subsequent cropping seasons.

5. PULSES FOR ECOLOGICAL SUSTAINABILITY

Pulses play an important role for sustainability in many ways. They are an important component of crop rotations, they require less fertilisers than other crops and they are a low carbon source of protein. Legumes are part of the rotational crops farmers can use to maintain soil fertility (Fig 2). Pulse crops are important components of production systems that are resilient to climate change. Pulses are able to convert nitrogen in the atmosphere into compounds in the soil. Some (types) of pulses are also able to free soil-bound phosphorous. Both nitrogen and phosphorus are essential nutrients for growing plants. Pulses can help to contribute to food security in a number of ways. Pulses are an inexpensive source of protein-a crucial component of any healthy diet, but especially in poorer areas where meat, dairy and fish are economically inaccessible (Kumar *et al.,* 2019).

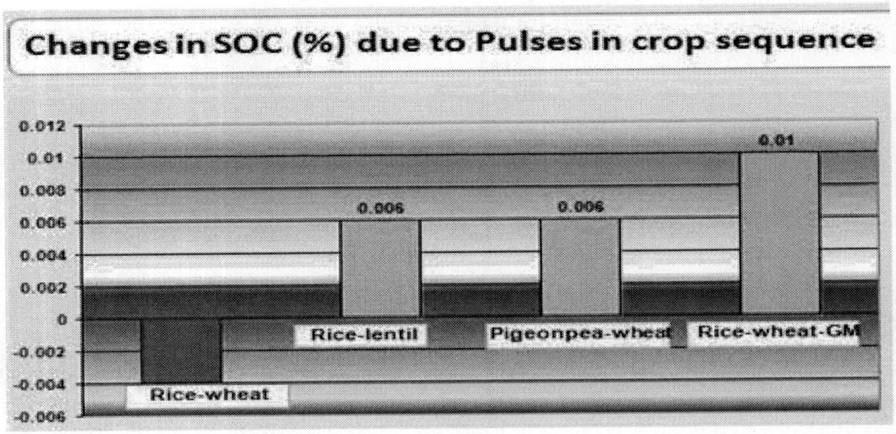

Fig. 2. Role of pulses in carbon sequestration

There, ecosystem services are grouped into four broad categories: provisioning, such as the production of food and water; regulating, such as the control of climate and disease; supporting, such as nutrient cycles (Fig 3) and oxygen production; and cultural, such as spiritual and recreational benefits.

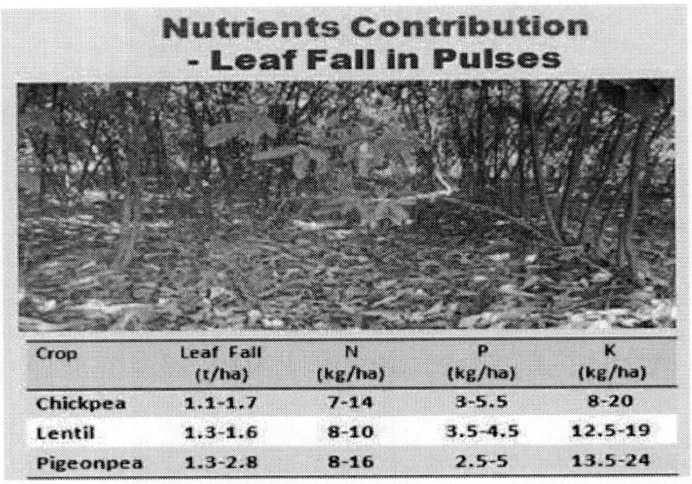

Crop	Leaf Fall (t/ha)	N (kg/ha)	P (kg/ha)	K (kg/ha)
Chickpea	1.1-1.7	7-14	3-5.5	8-20
Lentil	1.3-1.6	8-10	3.5-4.5	12.5-19
Pigeonpea	1.3-2.8	8-16	2.5-5	13.5-24

Fig. 3. Role of pulses in nutrient cycling in agro-ecosystem (Source: Kumar *et al.,* 2018)

6. PULSES AND BIODIVERSITY

Pulses are able to increase biodiversity as they are able to fix their own nitrogen into the soil, which increases soil fertility. It is estimated that there are hundreds of varieties of pulses, including many local varieties that are not exported or grown worldwide. The genetic diversity of these crops is an essential component for on-farm soil and pest management, especially for small-scale farmers. An important attribute of pulses is their ability to biologically fix nitrogen. These plants, in symbiosis with certain types of bacteria, namely *Rhizobium* and *Bradyrhizobium*, are able to convert atmospheric nitrogen into nitrogen compounds that can be used by plants, while also improving soil fertility. Some varieties of pulses are also able to free soil-bound phosphorous, and phosphorous also plays an important role in the nutrition of plants.

The presence of pulses in agro-ecosystems helps to maintain and/or increase vital microbial biomass and activity in the soil, so nourishing the development of those organisms that are responsible for promoting soil structure and nutrient availability. A high soil biodiversity provides ecosystems with not only greater resistance and resilience against disturbance and stress, but also improves the ability of ecosystems to suppress diseases. All these features are particularly important for mainstreaming soil health, which is the foundation of food security and health.

7. PULSES IN MULTIPLE CROPPING SYSTEMS AND BIODIVERSITY

Pulses cannot improve on-farm diversity per se. This means that if a farmer changes from cultivating only cereal species to cultivating only pulse species, on-farm diversity does not change. That said, pulses are a crucial component of multiple cropping systems, namely intercropping (Fig 4), crop rotation and agroforestry. These cropping systems have a higher species diversity than monocrop systems. Increasing species diversity of cropping systems could translate into not only a more efficient use of resources, namely light, water and nutrients, but also into higher outputs as yields are increased, and a lower risk of overall crop failure. The selection of which multiple cropping system to use is less important because the choice will be determined by the individual attributes of each agro-ecosystem. What is clear is that pulses should be an integral part of agro-ecosystems because they bring the agro-systems into balance. Additionally, cropping systems that are more resilient to climate change can be developed through the inclusion of local varieties, such as bambara beans, which are currently not well known or produced. In multiple cropping systems, services such as nutrient recycling and soil formation are improved through the pulses' abilities to fix nitrogen and free phosphorous and their capacity to increase soil biodiversity. At the same time, when used in multiple cropping systems, pulses also help to curb and control pests and diseases. Additionally, since pulses often promote higher rates of accumulation of soil carbon than cereals or grasses, they can contribute to improve the carbon sequestration of agro-ecosystems.

Fig. 4. Pulse based intercropping for crop diversification food sustenance

Good soil health is the foundation of food security and pulses help to increase soil microbial biomass and activity, thus improving soil biodiversity. Cultivating pulses in multiple cropping systems enriches agro-biodiversity, ensures resilience to climate change and improves ecosystem services. Pulses have a multiple role in promoting living organisms and ecological complexity to re-establish the natural good functioning of ecosystems

8. INDUSTRIAL USES OF PULSES AND PULSE PROTEINS

Alternative protein sources from vegetable origin, as exemplified by whole seed and/ or pulse protein isolates, have been investigated as food ingredients to answer to the increasing demands for innovative and functional food products, with the additional advantage of promoting health status. In fact, several functional foods proposed as having positive health benefits have been progressively included in a fast-growing global industry. Actually, the increasing number of people devoted to strict diets (e.g. lactose-free or gluten-free) or personal preferences (e.g. vegetarians, vegans) is leading to a rigorous selection of functional and specialty food products. Nowadays, food industry has been investing in R&D to address products to answer the consumer needs or trends. For instance, lactose-intolerant people can ingest calcium by drinking calcium fortified orange juice or vegetable drinks (rice, oat); soya-based products (tofu and related products) may be the alternative to animal protein in people with vegetarian diets. In this way, pulses (whole seed, flours, or derived products such as protein concentrates or protein isolates) represent functional foods, not only when consumed alone but also when mixed with other food products (e.g., cereals) and adequately cooked in order to increase their nutritional value and reduce their ANF content (Babault *et al.,* 2015). Nowadays, several scientific studies on pulse and/ or pulse related food products have been performed in collaboration with food industry, aiming to develop new food product formulations that might enhance human health and improve their technological functional properties.

9. FUTURE PROSPECTS OF PULSE PRODUCTION SUSTAINABILITY

Pulses are an important source of high-quality protein complementing cereal proteins for the substantial vegetarian population of India. Pulses can be produced with a minimum use of resources and hence become less costly even than animal proteins. Pulses meet their nitrogen requirements to a great extent by fixing atmospheric nitrogen in their root nodules. Pulses can provide a sustainable solution in rainfed area which occupies 67 per cent net sown area, contributing 44 per cent of foodgrains and supporting 40 per cent of the population. Development and cultivation of more draught tolerant varieties of pulses can save India's food and nutritional security (Fig 5).

Pulses are the most suitable crops to grow in water stressed regions. Pulses require less water, improve soil health, and suit the local micro climate for smart agriculture. Only forty-three gallons of water (one gallon equals 3.785 liters) are required to produce one pound of pulses, whereas wheat, rice, and meat require 660, 1,056, and 1,142 gallons of water, respectively. Therefore, in rainfed farming systems, the cultivation of pulses, considered as smart crops, can help address climate change in agriculture. In comparison with resource intensive crops such as wheat, soybean, paddy, and maize, pulses are more remunerative crops due to fewer input needs and high market value. In addition, pulses improve soil health by fixing atmospheric nitrogen and adding humus content to the soil, which improves the soil's biological, chemical, and physical properties.

The high return demand for promoting pulses in rainfed areas provides opportunities for small land holders to shift from subsistence farming to profitable farming. The BCR and ROI of pigeonpea are 3.82 and 2.82, whereas for millet the BCR and ROI are 2.41 and 1.41. The high BCR and ROI of pigeonpea indicate that this is a more profitable crop than millet. Traditionally, Indians have pulses, especially dais, as an important part of their daily diet, and pulses are the only sources of protein for more than half of the vegetarian population. Proteins are required to grow new cells and tissues, Claws are essential for all and especially fin growing children. India has a large vegetarian population, they can derive needed proteins from pulses and legumes.

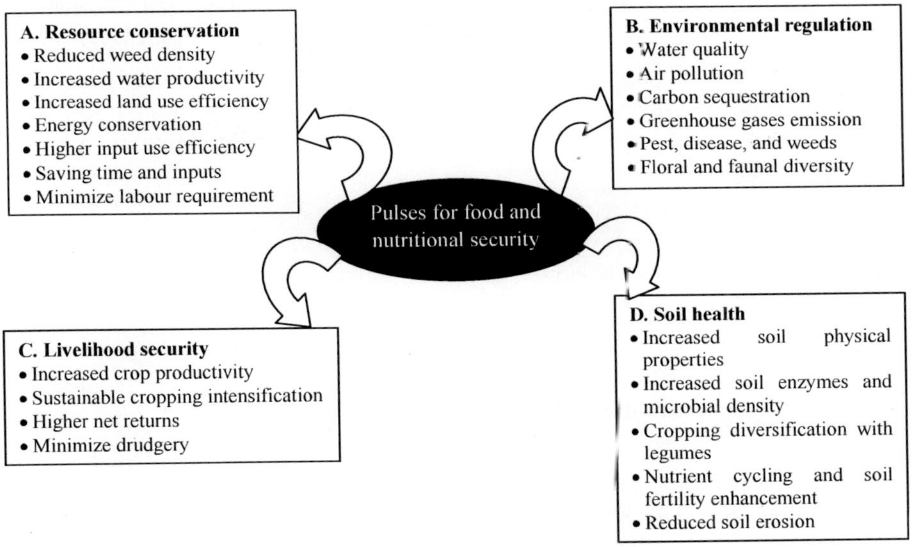

Fig. 5. Role of pulses in food security and agricultural sustainability

Pulses can be excellent sources of proteins in our diets. According to the Indian Market Research Bureau survey, which was conducted by interviewing 1,260 Indians, protein consumption in the diet of adult Indians (nine out of ten Indians) is less than their daily protein intake. The survey noted that 91 per cent of the vegetarians and 85 per cent of the non-vegetarians were protein deficient. The reason for this can be attribute to changes in eating habits and lifestyle. The tremendous increase in pulse prices also contributes significantly to his deficiency. Therefore, mass scale promotion and cultivation of pulses is needed to enhance the income and nutritional security in rural and urban India.

REFERENCES

Babault, N., Paizis, C., Deley, G., Guérin-Deremaux, L., Saniez, M., Lefranc-Millot, C., et al. (2015). Pea proteins oral supplementation promotes muscle thickness gains during resistance training: A double-blind, randomized, placebo-controlled clinical trial vs. Whey protein. *Journal of the International Society of Sports Nutrition* 12, 42–61.

Barać, M. B., Pešić, M. B., Stanojević, S. P., Kostić, A. Z., & Čabrilo, S. B. (2015). Techno-Functional properties of pea (Pisum sativum) protein isolates - a Review. *Acta Periodica Technologica* 46, 1–18.

Boye, J., Zare, F., & Pletch, A. (2010). Pulse proteins: Processing, characterization, functional properties and applications in food and feed. *Food Research International* 43, 414–431.

Day, L. (2013). Proteins from land plants - potential resources for human nutrition and food security. *Trends in Food Science & Technology* 32, 25–42.

FAO (2016). International year of pulses 2016. http://www.fao.org/pulses-2016/about/ en/, Accessed date: 27 February 2019.

Jones, O.G. (2016). Recent advances in the functionality of non-animal sourced proteins contributing to their use in meat analogs. *Current Opinion in Food Science* 7, 7–13.

Kumar N., Hazra K.K., Nath C.P., Praharaj C.S., Singh U. (2018). Grain Legumes for Resource Conservation and Agricultural Sustainability in South Asia. In: R. Meena, A. Das, G. Yadav, R. Lal (eds) Legumes for Soil Health and Sustainable Management. Springer, Singapore. 77-107 pages.

Kumar, N., Nath, C.P., Hazra, K.K., Das, K., Venkatesh, M.S., Singh, M.K., Singh, S.S., Praharaj, C.S. and Singh, N.P., 2019. Impact of zero-till residue management and crop diversification with legumes on soil aggregation and carbon sequestration. *Soil and Tillage Research* 189, pp.158-167.

Omosebi, M. O., Osundahunsi, O. F., & Fagbemi, T. N. (2018). Effect of extrusion on protein quality, antinutritional factors, and digestibility of complementary diet from quality protein maize and soybean protein concentrate. *Journal of Food Biochemistrye* 12508.

Pathak, A., Mahadik, K., Swami, M.B., Roy, P.K., Sharma, M., Mahadik, V.K. and Lundborg, C.S., 2017. Incidence and risk factors for surgical site infections in obstetric and gynecological surgeries from a teaching hospital in rural India. *Antimicrobial Resistance & Infection Control* 6(1), pp.1-8.

UN, United Nations (2016). World population prospects. The 2015 revision. https://esa.un.org/unpd/wpp/Graphs/Probabilistic/POP/TOT/27.02.19.

Venkateswarlu, B. and Shanker, A.K., 2012. Dryland agriculture: bringing resilience to crop production under changing climate. In Crop stress and its management: perspectives and strategies (pp. 19-44). Springer, Dordrecht.

World Bank. Information, Communication Technologies and infoDev (Program), 2012. Information and Communications for development 2012: Maximizing mobile. World Bank Publications.

World Health Organization, 2018. The state of food security and nutrition in the world 2018: building climate resilience for food security and nutrition. Food & Agriculture Org.

Zhong, L., Fang, Z., Wahlqvist, M.L., Wu, G., Hodgson, J.M., & Johnson, S. K. (2018). Seed coats of pulses as a food ingredient: Characterization, processing, and applications. *Trends in Food Science & Technology*, 80, 35–42.

CHAPTER 2

Climate change effects on grain legume production and its mitigation

Surendra Kumar Meena[1*], Banwari Lal[1], Vaibhav Kumar[2], Gurumurthy S[3] and Narendra Kumar[2]

1. *ICAR-Indian Institute of Pulses Research, Regional Centre, Bikaner-334004, Rajasthan, India*
2. *ICAR-Indian Institute of Pulses Research,Kanpur-208024 Uttar Pradesh, India*
3. *ICAR-National Institute of Abiotic Stress Management, Pune- 413115, Maharashtra, India*

*Corresponding author email: sdmeena84@gmail.com

Abstract

Global climate change phenomenon is a great threat to agriculture and allied activities. The resilience of pulse crops to present day weather extremes such as drought, elevated CO_2, high temperature and low irradiance are considered to predict adaptation to future climate change. Drought stress is the most important challenge in current situation and rising temperature worsening the drought stress management in pulse crops production. Developing new genotypes of pulses is one of the key options for adaptation of agriculture to climate change.

Keywords: Drought, Elevated CO_2, Heat, Global dimming, Global warming

1. INTRODUCTION

According to IPCC, global warming is the major threat to socio-economic development. Climatic conditions vary tremendously worldwide, including large diurnal ranges in temperature, frequent strong winds, highly variable and unpredictable precipitation (https://www.ipcc.ch). Global climate change phenomenon is a great threat to agriculture and allied activities. Climate change associated with environmental stresses is the new form of risk and uncertainties for pulses production as well as global food security. By the end of this century, the concentration of atmospheric carbon dioxide is projected to increase to 900 ppm from the present level of 400 ppm, whereas the global temperature will rise by 2-4°C which may result in 10-40% reduction in crop production in South Asia (IPCC, 2001). Climate change has the impact on food production, food security and food safety globally. The atmospheric greenhouse gases associated with agriculture and its allied sectors are methane, carbon dioxide and nitrous oxide. These gases will increase the temperature and rising temperature more than 2.5°C may enhance the crop production in the cooler regions of the earth but affect negatively in the warmer tropical regions. Overall, increased temperature has negative effects on crop production (IPCC, 2001).

The severity of climate change on agriculture depends on the increased magnitude of temperature. Drought stress is the most important challenge in current situation and rising temperature worsening the drought stress management in pulse crop production. Elevated CO_2 and increasing temperature will change the dynamics of insects, diseases and weeds worldwide and these are the new challenges in pulse breeding programmes and crop management. This chapter presented the current scenario of climate change and its impact on pulses production and their mitigation approaches for nutritional and food security. It also provides an overview of the relationship between pulse production and climate change and suggests the possible options for enhancing the pulse productivity.

2. PHYSIOLOGICAL RESPONSES OF PULSE CROPS UNDER CLIMATE CHANGE

This chapter focuses on primary four factors, viz., increasing temperature (global warming), elevated CO_2, drought and low light intensity (global dimming) that have major impact on pulses production in the coming years. The physiological impacts of these factors on pulses production is covered in this section.

2.1 Increasing temperature (Global warming)

Heat stress is described as the rise in temperature beyond a threshold level for a short period of time which can lead to irreversible damage to plant growth and development (Wahid *et al.,* 2007). High temperature negatively affects germination, seedling establishment, gaseous exchange through leaves, membrane stabilization and reproductive organs (pollen viability and germination), primary and secondary metabolites, photosynthate and nutrient movement (Wahid *et al.,* 2007). Based on the germination sensitivity to temperatures and frost, pulse crops can be classified as cool-season chickpea (*Cicer arietinum* L.), lentil (*Lens culinaris* Medik) and field pea (*Pisum sativum* L.) and warm season cowpea (*Vigna unguiculata* L.), pigeonpea (*Cajanus cajan.* L. Millsp), mungbean (*Vigna radiata* L. Wilczek) and urdbean (*Vigna mungo* L.) crops (Miller *et al.,* 2002).

The growth and development of pulses is highly dependent on high temperature due to the global climate change (Vadez *et al.,* 2012). Heat stress severely affects seed germination and seedling establishment in different pulse crops (Hasanuzzaman *et al.,* 2013). At vegetative growth, heat stress results in water loss from cells, reduced cell size and growth, and hence reduced leaf area and biomass. Heat stress during reproductive and seed filling stages can cause severe loss in the quality and productivity of grain legumes. Temperature fluctuations and heat stress during seed filling drastically reduce yield in legume crops. Summerfield et al (1984) observed drastic reduction in chickpea seed yields with greater exposure to hot days (30-35 ºC), during the reproductive period. The reproductive failures in mungbean might be due to the impaired sucrose metabolism in leaves and anthers which inhibit the pollen and stigma function under heat stress conditions. Heat stress can also reduce the seed germination by inhibiting the supply of assimilates to seeds at development and maturity stage (Larmure and Munier-Jolain, 2019). Seed formation and grain filling is the completion of growth and development in crop plants, which involves transport and remobilization of photosynthates, nutrients and other metabolites in the seeds, sink limitation of seed N accumulation can be affected by high temperatures (Larmure and Munier-Jolain, 2019).

High-temperature stress reduces seed size due to insufficient accumulation of photosynthates during seed filling (Sita *et al.*, 2017). Heat stress decreases the accumulation of carbohydrates in pollen grains and stigmatic tissue by changing partitioning of the assimilates and the proportion between symplastic and apoplastic loading of the phloem, which affects pollen viability (Kaushal *et al.*, 2013). Pollen sterility in pigeonpea (Basu *et al.*, 2016) and lentil (Sita *et al.*, 2017) has been recorded due to high temperature. Heat stress decreases the activity of sucrose synthase and vacuolarinvertases in developing pollen grains; as a result, sucrose and starch turnover is impaired and accumulation of soluble carbohydrates is reduced in several pulse crops (Kaushal *et al.*, 2013).

2.2 Drought

According to the IPCC (Intergovernmental Panel on Climate Change) report, a serious threat to agriculture is water deficit. Water stress prone agricultural regions can undergo yield loss up to 50% or more; up to 45% of world agricultural area is affected by water deficit impacting the life of 38% of population globally (IPCC, 2001). Drought and heat stress influence a broad variety of physiological processes by changing the plant water relationship affecting crop growth and development (Wahid *et al.*, 2007). Drought is the most widespread abiotic stress in legume crops and its effect on yield and its attributes of legume crops was well reported (Turner *et al.*, 2001). Drought stress has adverse effects on total biomass, pod number, seed number, seed weight and quality, seed yield/plant in moth bean (Garg *et al.*, 2004), mungbean (Kumar and Sharma, 2009) and pigeonpea (Singh *et al.*, 2013). The reduction in relative water content (RWC) under water deficit conditions is reported in moth bean (Garg *et al.*, 2004), in mungbean (Kumar and Sharma, 2009) and in pigeonpea (Singh *et al.*, 2013).

Reduction in leaf expansion can be considered as a first line of defence against drought to maintain water potential of cells. Plants reduce their leaf area, leaf area index and canopy size to cope up with moisture stress (Blum, 2005). Reduction in leaf area due to drought stress was frequently reported in almost every pulse crop. Specific leaf area was associated with relative seedling growth rate and net assimilation rate that decreases under drought conditions. Flowering stage is more sensitive to soil moisture stress and resulted in low yield of pulse crops. Several pulses (chickpea, cowpea, common bean, pigeonpea) maintain cellular water content and turgor by lowering stomatal conductance, while in other beans (pea, faba bean, mungbean), it is achieved by lowering the osmotic potential in response to drought (Amede *et al.*, 2003). Pulses are generally most susceptible to terminal drought stress during the reproductive stage (Fang *et al.*, 2010). Drought stress can delay or completely inhibit flowering during the initiation of reproductive development. Drastic impact on seed yield and harvest index of chickpea (Fang *et al.*, 2010) and mungbean (Kumar and Sharma, 2009) have been reported due to drought stress at the reproductive stage.

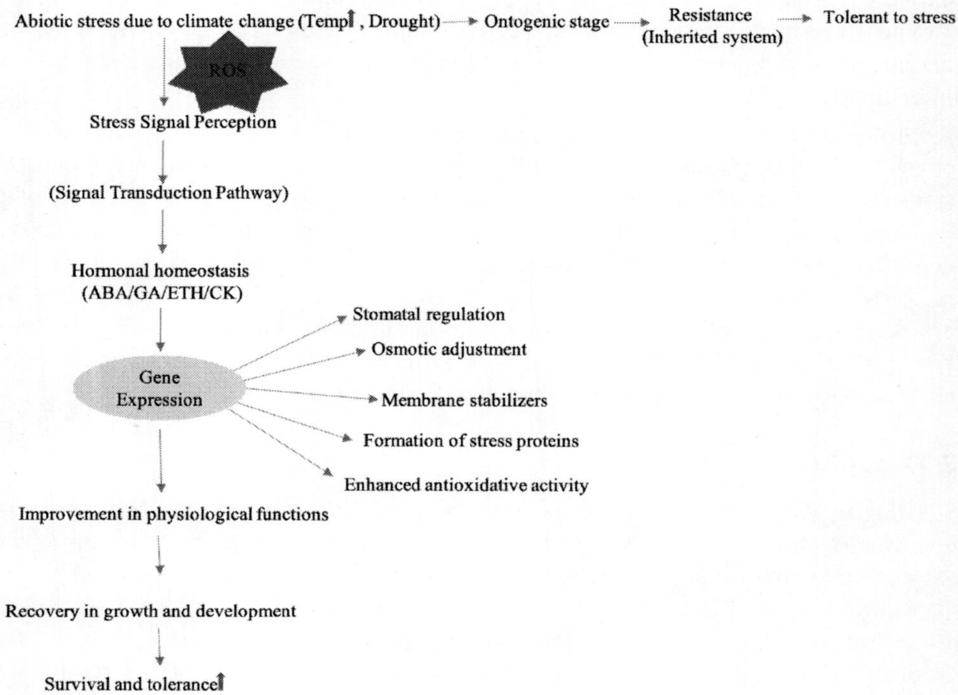

Fig. 1. Effects of climate change on legume crops and their responses.

2.3 Elevated CO_2

It is projected that by the middle of this century, the CO_2 would surpass 550 µmol/mol and over 700-900 µmol/mol by the end of the 21st century (*https://www.ipcc.ch*). Increasing CO_2 level enhances plant growth and affects yield quality. When plants grew at elevated CO_2 for weeks or months, photosynthetic capacity may be decreased. The phenomenon is widely known as photosynthetic acclimation. The source-sink balance is crucial in the occurrence and extent of the acclimation to the elevated CO_2. In legumes, nodulation largely represents plant sink capacity and therefore legumes do not acclimate to elevated CO_2 in the absence of growth-limiting factors (nodule sink strength is strong enough to avoid leaf carbon accumulation caused by elevated CO_2). Increasing CO_2 enhances the photosynthesis and inhibits the photorespiration which increases the biomass and yield of crop plants. Several studies have shown that elevated CO_2 resulted in increased growth, diameter, volume and surface area of roots in different crops (Pandey *et al.,* 2015). Many leguminous crops such as mungbean, chickpea, pigeonpea, broad bean, blackgram and frenchbean have shown increase in seed as well as biomass yield under elevated CO_2. Increase in photosynthesis under elevated CO_2 was supposed to be the reason for higher biomass accumulation in these pulses (Pandey *et al.,* 2015).

Higher metabolic activity or storage sink capacity is usually required to sustain photosynthetic response to elevated CO_2 in legumes (Sharma-Natu *et al.*, 2004). A large proportion of potential sinks remain unrealized in mungbean, resulting in shedding of flowers and immature fruits, this way the source limits the sink yields in mungbean (Mitra and Ghildiyal, 1988). Legumes may have advantage over non-legumes when grown under increased CO_2 concentration because of their ability to fix atmospheric nitrogen (N) (De-Graaff *et al.,* 2006). Elevated CO_2 increased nitrogenase activity either directly or indirectly by increasing the rate of photosynthesis, whereas it had no effect on N-fixation in the absence of added non-N nutrients (*e.g.*, P, Mo, and K), and a stimulative effect on N-fixation if non-N nutrients were available (De-Graaff *et al.,* 2006).

2.4 Low light intensity (Global dimming)

Climate change is associated with global warming as well as global dimming. Reduction in the solar radiation reaching the earth surface is known as global dimming. Global dimming is caused due to increased cloud thickness, cloud cover and suspended particulate matter in the air (aerosol). Solar radiation is the primary energy source for plants which governs the plant growth and development. It plays an important role in many plant physiological and bio-chemical processes like chlorophyll synthesis, enzyme activation, photosynthesis that regulating the growth and development of plants. Long-term observations of the global surface solar radiation in recent decades have shown a widespread declining trend in many parts of the world (Liepert, 2002). Stanhill and Cohen (2001) reported averaged globally reduction of 0.51W/m^2, or 2.7 per cent decade, since last 50 years. India also showed a continuous and steady light dimming since 1960s. Although, crop growth and development is adversely affected by low irradiance due to global dimming but recent updates on growth and physiology of legume pulses in response to low irradiances is meagre. Experiments conducted in 1970-80s have shown that reduced irradiance caused the abortion of reproductive structures. Reduced levels of solar irradiance during the growing season greatly impair growth, development, physio-biochemical processes and yield in crop plants. Decreased plant height is recorded in chickpea by which plants allocate photosynthates to stem and leaves to compensate less solar radiation (Mina *et al.,* 2015).

3. MITIGATION STRATEGIES

3.1 Adopting pulse based cropping system for sustained yield

Leguminous crops are climate smart crops as they have great adaptability to weather change thus contribute to mitigate the deleterious effects of climate change. Pulses are hardier than other crops and improve soil fertility through symbiotic nitrogen fixation. It has been observed that inclusion of pulses in cereal based cropping systems enhances soil health, productivity and sustainability in long run. Pulses also help in food security and improving farmer income.

3.2 Cultivation of neglected and underutilized grain legumes

The available information of neglected and underutilized pulse crops is limited in literature. These crops are adapted to wide range ecological niches, cultivated with low inputs and tolerant to biotic and abiotic stresses. Generally, neglected crops are defined as drought tolerant and could therefore prove beneficial in mitigating the adverse effects of climate change. These grain legumes (rice bean, lathyrus, cowpea and bambara groundnut) mostly suitable to semi-arid and arid regions. These crops have a strong well-developed tap root system which proves beneficial in the event of drought and nutrient mining.

3.3 Breeding strategies for development of drought and heat tolerance in pulses

To mitigate the deleterious impact of climate change on agriculture, plant breeders are required to accelerate efforts to identify and develop the genotypes that can withstand or able to take advantage/ resilient to/ of climate disturbances in respect of high temperature, erratic rainfall, drought stress and increased CO_2 level. Developing new genotypes of plants is one of the key options for adaptation of agriculture to climate change. To identify the donors or desirable traits, diversified germplasm with landraces and wild accessions/ species should be evaluated and characterized for tolerance to various abiotic and biotic stresses under controlled conditions.

3.3.1 Incorporation of drought tolerance

Drought is the most challenging abiotic stress in pulses cultivation. Therefore, development of tolerant varieties to moisture stress is a great deal of the present scenario under climate change.

Greater early vigour is related to enhance N and P uptake and water use efficiency (Ryan *et al.,* 2015). Various reports suggest that plant vigour or good ground cover has an effect on radiation use efficiency and contributes to the reduction of soil evaporation thus helps to increase soil water availability for transpiration. Incorporation of early vigour trait which is an integration of several physiological factors like germination, water uptake, good root growth and shoot development and leaf area will benefit the breeding programme that is targeted for stress prone areas.

Root architecture also plays an important role in drought tolerance in several cultivated crop species. Although primary root remains unaffected by drought stress, lateral root growth is significantly reduced (Basu *et al.,* 2016). High prolific root system not only influences transpiration through proper utilization of soil moisture but also improves biomass and harvest index under terminal drought conditions (Zaman-Allah *et al.,* 2011), however a deeper root system alone had not been acceptable trait for higher grain yields. Greater root length density increased water storage capacity of the root zone, and a deeper root system is associated with more water uptake from the soil with better crop performance under drought conditions. The most productive genotypes under water deficit is the higher root density in deeper soil layers, suggesting the role of vigorous root growth in soil water uptake (Zaman-Allah *et al.,* 2011). A deeper rooting system would also be a promising trait to improve soil water uptake from the subsoil, thereby improving drought tolerance in pigeonpea (Ali *et al.,* 2005).

It is therefore imperative to develop crop varieties that would have efficient root system for extracting water and for absorption of nutrients from deeper layers of the soil. For example, in case of chickpea, out of more than 1500 genotypes when subjected to field screening for drought tolerance, only one genotype ICC 4958 had higher root mass and it gave the best performance at many locations for coping with terminal drought (Ali *et al.,* 2005). Another possibility for regulating water flow at the root level is to use aquaporins to modulate water conductance in the root. Aquaporins play a major role in controlling water movement by regulating the hydraulic conductance of roots. Understanding the interaction between aquaporins and abscisic acid in regulating root hydraulic conductance and the rate of plant water use may be an important research priority in the adaptation of pulse crops to moisture deficit stress expected under climate change. Drought tolerance includes osmotic adjustment (accumulation of metabolites, osmoprotection *e.g.*, proline) and the antioxidant defence systems (Farooq *et al.,* 2009). Blum (2005) gave a detailed account of increasing evidence suggesting a relationship between high osmotic adjustment and maintenance of biomass and yield under stress. Lentil and chickpea have considerable potential for drought resistance through osmotic adjustment (Turner *et al.,* 2001). Higher relative water content and enhanced accumulated proline content were observed in high yielding genotypes under water deficit conditions.

Seed formation and grain filling showed more sensitivity to moisture stress, hence dry matter partitioning becomes the most potential trait for drought screening in mungbean and pigeonpea (Kumar and Sharma, 2009). The essence of **drought avoidance** is to reduce water loss while enhancing or maintaining uptake by the roots. Drought avoidance involves crop responses such as stomatal regulation, enhanced capture of soil water through an extensive and prolific root system (Kavar *et al.,* 2008). Several root characteristics such as biomass, length, depth and thickness (volume) are thought to contribute to final yield under drought stress (Kavar *et al.,* 2008) due to improved water capture. Blum (2005) also reported that drought avoidance is associated with reduced season duration due to reduced leaf number; reduced season duration is also characteristic of drought escape, suggesting that the mechanisms do not work in isolation.

Drought escape is an important phenological characteristic at sites with terminal drought (Berger *et al.,* 2004). High-yielding genotypes generally flower early, produce pod early, and have a relatively long flowering period. However, under severe drought stress high-yielding genotypes have a high degree of biomass translocation from leaves to stems to pods. Berger et al (2004) reported that physiological processes and phenological development depends on the type and level of the water scarcity in chickpea which play a major role in adaption of chickpea to water deficit conditions.

3.3.2 Incorporation of heat tolerance

Increasing temperature will affect the rate of crop growth and development as the accumulation of thermal time increases and accelerates the phenological development. Therefore, the study will be needed on the understanding of the key phenological mechanisms under increasing temperature. There is a need of the hour to develop an improved cultivars that is associated with better balance of crop duration with available soil water and maximize light capture and also capable of setting seed at supra optimal temperature. As photosynthesis and reproductive development are the most sensitive physiological processes to stress, a donor for heat tolerance will be usually considered by higher photosynthetic rates reflected in stay-green leaves, better osmotic adjustment, increased activities of antioxidative enzymes, increased membrane-thermostability and successful seed set under heat stress (Sita *et al.,* 2017). Irrespective of the screening method, the main target for plant breeders is to establish an appropriate collection of thermotolerance markers that can be used further to add heat tolerance to various crop species. Screening and selection of kharif pulses (greengram, blackgram and cowpea) and rabi pulses (chickpea, lentil, pea), which can retain the maximum number of flowers and produce productive pods during high temperature (>40 °C for kharif pulses; 33-35°C for rabi pulses). Pollen viability and seed-set as measures of reproductive success under heat stress are also candidate traits for selection in breeding programs, and generally there is a strong correlation between pollen production and viability, anther dehiscence, and seed-set.

3.3.3 Incorporation of low irradiance/light tolerance

Development of photo-insensitive crop variety which can perform better under low light intensity and are efficiently able to harvest as much as light energy as it reaches to crop canopy. Increased chlorophyll content was a mechanism to improve light interception in order to sustain normal growth and development of the crop under shading conditions. This increased content of the chlorophyll was to enhance efficiency of the PAR absorption indicating adaptation of crops to the low light intensity (Cartechini and Palliotti, 1995). In order to avoid low light intensity, plants increase overall light absorption for photosynthesis and reduce reflected and transmitted light from the leaves. Plants at low irradiance increased their leaf area ratio by producing leaves of larger specific leaf area for maintaining their relative growth rate. Generally, adaptive responses of plant to low irradiance are increase of chlorophyll content, leaf area ratio, leaf to stem mass and stem length. On the other hand, the adaptive responses also include decreased leaf thickness, chlorophyll a to b ratio and root growth relative to shoot growth.

4. CONCLUSION

In present climate change scenario, climate smart agriculture has become extremely relevant solution that promises to solve both adaptation and mitigation issues. Among the foodgrains crops, pulses play an important role in the sustainable production system due to their inherent capacity to maintain soil fertility, in addition to their role as rich source of protein in the human food and animal feed. Therefore, pulses have a key role in achieving food and nutritional security while addressing effects of climate change. Improved tolerant cultivar with the better crop practices and management, pulses plays an important role in climate change mitigation.

REFERENCE

Ali, Y.M., Johansen, C., Krishnamurthy, L. and Hamid, A. (2005). Genotypic variation in root systems of chickpea (*Cicer arietinum* L.) across environments. *Journal of Agronomy and Crop Science* 191(6), 464-472.

Amede, T., Schubert, S. and Stahr, K. (2003). Mechanisms of drought resistance in grain legumes I: Osmotic adjustment. *SINET: Ethiopian Journal of Science 26*(1), 37-46.

Basu, P.S., Singh, U., Kumar, A., Praharaj, C.S., and Shivran, R.K. (2016). Climate change and its mitigation strategies in pulses production. *Indian Journal of Agronomy* 61(Special issue), S71-S82.

Berger, J.D., Turner, N.C., Siddique, K.H.M., Knights, E.J., Brinsmead, R.B., Mock, I., Edmondson, C. and Khan, T.N. (2004). Genotype by environment studies across

Australia reveal the importance of phenology for chickpea (*Cicer arietinum* L.) improvement. *Australian Journal of Agricultural Research 55*(10), 1071-1084.

Blum, A. (2005). Drought resistance, water-use efficiency, and yield potential—are they compatible, dissonant, or mutually exclusive? *Australian Journal of Agricultural Research* 56(11), 1159-1168.

Cartechini, A. and Palliotti, A. (1995). Effect of shading on vine morphology and productivity and leaf gas exchange characteristics in grapevines in the field. *American Journal of Enology and Viticulture* 46(2), 227-234.

De-Graaff, M.A., Van Groenigen, K.J., Six, J., Hungate, B. and Van Kessel, C. (2006). Interactions between plant growth and soil nutrient cycling under elevated CO_2: a meta-analysis. *Global Change Biology 12*(11), 2077-2091.

Fang, X., Turner, N.C., Yan, G., Li, F. and Siddique, K.H. (2010). Flower numbers, pod production, pollen viability, and pistil function are reduced and flower and pod abortion increased in chickpea (*Cicer arietinum* L.) under terminal drought. *Journal of Experimental Botany* 61(2), 335-345.

Farooq, M., Wahid, A., Kobayashi, N., Fujita, D.B.S.M.A. and Basra, S.M.A. (2009). Plant drought stress: effects, mechanisms and management. *Agronomy for Sustainable Agriculture* 29, 185-212.

Garg, B.K., Burman, U. and Kathju, S. (2004). The influence of phosphorus nutrition on the physiological response of moth bean genotypes to drought. *Journal of Plant Nutrition and Soil Science* 167(4), 503-508.

Hasanuzzaman, M., Nahar, K., Alam, M., Roychowdhury, R. and Fujita, M. (2013). Physiological, biochemical, and molecular mechanisms of heat stress tolerance in plants. *International Journal of Molecular Sciences* 14(5), 9643-9684.

IPCC. (2001). Special Report on Emissions Scenarios. *A special report of Intergovernmental Panel on Climate Change (IPCC) Working Group III*. Cambridge, UK: Cambridge University Press.

Kaushal, N., Awasthi, R., Gupta, K., Gaur, P., Siddique, K. H. and Nayyar, H. (2013). Heat-stress-induced reproductive failures in chickpea (*Cicer arietinum*) are associated with impaired sucrose metabolism in leaves and anthers. *Functional Plant Biology 40*(12), 1334-1349.

Kavar, T., Maras, M., Kidric, M., Sustar-Vozlic, J. and Meglic, V. (2008). Identification of genes involved in the response of leaves of *Phaseolus vulgaris* to drought stress. *Molecular Breeding* 21(2), 159-172.

Kumar, A. and Sharma, K.D. (2009). Physiological responses and dry matter partitioning of summer mungbean (*Vigna radiata* L.) genotypes subjected to drought conditions. *Journal of Agronomy and Crop Science 195*(4), 270-277.

Larmure, A. and Munier-Jolain, N.G. (2019). High temperatures during the seed-filling period decrease seed nitrogen amount in pea (*Pisum sativum* L.): Evidence for a sink limitation. *Frontiers in Plant Science 10*, 1608.

Liepert, B.G. (2002). Observed reductions of surface solar radiation at sites in the United States and worldwide from 1961 to 1990. *Geophysical Research Letters* 29(10), 61-1.

Miller, P.R., McConkey, B.G., Clayton, G.W., Brandt, S.A., Staricka, J.A., Johnston, A.M., Lafond, G.P., Schatz, B.G., Baltensperger, D.D. and Neill, K.E. (2002). Pulse crop adaptation in the northern great plains. *Agronomy Journal* 94(2), 261-272.

Mina, U., Singh, S.D., Singh, B. and Khaund, M. (2015). Response of wheat and chickpea cultivars to reduced levels of solar irradiance. *Journal of Agrometeorology* 17(2), 165.

Mitra, S. and Ghildiyal, M.C. (1988). Photosynthesis and assimilate partitioning in mungbean in response to source-sink alteration. *Journal of Agronomy and Crop Science* 160(5), 303-308.

Pandey, R., Zinta, G., AbdElgawad, H., Ahmad, A., Jain, V. and Janssens, I.A. (2015). Physiological and molecular alterations in plants exposed to high (CO_2) under phosphorus stress. *Biotechnology Advances* 33(3-4), 303-316.

Ryan, P.R., Liao, M., Delhaize, E., Rebetzke, G.J., Weligama, C., Spielmeyer, W. and James, R.A. (2015). Early vigour improves phosphate uptake in wheat. *Journal of Experimental Botany* 66(22), 7089-7100.

Sharma-Natu, P., Pandurangam, V. and Ghildiyal, M.C. (2004). Photosynthetic acclimation and productivity of mungbean cultivars under elevated CO_2 concentration. *Journal of Agronomy and Crop Science* 190(2), 81-85.

Singh, S.D., Chakrabarti, B., Muralikrishna, K.S., Chaturvedi, A.K., Kumar, V., Mishra, S. and Harit, R. (2013). Yield response of important field crops to elevated air temperature and CO_2. *Indian Journal of Agricultural Sciences* 83(10), 1009-12.

Sita, K., Sehgal, A., Kumar, J., Kumar, S., Singh, S., Siddique, K.H. and Nayyar, H. (2017). Identification of high-temperature tolerant lentil (*Lens culinaris* Medik.) genotypes through leaf and pollen traits. *Frontiers in Plant Science* 8, 744.

Stanhill, G. and Cohen, S. (2001). Global dimming: a review of the evidence for a widespread and significant reduction in global radiation with discussion of its probable causes and possible agricultural consequences. *Agricultural and Forest Meteorology* 107(4), 255-278.

Summerfield, R.J., Hadley, P., Roberts, E.H., Minchin, F.R. and Rawsthorne, S. (1984). Sensitivity of chickpeas (*Cicer arietinum*) to hot temperatures during the reproductive period. *Experimental Agriculture* 20(1), 77-93.

Turner, N.C., Wright, G.C. and Siddique, K.H.M. (2001). Adaptation of grain legumes (pulses) to water-limited environments. *Advances in Agronomy* 71, 193-231.

Vadez, V., Berger, J.D., Warkentin, T., Asseng, S., Ratnakumar, P., Rao, K.P.C., Gaur, P.M., Munier-Jolain, N., Larmure, A., Voisin, A.S. and Sharma, H.C. (2012). Adaptation of grain legumes to climate change: a review. *Agronomy for Sustainable Development* 32(1), 31-44.

Wahid, A., Gelani, S., Ashraf, M. and Foolad, M.R. (2007). Heat tolerance in plants: an overview. *Environmental and Experimental Botany* 61(3), 199-223.

Zaman-Allah, M., Jenkinson, D.M. and Vadez, V. (2011). A conservative pattern of water use, rather than deep or profuse rooting, is critical for the terminal drought tolerance of chickpea. *Journal of Experimental Botany* 62(12), 4239-4252.

CHAPTER 3

Agronomic interventions for drought stress mitigation in pulses

Kancheti Mrunalini[1*], Biswaranjan Behera[2], Narendra Kumar[1] and C.S. Praharaj[1]

1. *ICAR-Indian Institute of Pulses Research, Kanpur, Uttar Pradesh*
2. *ICAR-Indian Institute of Water Management, Bhubaneswar, Odisha*

*Corresponding author email: mrunalinikancheti@gmail.com

Abstract

Grain legumes grown under marginal and sub-marginal lands are often subjected to low soil moisture stress or drought. This stress reduces crop productivity by severely affecting crop growth processes. This chapter reviews the plant growth processes under the drought stress and suggests possible agronomic interventions to mitigate it. The plant morphology, reproductive growth, nitrogen fixation, content of scavenging enzymes and crop productivity vary significantly under low moisture stress. However, the extent of changes varies among species, growing seasons and locations. The field water use efficiency can be enhanced by modifying the land configuration, conserving moisture in-situ, providing nutrients optimally, managing weeds and using moisture conserving chemicals. Contingency measures for drought stress help in preventing the complete crop failure of pulse crops during the rainy season. Harvesting of rainwater during monsoon season helps in providing supplemental or lifesaving irrigation to winter season legume crops. Hence, in rainfed areas, growing pulse crops with suitable agronomic interventions will increase crop productivity by offsetting the low soil moisture stress.

Keywords: Contingency measure, Dryland, Life-saving irrigation, Paira cropping, Rainfed, Rice fallow, Seed priming

1. INTRODUCTION

Drought stress imposes perpetual challenges for global agricultural productivity through its effects on agricultural crop production and its intensity and severity are predicted to escalate in near future. Grain legumes are characterized by high sensitivity to drought stress, particularly at vegetative and reproductive stages. Mostly their cultivation is restricted to marginal and drylands with unpredictable rainfall patterns and therefore, pulse crops are frequently subjected to low moisture stress (Table 1). The impact of low moisture stress (drought) on pulses production varies with the soil properties, agro-ecological regions and timing of drought (Pradhan *et al.*, 2019). This chapter aims at characterisation of pulse crop responses to drought occurrence; its impact on yield loss and more emphasis is given on agronomic interventions for mitigating yield loss due to low moisture stress, which is of prime importance in planning contingency and holistic measures to enhance sustainable resilience of pulses crop production systems.

Table 1. Total share of pulses area under rainfed conditions (2015-16)

Crops	Area (M ha)	Production (Mt)	Area under rainfed (%)
Gram	8.40	7.06	61.4
Redgram (Arhar)	3.96	2.56	96.2
Total pulses	24.91	16.32	80.9

Source: Directorate of Economics and Statistics (DAC&FW, 2019)

2. OCCURRENCE OF DROUGHT DURING CROP SEASON AND THEIR CONSEQUENCES IN PULSES

In India, pulses are mainly grown under two distinct scenarios in which moisture stress can occur during the crop growth period. In the first scenario, pulse crops are grown in *kharif* season mostly under rainfed condition during which moisture stress can occur due to erratic rainfall. In the second scenario, pulse crops are grown in *rabi* season, the available residual soil moisture may not be sufficient for optimum crop growth and development. In these contexts, contingency measures for minimising the effect of soil moisture stress can play a major role in preventing the complete crop failure. The consequences of moisture stress at different growth stages of pulses have been depicted in Figure 1.

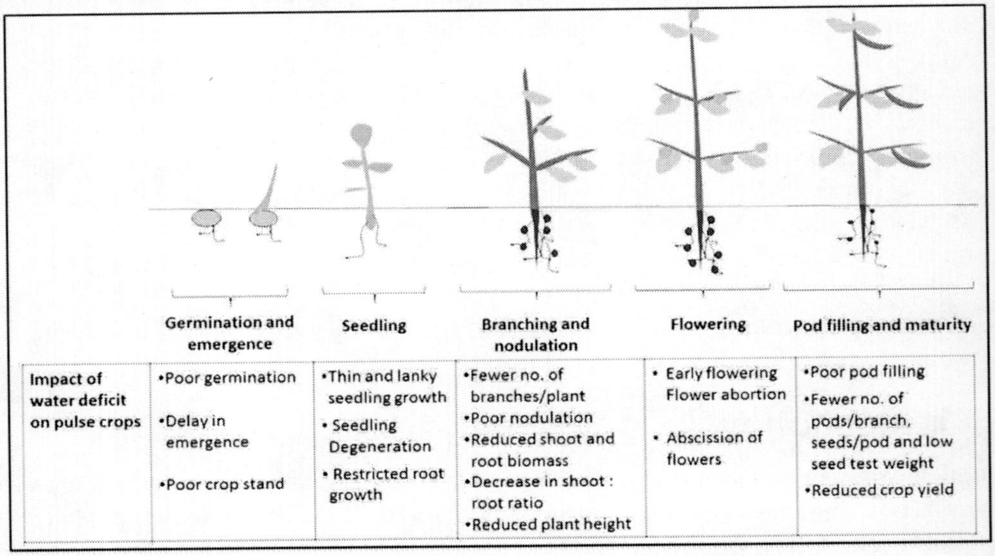

Fig. 1. Consequences of moisture stress at different growth stages of pulses

3. EFFECT OF DROUGHT STRESS

3.1 Morpho-physiological processes

Primary effect of drought stress at the start of the growing season is majorly on crop establishment due to lack of sufficient moisture for the seed to germinate (water deficit hinders the absorption and enzyme activation processes responsible for the hydrolysis of reserved food material). Drought stress immediately after sowing in pulses resulted in impaired germination (Harris *et al.*, 2002; Okcu *et al.*, 2005) and poor seedling /crop stand (Kaya *et al.*, 2006).

In general low moisture stress shows reduced crop growth and wilting. The leaf is a vital organ in the plant which acts as a kitchen in providing photosynthates to other plant organs through photosynthesis (Pradhan *et al.*, 2019). Drought stress during plant growth and development have a significant negative influence on photosynthetically active leaf area which further affect new leaf emergence (number of leaves) and leaf expansion process (size of leaves) (Lee *et al.*, 2012; Turner *et al.*, 2012; Esmaeilzade-Moridani *et al.*, 2015; Marquez-Garcia *et al.*, 2015) due to reduction in cell expansion and cell division and shrinkage/wilting of previously formed leaves (Burling *et al.*, 2013; Scoffoni *et al.*, 2014).

During early phase of drought, younger leaves start feeding on older leaves (remobilization of nutrients) owing to the production of abscisic acid (due to altered apoplastic pH), responsible for stomata closure which further hinders the photosynthesis process (absence of substrate (CO_2) and subsequent ethylene production in plant leads to senescence of older leaves. However, such leaf senescence is unusual in severe drought and is characterized by a reduction in cell expansion and cell division, degradation of chlorophyll and protein sources, limitation in sink size and activity, and reduced nitrogen remobilization (Pradhan *et al.*, 2019). Especially in legumes, vegetative storage proteins might accumulate in senescing leaves during drought stress when there is a limitation in nitrogen export capacity (Lee *et al.*, 2014).

In general, beyond a critical point leaf senescence turns out to be irreversible and this causes a reduction in total leaf biomass. Water deficit significantly reduced leaf growth in urdbean (Manivannan *et al.*, 2007a) and reduced total biomass in common bean and greengram (Webber *et al.*, 2006). The shoot growth of a plant is more sensitive to moisture stress than the root growth (Pradhan *et al.*, 2019). Besides, more information on the effect of drought stress is morpho-physiological processes in pulses is given in Table 1.

3.2 Reproductive growth

The grain legumes are relatively more tolerant to moisture stress during the vegetative stage even though they may experience retarded cell division, elongation and differentiation (Farooq *et al.*, 2009). The availability of adequate soil moisture during critical stages (flowering and seed formation) is very crucial in pulses, as these critical stages ultimately influence the yield (Moradi *et al.*, 2008). Drought stress results in the failure of pollen development (Sheoran and Saini, 1996) and further it enhances early flowering and maturity in legumes (Nam *et al.* 2001).

Water deficit during the reproductive period in pulses can potentially reduce flowering by 50% along with reduction in total pods located on the branches rather than on the main stem in common bean (Nu˜nez Barrios *et al.*, 2005). Drought stress coinciding with the reproductive stage (flowering) reduces seed yield in pigeonpea by 40-55% (Nam *et al.*, 2001), chickpea by 45-69% (Nayyar *et al.*, 2006), common beans by 58-87% (Martínez *et al.*, 2007) and in cowpea by 60-71% (Ogbonnaya *et al.*, 2003).

3.3 Root system

A prolific root system is vital for accelerated plant growth during early phases to tap water from the top (shallow) soil profile which is otherwise lost easily by evaporation (Johansen *et al.*, 1994). For improved productivity of pulses under drought stress conditions, the selection of deep and extensive root system plays an important role in optimizing the ability of crops to acquire water (Subbarao *et al.*, 1995). Increased root to shoot ratio was related to abscisic acid concentration of roots and shoots under drought conditions (Sharp and LeNoble, 2002; Manivannan *et al.*, 2007b). Different soil textural groups and their corresponding water availabilities, xylem vessel size and number have a significant influence on pulses crop adaptability and are determined through changes in root morphology and anatomy. For instance, chickpea is considered to be more adaptive in fine-textured soils (fine-textured soils restrict lateral movement of water and facilitate better soil water absorption than coarse-textured soils) than soybean, cowpea, or common bean due to the presence of dense roots and high root thickening properties i.e. roots become thicker with an increase in water-deficit stress and the existence of thin xylem vessels compared to other pulse crops which have broader xylem vessels. However, in terms of yield chickpea does not perform better when compared to soybean and common bean due to differences in root morphology and anatomy. Conversely, soybean shows a lower reduction in yields under drought stress in comparison to common bean despite their similarities in root characteristics or drought sensitivity (Purushothaman *et al.*, 2013) which indicates enhancing yield performance does not depend on merely a single trait.

3.4 Nitrogen fixation

Water deficit stress impairs biological nitrogen (N_2) fixing ability through a decrease in oxygen diffusion to pulses root nodules (Pankhurst and Sprent, 1976; Davey and Simpson, 1990) or increased oxygen diffusion resistance to root microbes which results in a reduction in nitrogenase activity (Table 1) that significantly reduce biosynthesis of proteins (primary reserve in grain legumes) and further leads to a reduction in grain yields (Purcell and King, 1996). However, the reduction in nitrogenase activity was primarily due to a reduction in nodule number and weight (Ramos *et al.*, 1999).

3.5 Nutrient uptake

Low moisture stress negatively affects the nutrient uptake (Samarah *et al.*, 2004; Gunes *et al.*, 2006) and drastically diminishes nitrogen fixation in legumes like soybean (Serraj, 2003) and pea (Gonzalez *et al.*, 2001). Further, these effects eventually reduce assimilate production and mobilization to the sink (grains) in different crops (Leport *et al.*, 2006; Mafakheri *et al.*, 2010; Zare, 2012).

Drought stress is known to reduce the accumulation of nutrients in developing grains. For instance, common bean showed low Fe, Zn, P and N concentrations under drought stress which it further directly correlate with a reduction in total protein content (Ghanbari *et al.*, 2013); in chickpea drought stress during pod filling resulted in a reduction in protein and amino acids in the seeds and this impact is more in '*Kabuli*' than '*Desi*' chickpea types (Nayyar *et al.*, 2006). Similarly, reduction in seed protein content is reported in mungbean (Vamsi Krishna *et al.*, 2018) and lentil and lathyrus (Talukdar, 2013). On the other hand, increase in leaf proline content is observed with the increase in drought stress in chickpea (Mafakheri *et al.*, 2010), mungbean (Vamsi Krishna *et al.*, 2018) and in lentil and lathyrus (Talukdar, 2013).

3.6 Crop yield

Yield and yield attributes are mainly dependent on the availability of moisture at critical growth stages of the crop. Pulses, in general, were cultivated in marginal lands, arid to semi-arid environments, and drylands [which are frequently exposed to low moisture stress (drought stress) where the yields are uncertain and highly variable (Leport *et al.*, 1999; Mafakheri *et al.*, 2010; Baroowa *et al.*, 2012; Vanaja *et al.*, 2015; Suresh *et al.*, 2016). In comparison to other crop groups (cereal crops/tuber crops), the variation in yield reduction in pulses varies between species, some species like field pea, mungbean exhibit lesser variation in yield loss when compared to other pulses like lentil, pigeonpea and chickpea (Table 1). Among three crop groups (cereals, legumes, and tuber crops), greater variability of yield reduction is observed with the legume group (20-85%) and tuber group (25-70%) when compared to the cereal group (25-40%), excluding anaerobic rice. The cereal group is reported to exhibit lower variation in yield reduction compared to the other two crops (legumes and tuber crops) groups (Pradhan *et al.*, 2019).

Table 1. Impact on drought stress on growth and yield parameters in pulse crops

Crop	Location	Parameter	% variation in water stress vs irrigated treatments	References
Pigeonpea	Gujarat	Seed yield	-50 to -70	Suresh *et al.*, 2016
		RWC	-11 to -22	
	ICRISAT, Hyderabad	Total drymatter production	-3.7 to -37	Nam *et al.*, 2001
		Seed yield	-52 to -66	
	CRIDA, Hyderabad	Root dry weight	-30	Vanaja *et al.* 2015
		Seed yield	-50.5	
		Harvest index	-21	
Blackgram	CRIDA, Hyderabad	Photosynthetic rate	-42	Gurumurthy *et al.*, 2019
		Root dry weight	-34	
		Peroxidase activity	+13	
		Superoxide dismutase	+7.9	
	Tirupati, AP	Total drymatter production	-38	Yohan *et al.*, 2017
		Seed yield	-15	
		Harvest index	-7.2	
	CRIDA, Hyderabad	Net photosynthetic rate	-0.7 to -55.6	Anitha *et al.*, 2015
		Transpiration rate	-9.3 to -52.6	
		Seed yield	-3.8 to -58.5	
	Assam	Seed yield	-25.5	Baroowa *et al.*, 2012

Mungbean	Allahabad, UP	Relative water content	-16.2	Vamsi Krishna *et al.*, 2018
		Total chlorophyll content	-14.2	
		Protein content	-27.8	
		Proline content	+67.4	
	Bangladesh	Flower and pod abscission	+ 33	Uddin *et al.*, 2013
		NAR	-72	
		Root dry weight	-62	
		Root/shoot ratio	-12	
	Assam	Seed yield	-16	Baroowa *et al.*, 2012
	Srilanka	Seed yield	-28	Ranawake *et al.*, 2011
Chickpea	Iran	Shoot height	-25	Mafakheri *et al.*, 2010
		Proline content	+40 to +50	
		Seed yield	-60 to -66	
		Seed yield	-54	Shamsi *et al.*, 2010
	Tunisia	Individual nodule biomass	-18.6	Labidi *et al.*, 2009
		Specific N fixation	-24	
		Leghaemoglobin	-38	
	Western Australia	Seed yield	-47.5	Leport *et al.*, 1999
		Pod abortion on primary branches	+186	Leport *et al.*, 2006
		Pod abortion on secondary branches	+217	
		Flower abortion	+278	Fang *et al.*, 2010
		Pod abortion	+8.7	
		Seed yield	-84.9	
	Pakistan	Days to flowering	-10.5	Hussain *et al.*, 2015
		Seed yield	-15	
		100-seed weight	-8.2	
	Iran	Superoxide Dismutase	+59	Mohammadi *et al.*, 2011
		Catalase	+73	
		Glutathione Peroxidase	+50	
		Malondialdehyde	+85	
	Ludhiana, India	Dry root weight	-31	Randhawa *et al.*, 2014
		LAI	-44.5	
		Total number of pods	-22	
	ICRISAT, Hyderabad	Days to 50% flowering	-9	Ramamoorthy *et al.*, 2016
		Days to maturity	-18.3	
		Seed yield	-4	

Lentil	West Bengal	Days to maturity	-48	Talukdar, 2013
		Relative water content	-59	
		Net photosynthesis rate	-50	
		Leaf proline content	-33.5	
		Seed protein content	-55	
		Seed yield	-70	
Field pea	New Zealand	Total dry weight	-17	Martin & Jamieson, 1996
		No. of pods/m^2	-30	
	England	Plant height	-22	Grzesiak *et al.*, 1997
		Shoot dry matter	-18 to -26	
		Root dry matter	-10 to -22	
		Shoot/root ratio	+1.4 to -18	
Field bean	England	Plant height	-21	Grzesiak *et al.*, 1997
		Shoot dry matter	-22	
		Root dry matter	-10 to -20	
		Shoot/root ratio	-3 to -10	
Common bean	UK	Nodule dry weight	-28.5	Ramos *et al.*, 1999
		Nitrogenase activity	-42.2	
		O_2 diffusion resistance	+33	
Lathyrus	Western Australia	Total drymatter	-60	Gusmao *et al.*, 2012
		Flower abortion	+73	
		Seed yield	-87	
	West Bengal	Days to maturity	-11	Talukdar, 2013
		Relative water content	-22.2	
		Net photosynthesis rate	-10	
		Leaf proline content	+52.3	
		Seed protein content	-24	
		Seed yield	-12.5	

4. AGRONOMIC MEASURES FOR DROUGHT STRESS MITIGATION

4.1 Land modification practices

Field water use efficiency can be enhanced by modifying the land configuration. These practices also conserve soil moisture that can meet crop water need during the growing season.

4.1.1 Tillage

Deep tillage breaks soil crust, makes the soil more porous, and improves soil hydraulic properties. During pre-monsoon rainfall, it helps in recharging groundwater by increasing water infiltration to soil. The continuity of a slope on graded land can be broken by ploughing across the slope. This not only reduces runoff losses but also conserves soil and water. The summer ploughing uproots weeds that are desiccated later under the hot sun and prevents moisture depletion during the fallow period. Light harrowing can be done to break the capillarity on surface soil and prevent evaporation loss when there is no crop in the field. Jat *et al.* (2013) reported a 16-21% increase in seed yield of blackgram under deep tillage as compared to shallow tillage. Growing blackgram on a raised-bed of 40 cm width with summer deep ploughing resulted in seed yield of 1239 kg/ha (82% to 118% increase as compared to other conservation practices) and water use efficiency of 2.01 kg/ha/mm.

4.1.2 Broad Bed and Furrow (BBF)

The BBF method controls surface drainage by forming beds and furrows. This system consists of broad beds about 100-120 cm wide separated by sunken furrows about 50 cm wide, which can be modified according to numbers and spacing of crop rows (FAO, 2020). In the BBF system, good contact between water and the soil in broad beds results in vertical infiltration in broad beds, and both vertical and horizontal infiltration (in the form of seepage) in furrows. In India, it is mostly practiced in deep vertisols to store moisture in the soil profile up to 25 cm. The stored moisture acts as a contingency source during mid and late-season dry spells to support plant growth and development. In the BBF method, excess runoff water can be safely disposed of and stored *ex-situ,* which can be used as lifesaving irrigation during moisture stress. The judicious use of water in BBF method improves crop water use efficiency. Choudhary (2016) reported 40.6% increase in water use efficiency of frenchbean under BBF method as compared to flat-bed method of planting. The major limitation in vertisols that leaves these soil uncropped during the rainy season is a narrow range of optimum moisture for its workability. To avoid difficulty in sowing after the commencement of rain in vertisols, dry seeding of pulses can be done in BBF after tilling the raised beds before the rains. The BBF planting and post-monsoon irrigation at 0.4 IW: CPE ratio increased the yield of pigeonpea by 10% over flat-bed planting without any irrigation (Kantwa *et al.*, 2005). Similarly, Kumar *et al.* (2012) reported 16% increase in pigeonpea equivalent yield in a pigeonpea + mungbean intercropping under BBF planting as compared to flat-bed planting.

4.1.3 Laser land levelling

The slope and gravity in a field determine the path of water flow on the soil surface. Uniform water distribution in the field is a pre-requisite for better soil, water and plant interaction. The traditional land levelling practice using bullock drawn plank leads to less input use efficiency. Whereas, laser-land levelling improves input use efficiency and results in better crop stand and grain yield per unit quantity of water (Praharaj *et al.*, 2016). Moisture conservation practices like sowing under surface residue condition using zero-till seed drill on a laser levelled field resulted in better seed placement, germination and uniform distribution of irrigation water and plant nutrients as compared to unlevelled field (Sankaranarayanan *et al.*, 2010; Praharaj *et al.*, 2015). Laser levelling is a process of smoothening the land surface (± 2 cm) from its average elevation using laser-equipped drag buckets to achieve precision in land levelling (Jat *et al.*, 2006). A perfectly level field conserves water by reducing runoff and allowing uniform distribution of water. Laser land levelling can reduce water use by 20-30% and increase crop yields by 10-20%.

4.1.4 Tied ridging

A series of ridges and furrows are made and after which ridges are tied by damming furrows with small mounds. This converts furrows into small compartments that prevent runoff and increase infiltration during rainfall. The tied ridging at 4 weeks after planting resulted in grain yield increases of 79, 31, and 96% for fababean, lentil, and field pea, respectively, compared with flat planting due to better soil water availability during the growing season (Tesfahunegn and Wortmann, 2008).

4.2 Nutrient management

In general for pulses crops, all the recommended fertilizers should be applied as basal application i.e. 15-20 kg N/ha and 20-40 kg P_2O_5/ha and 20-40 kg K_2O/ha in drylands areas. Foliar application of potassium and urea can up regulate the physiological processes of plants during drought stress which further aid in enhancing crop yields (Raza *et al.* 2013; Yadav *et al.*, 2019). The foliar spray of 2% DAP + 40 ppm NAA + 0.5% $ZnSO_4$ + 1% $FeSO_4$ at pre-flowering and flowering improved reproductive efficiency and grain yield of rice fallow pulses due to reduction in flower dropping. Das and Jana (2005) reported a significantly higher seed yield of pulses with application of 2% urea spray over basal dose of fertilizer application. The use of drip-fertigation can supply both water and nutrients to the root zone thereby enhancing water and nutrient use efficiencies in drylands (Ramamurthy *et al.*, 2009). Urea can be applied as 2% foliar spray at 50% flowering to cope up terminal drought and post-flowering nodule degeneration in pulses like chickpea (Ahlawat *et al.*, 2007).

4.3 Irrigation management

In drylands, water is the major limiting source for crop production. Supplemental irrigation is given to sustain during drought periods and take the advantage of subsequent rains. Supplemental irrigation is a highly efficient practice for enhancing the productivity of pulses grown in drylands. In general for all the pulse crops, flowering and pod development are considered critical stages for irrigation, supplemental irrigation given at these stages during moisture deficit stress provide insurance against crop yield losses. With available scarce water resources in drylands, it is advisable to adapt micro-irrigation technologies like sprinklers and drip irrigation without conveyance losses. For instance, sprinkler irrigation with 60 mm water given that three crop growth stages (sowing, branching, and pod-formation stages) in chickpea showed 44% saving in water use (Chandegara and Yadavendra, 1998), whereas in mungbean, 39.7% yield increase and 49.8% water saving has been reported. Similarly, in pigeonpea drip irrigation showed a 20% increase in seed yield (Mahalakshmi *et al.*, 2011; Praharaj and Kumar, 2012).

4.4 Weed management

Weeds compete with the crop for different growth resources like moisture and nutrients and reduce the crop yields by 30 to 60% in drylands. Vigorous weed competition is observed in pulse crops with slow initial growth at early growth stages. Therefore, the weed control especially during the early stages of crop growth reduces the impact of a dry spell by soil moisture conservation leading to higher fertilizer use efficiency and higher yields. In these situations, the use of post-emergence herbicides in combination with pre-emergence herbicides could be a better alternative for effective weed management in pulses. For instance, in *kharif* pulses use of pendimethalin 1.0 kg/ha (PE) followed by imazethapyr 100 g/ha (PoE), and in rabi pulses use of pendimethalin 1.0 kg/ha (PE) followed by quizalofop-ethyl 50 g/ha (PoE) offers best weed management (Kumar *et al.*, 2013; Kumar *et al.*, 2015). Along with chemical weed control, other alternatives for weed management like residue management as mulch, crop rotation, intercropping, solarisation and integrated weed management practices play a crucial role in achieving effective weed management in drylands. In dryland areas, during fallow period leguminous crops like *Sesbania* can be grown and knocked down by herbicides like 2,4-D to suppress weeds, conserve moisture and increase soil fertility (Behera *et al.*, 2018). An integrated weed management practice including quizalofop-ethyl @ 0.04 kg ha^{-1} as PoE at 20-25 DAS + 1 hand weeding + 1 cultural practice at 40-45 DAS led to 60% yield increase in cowpea as compared to unweeded control (Kumar and Singh, 2017).

4.5 Soil mulching

Various materials are used to cover the soil surface which modify the physical, chemical and biological properties of soil. The organic mulches like crop residues increase infiltration by increasing the opportunity time and surface roughness. Such mulches also moderate soil temperature by increasing it during winter (by entrapping terrestrial radiation) and reducing it during summer (by reflecting solar radiation due to the albedo effect). Chemicals like hexadecanol act as a chemical mulch to reduce evaporation when it is mixed with soil. Dust mulching is a practice of stirring dry surface soil of 5-7 cm depth. It reduces evaporation loss of soil moisture by breaking the pore continuity. In deep black soils, vertical mulching can be done by filling organic materials in a trench of 40 cm depth and 15 cm width. This type of mulch increases the available soil moisture by 4-5 cm by increasing the infiltration of water to the soil. Under deficit rainfall condition, soil mulching resulted in significantly higher seed yield and water use efficiency in chickpea as compared to no mulching (Mishra *et al.*,2012). Sharma *et al.* (2009) reported beneficial effects of straw, polythene and dust mulching on soil physical properties, crop productivity and water use efficiency.

4.6 Use of chemicals

Chemicals like anti-transpirants, super absorbent polymers and seed hardeners can be used to manage moisture stress in crop field.

4.6.1 Anti-transpirants

According to Waggoner (1966),

$$\text{Water loss through transpiration} = \frac{\Delta e}{Rs \times Ra}$$

Where Δe = water vapour deficit between leaf cell and atmosphere, Rs= Stomatal resistance, and Ra = Boundary layer resistance.

Therefore, the transpiration can be reduced by either decreasing Δe or increasing Rs and Ra. The stomatal closing/growth retardant type anti-transpirants like cycocel, PMA, 2,4-D, *etc.* increases stomatal resistance, whereas film forming anti-transpirants like mobileaf increases boundary layer resistance. The reflecting type anti-transpirants like kaolin reduces the water vapour deficit by reducing solar radiation incidence due to the albedo effect. Ansari and Rana (2012) reported 8% reduction in consumptive use of pigeonpea due to application of PMA @ 320 ppm. The moisture use efficiency was also improved by application of cycocel at 200 ppm concentration due to reduction in the water loss through transpiration (Ansari and Rana, 2012).

4.6.2 Super absorbent polymers (SAP)

The SAP is a soil conditioner that can absorb water >100 times of its weight and hence increases the water holding capacity of the soil. Examples: Aquasorb, Hydrosorb, Skygel, Agragel, Jalashakti, Pusa hydrogel, *etc.* The Jalashakti is a starch-based free-flowing granular product that absorbs water 150-200 times of its weight. Before crop sowing, Jalashakti is mixed @ 5 kg/ha in a soil of 10-15 cm depth in the plant root zone. The application of Jalashakti @ 5 kg/ha increased seed yield of chickpea by 17% as compared to no mulching (Dhar *et al.*, 2008). The limitations in the use of Jalashakti are its stickiness, instability at high temperature and cost ineffectiveness. Another popular SAP is Pusa hydrogel which is a semi-synthetic cellulose-based substance having stability at high temperature and no stickiness. The Pusa hydrogel absorbs water upto 350 times its weight when applied at 2.5-5 kg/ha.

4.6.3 Seed primers/hardeners

Seed priming is a technique of soaking seeds in water or low osmotic potential solution to begin germination related metabolic activities without initiating radicle emergence. It has been reported that during drought stress conditions, hydro-priming (soaking seeds in water) enhances the germination and seedling growth by 3-4 times as compared to non-priming (Maurya *et al.*, 2020). The seed primers/hardeners do physiological preconditioning of crop seeds to withstand soil moisture stress under rainfed condition. Seed hardening is done by soaking the seeds in water or suitable chemicals and then drying them under shade. The alternate soaking and drying practice makes the protoplast flexible by expanding and contracting it. This helps the seeds to withstand moisture stress just after sowing. Some seed hardeners are 100 ppm $ZnSO_4$ (for pigeonpea and cowpea), 1% KH_2PO_4 (for chickpea), 50% ash + 3% gum acacia (for greengram and blackgram), and 100 ppm $MnSO_4$ (for greengram). Khan *et al.* (2008) recorded 12% yield increase in mungbean due to seed treatment with water and polyethylene glycol. The normal development of nodules in primed plants occurs due to higher sucrose synthesis activity in it. The enhanced phenology in mungbean due to primed seed is associated with faster emergence and reduced germination imbibition periods (Maurya *et al.*, 2020).

5. SEASONAL DROUGHT STRESS MANAGEMENT

5.1 Growing pulses in the rainy season

In India, more than 50% of the total rainfall occurs in few events. The unevenness in rainfall distribution within a season can result in moisture stress during crop growth periods. In kharif season, delayed onset of monsoon, prolonged dry spell, and early withdrawal of monsoon can lead to complete crop failure, which can otherwise be prevented by following contingency measures (Figure 2).

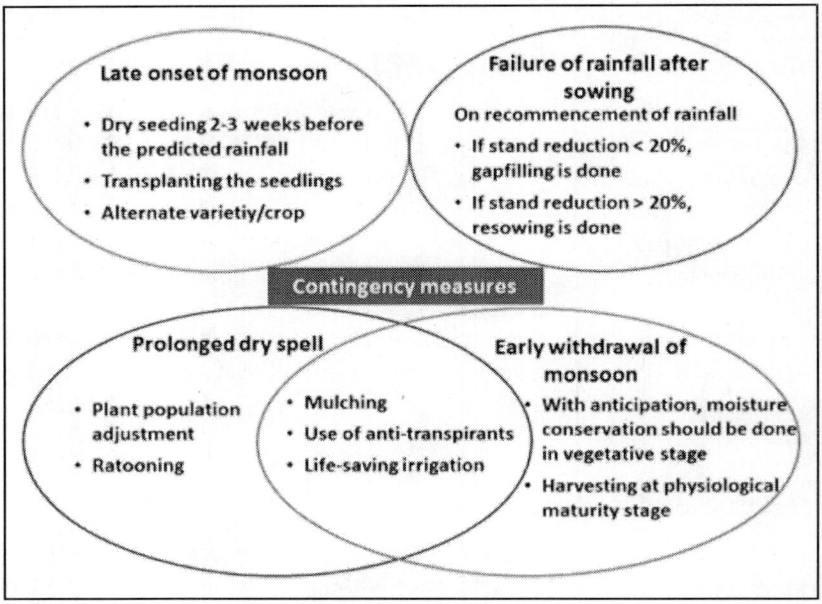

Fig. 2. Contingency measures to manage moisture stress during erratic rainfall

5.1.1 Delayed onset of monsoon

The late onset of monsoon, among different erratic behaviours, causes a maximum reduction in crop productivity. Time of sowing is a major non-monetary input that significantly affects the crop yield. It has been seen that delayed sowing of *kharif* pulses after 2nd fortnight of July significantly affects the crop productivity. Arunkumar *et al.* (2018) reported a 37% yield decline in 1st August sown and a 56% yield decline in 20th August sown pigeonpea crop as compared to that of 10th July sown. The seeding of crop seeds in dry soil 2-3 weeks before the predicted rainfall can result in quick germination after the wetting event. In the case of limited irrigation availability, growing seedlings of pulses like pigeonpea in a nursery and transplanting them in the main field after the onset of monsoon can avoid moisture stress in the main field. After the commencement of monsoon, if rainfall fails immediately after sowing, gap filling should be practiced when stand reduction is <20%, whereas, re-sowing should be done when stand reduction >20% after recommencement of rainfall. If the onset of monsoon is delayed considerably, low water requiring *kharif* pulses like horsegram, greengram, *etc.* can be grown to avoid severe yield loss.

5.1.2 Prolonged dry spell during cropping season

A dry spell with no rainfall event for > 10 days (in sandy soil) or 15 days (in clay soil) can severely affect the crop growth. The harvesting and storing of rainwater can be useful in supplying live-saving irrigation during critical growth stages. This supplementary irrigation especially during long dry spells after rainy months could alleviate moisture stress in growing crops (Praharaj *et al.*, 2016). For instance, Praharaj *et al.* (2017) reported a 20% yield increase in pigeonpea with supplemental irrigation at branching as compared to that of the rainfed crop. Mulching the soil surface with suitable materials (organic/inorganic) can prolong the effect of supplemental rainfall and irrigation by reducing evaporation and runoff losses. The anti-transpirants like cycocel, kaolin and mobileaf can be used to reduce transpiration loss and conserve water at cellular level. Despite all these measures, if a dry spell persists and plants don't regain their normal growth and development, the plant population can be reduced to half by removing every alternate row.

5.1.3 Early withdrawal of rainfall

With the anticipation of early withdrawal of monsoon rains during *kharif* season, early moisture conservation practices both *in-situ* and *ex-situ* should be practiced. The *ex-situ* stored water in farm/village pond can be utilized for saving crops by providing lifesaving irrigation. The crop can be harvested at the physiological maturity stage to avoid shattering of pods.

5.2 Growing pulses in the winter season

In rainfed areas, when the quantity of rainfall in the rainy season is sufficient enough to grow major cereal crops, the pulses are generally grown as *paira/utera* crops with residual soil moisture in the *rabi* season. In *paira* cropping, *rabi* pulse crops are sown before the harvest of *kharif* cereal crops like rice. It is widely practiced in eastern India where soil remains saturated during the harvesting of rice. High soil moisture content delays the land preparation for sowing *rabi* crops by 15-20 days that lead to moisture stress during the later phase of the *rabi* crop. Conventionally, farmers don't take any crops under such conditions and keep the land fallow during the *rabi* season. In some areas *paira* cropping of grasspea and lentil is followed under rice fallow situation after harvesting of rice on residual soil moisture, but chickpea cultivation also possible in low to medium land ecosystem (Bitew and Asargrew, 2014). The management practice under *paira* cropping is depicted in the Figure 3.

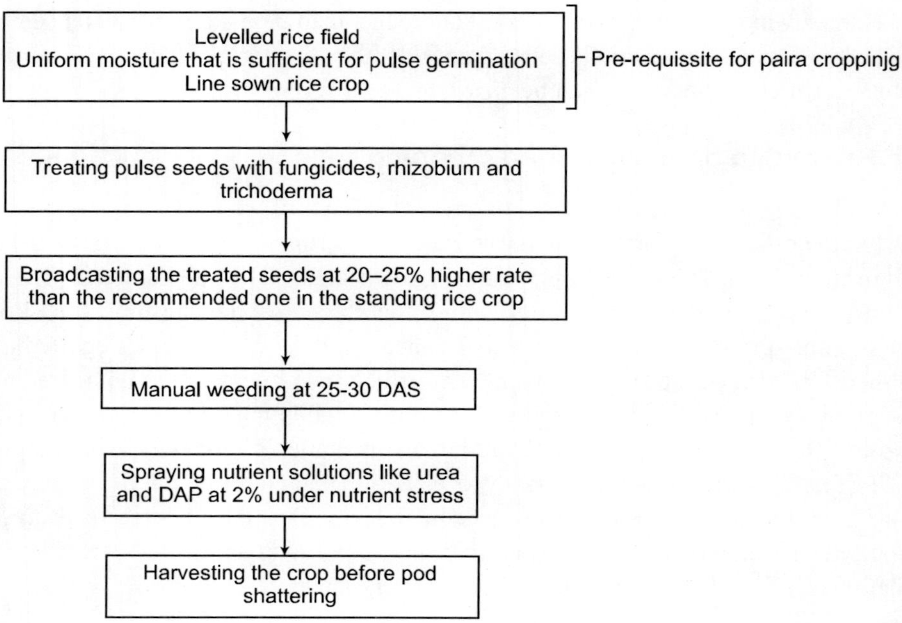

Fig. 3. Agronomic practices for *paira* cropping of pulses in rice fallow of eastern India

When the quantity of rainfall in *kharif* season is not sufficient to meet the water requirement of any major food crops, rainwater is conserved by keeping the land fallow and pulse crops are grown with the conserved water during the *rabi* season. Deep summer ploughing, rainwater harvesting, soil mulching, groundwater recharging, *etc*. can be practiced to conserve water in the soil. An excellent traditional water harvesting and runoff farming system, known as 'Haveli' cultivation, can also be useful. In the Haveli system, cultivated fields are levelled and surrounded by large bunds and kept fallow during the *kharif* season. The standing water in the field during monsoon not only recharges the groundwater but also suppresses weeds. After drainage of the impounded water in October, the *rabi* crops are sown in moist soil with a minimum tillage operation. This system provides sufficient soil moisture to meet the crop water need of pulses like lentil, chickpea, *etc*. without any rainfall or irrigation from any sources between sowing and harvest (Shrivastava *et al*., 2018). This practice has been followed traditionally since long back in history in central India particularly in the Kymore Plateau and Satpura hills Agro-climatic zone of Madhya Pradesh. Due to this practice, the regions could maintain a huge reverse of groundwater bellow alluvium and produced quality products in pulse (Awasthi *et al*., 2011).

6. CONCLUSION

The soil moisture led yield reduction depends on the phenological stage of occurrence of drought stress and differs with genotypes of pulse crops. In general, yield reduction is higher when moisture deficit occurs during reproductive stage in comparison to vegetative stages. Agronomic interventions in managing low moisture stress could be a promising way to reduce yield loss in rainfed areas. The practices like land modification, selection of crop, varieties and maintenance of plant stand as per water availability, mulching, supplemental irrigation, weed and nutrient management, use of chemicals *etc.* can address the moisture stress due to aberrant weather during *kharif* season. Pulses can also be grown on residual soil moisture in winter season by adapting various moisture conservation practices. In rainfed areas, growing pulse crops with suitable agronomic interventions will contribute national food basket by increasing the cropping intensity.

7. FUTURE THRUST

In view to enhance resilience in farming community to drought risks, emphasis should be made in crop improvement programs to develop short/early duration cultivars which can escape terminal drought stress and yield better in water stress condition. These early/short duration cultivars have vast potential to have place in cropping systems with narrow sowing windows and in new production niches. Cultivar with good biological nitrogen fixation ability even in water stress conditions allows the crop to be a component in sustainable cropping systems. Further, these drought tolerant cultivars offers best cost effective long-term solutions against adverse drought effects and returns per investment in these breeding strategies for drought tolerance are higher compared to other drought management strategies. In addition to cultivars, crop management strategies need to developed which include in-situ soil moisture storage, farm or community ponds for storing runoff water to give life saving irrigation during drought, foliar nutrition, crop contingency planning and identification of climate resilience cropping systems

REFERENCES

Ahlawat, I.P.S., Gangaiah, B. and Zahid, M.A. (2007). Nutrient management in chickpea. In: Chickpea breeding and management (Yadav S.S., Redden R., Chen W., Sharma B., eds). CAB International, Wallingford, Oxon, United Kingdom. pp. 213-232.

Anitha, Y., Vanaja, M., Sunitha, V., Sowmya, P., Sathish, P. and Kumar, G.V. (2015). Performance of blackgram genotypes under moisture deficit stress-Variability in physiological and yield contributing attributes. *International Journal of Current Science* 17, 82-89.

Ansari, M.A. and Rana, K.S. (2012). Effect of transpiration suppressants and nutrients on productivity and moisture-use efficiency of pearlmillet (*Pennisetum glaucum*)–pigeonpea (*Cajanus cajan*) intercropping system under rainfed conditions. *Indian Journal of Agricultural Sciences* 82(8), 676–680.

Arunkumar, M.M. and Meena, M.K. (2018). Phenology and productive performance of pigeon pea as influenced by date of sowing. *Journal of Pharmacognosy and Phytochemistry* 7(5), 266-268.

Awasthi M.K., Nema R.K., Shrivastav R.N. and Tiwari Y.K. (2011). Proceeding of National seminar on Re-structuering of irrigated Agriculture. *Status and Strategies* 15-17March 2011.pp. 197-203.

Baroowa, B., Gogoi, N., Paul, S. And Sarma, B. (2012). Morphological responses of pulse (*Vigna* spp.) crops to soil water deficit. *Journal of Agricultural Sciences* 57(1), 31-40.

Behera, B., Das, T. K., Ghosh, S., Kaur, R. and Singh, R. (2018). Brown manuring in maize (*Zea mays*): Effects on weed interference and crop productivity. *Indian Journal of Agronomy* 63(4), 524-527.

Bürling, K., Cerovic, Z.G., Cornic, G., Ducruet, J.M.,Noga, G. and Hunsche, M. (2013). Fluorescence-based sensing of drought-induced stress in the vegetative phase of four contrasting wheat genotypes. *Environmental and Experimental Botany* 89, 51-59.

Choudhary, V.K. (2016) Response of land configuration and mulches on maize–frenchbean–toria cropping system. *Agronomy Journal* 108(5), 2147-2157.

DAC & FW. (2019). Agriculture Statistics at a Glance.Directorate of Economics and Statistics, *Department of Agriculture and Cooperation, Ministry of Agriculture, Govt. of India.*https://eands.dacnet.nic.in/PDF/At%20a%20Glance%202019%20 Eng.pdf.

Das, S.K. and Jana, K. (2015). Effect of foliar spray of water soluble fertilizer at pre flowering stage on yield of pulses. *Agricultural Science Digest-A Research Journal* 35(4), 275-279.

Davey, A.G. and Simpson, R.J. (1990). Nitrogen fixation by subterranean clover at varying stages of nodule dehydration: I. carbohydrate status and short-term recovery of nodulated root respiration. *Journal of Experimental Botany* 41(9), 1175-1187.

Dhar, S., Das, S.K., Kumar, S. and Singh, J.B. (2008). Effect of tillage and soil moisture conservation practices on crop yields of chickpea (*Cicer arietinum*) and soil properties under rainfed conditions. *Indian Journal of Agricultural Sciences* 78(12), 1042-1053.

Esmaeilzade–Moridani, M., Kamkar, B., Galeshi, S., Ghaderi-Far, F. and Da Silva, J.A.T. (2015). Leaf expansion and transpiration responses of millet species to soil water deficit. *Pedosphere* 25(6), 834-843.

Fang, X., Turner, N.C., Yan, G., Li, F. And Siddique, K.H. (2010). Flower numbers, pod production, pollen viability, and pistil function are reduced and flower and pod abortion increased in chickpea (*Cicer arietinum* L.) under terminal drought. *Journal of Experimental Botany* 61(2), 335-345.

FAO. (2020). Water conservation. http://www.fao.org/3/T0321E/t0321e-11.htm assessed on 14.09.2020.

Farooq, M., Wahid, A. and Kobayashi D. Fujita S.M.A. Basra, N. (2009). Plant drought stress: effects, mechanisms and management. *Agronomy for Sustainable Development* 29(1), 185-212.

Ganapathy, M., Baradhan, G. and Ramesh, N. (2008). Effect of foliar nutrition on reproductive efficiency and grain yield of rice fallow pulses. *Legume Research-An International Journal* 31(2), 142-144.

Ghanbari, A.A., Shakiba, M.R., Toorchi, M. and Choukan, R. (2013). Nitrogen changes in the leaves and accumulation of some minerals in the seeds of red, white and chitti beans (*Phaseolus vulgaris*) under water deficit conditions. *Australian Journal of Crop Science* 7(5), 706.

González, E., Gálvez, L., Royuela, M., Aparicio-Tejo, P. and Arrese-Igor, C. (2001).

Insights into the regulation of nitrogen fixation in pea nodules: lessons from drought, abscisic acid and increased photoassimilate availability. *Agronomie* 21(6-7), 607-613.

Grzesiak, S., Iijima, M., Kono, Y. and Yamauchi, A. (1997). Differences in drought tolerance between cultivars of field bean and field pea. A comparison of drought-resistant and drought-sensitive cultivars. *ActaPhysiologiaePlantarum* 19(3), 349-357.

Gunes, A., Cicek, N., Inal, A., Alpaslan, M., Eraslan, F., Guneri, E. and Guzelordu, T. (2006). Genotypic response of chickpea (*Cicer arietinum* L.) cultivars to drought stress implemented at pre-and post-anthesis stages and its relations with nutrient uptake and efficiency. *Plant Soil and Environment* 52(8), 368.

Gurumurthy, S., Sarkar, B., Vanaja, M., Lakshmi, J., Yadav, S.K. and Maheswari, M. (2019). Morpho-physiological and biochemical changes in black gram (*Vigna mungo* L. Hepper) genotypes under drought stress at flowering stage. *ActaPhysiologiaePlantarum* 41(3), 42.

Gusmao, M., Siddique, K.H.M., Flower, K., Nesbitt, H. and Veneklaas, E.J. (2012). Water deficit during the reproductive period of grass pea (*Lathyrus sativus* L.) reduced grain yield but maintained seed size. *Journal of Agronomy and Crop Science*, 198(6), 430-441.

Harris, D., Tripathi, R.S. and Joshi, A. (2002). On-farm seed priming to improve crop establishment and yield in dry direct-seeded rice. In: S Pandey, M Mortimer, L Wade, TP Tuong, K Lopez, and B Hardy (ed), Direct seeding: Research strategies and opportunities. International Rice Research Institute, Manila, Philippines. pp. 231-240.

Hussain, N., Aslam, M., Ghaffar, A., Irshad, M. and Din, N.U. (2015). Chickpea genotypes evaluation for morpho-yield traits under water stress conditions. *Journal of Animal & Plant Sciences* 25(1), 206-211.

Jat, M.L., Chandna, P., Gupta, R., Sharma, S.K. and Gill, M.A. (2006). Laser land leveling: A precursor technology for resource conservation. *Rice-Wheat Consortium Technical Bulletin Series* 7, 48.

Jat, M.L., Sharma, S.K., Kothari, A.K., Sharma, R.K., Pareek, K. And Kumari, M. (2013). Moisture conservation practices in blackgram for pulse security in semi-arid tropics. *Indian Journal of Soil Conservation* 41(2), 158-162.

Johansen, C., Baldev, B., Brouwer, J. B., Erskine, W., Jermyn, W. A., Li-Juan, L. and Silim, S.N. (1994) Biotic and abiotic stresses constraining productivity of cool season food legumes in Asia, Africa and Oceania. In: F.J. Muehlbauer & K.J. Kaiser (Eds.), Expanding the Production and Use of Cool Season Food Legumes, pp. 175–194. Kluwer Academic Publishers, Dordrecht, The Netherlands.

Kantwa, S.R., Ahlawat, I.P.S. and Gangaiah, B. (2005). Effect of land configuration, post-monsoon irrigation and phosphorus on performance of sole and intercropped pigeonpea (*Cajanus cajan*). *Indian Journal of Agronomy* 50(4), 278-280.

Kaya, M.D., Okçu, G., Atak, M., Cıkılı, Y. and Kolsarıcı, Ö. (2006). Seed treatments to overcome salt and drought stress during germination in sunflower (*Helianthus annuus* L.). *European Journal of Agronomy* 24(4), 291-295.

Khan, A., Khalil, S.K., Khan, A.Z., Marwat, K.B. and Afzal, A. (2008). The role of seed priming in the semi-arid area for mung bean phenology and yield. *Pakistan Journal Botany* 40(6), 2471-2480.

Kumar, P. and Singh, R. (2017). Integrated weed management in cowpea (*Vigna unguiculata* (L.) Wasp.) underrainfed conditions. *International Journal of Current Microbiology and Applied Sciences* 6(3), 97-101.

Kumar, P., Rana, K.S. and Rana, D.S. (2012). Effect of planting systems and phosphorus with bio-fertilizers on the performance of sole and intercropped pigeonpea (*Cajanus cajan*) under rainfed conditions. *Indian Journal of Agronomy* 57(2), 127-132.

Labidi, N., Mahmoudi, H., Dorsaf, M., Slama, I. and Abdelly, C. (2009). Assessment of intervarietal differences in drought tolerance in chickpea using both nodule and plant traits as indicators. *Journal of Plant Breeding and Crop Science* 1(4), 080-086.

Lee, B.R., Lee, D.G., Avice, J.C. and Kim, T.H. (2014). Characterization of vegetative storage protein (VSP) and low molecular proteins induced by water deficit in stolon of white clover. *Biochemical and Biophysical Research Communications* 443(1), 229-233.

Lee, S., Seo, P.J., Lee, H.J. and Park, C.M. (2012). A NAC transcription factor NTL4 promotes reactive oxygen species production during drought-induced leaf senescence in Arabidopsis. *The Plant Journal* 70(5), 831-844.

Leport, L., Turner, N.C., Davies, S.L. and Siddique, K.H.M. (2006). Variation in pod production and abortion among chickpea cultivars under terminal drought. *European Journal of Agronomy* 24(3), 236-246.

Mafakheri, A., Siosemardeh, A.F., Bahramnejad, B., Struik, P.C. and Sohrabi, Y. (2010). Effect of drought stress on yield, proline and chlorophyll contents in three chickpea cultivars. *Australian Journal of Crop Science* 4(8), 580.

Manivannan, P., Jaleel, C.A., Kishorekumar, A., Sankar, B., Somasundaram, R., Sridharan, R. and Panneerselvam, R. (2007a). Changes in antioxidant metabolism of *Vigna unguiculata* (L.) Walp. by propiconazole under water deficit stress. *Colloids and Surfaces B:Biointerfaces* 57(1), 69-74.

Manivannan, P., Jaleel, C.A., Sankar, B., Kishorekumar, A., Somasundaram, R., Lakshmanan, G.A. and Panneerselvam, R. (2007b). Growth, biochemical modifications and proline metabolism in *Helianthus annuus* L. as induced by drought stress. *Colloids and Surfaces B:Biointerfaces* 59(2), 141-149.

Marquez-Garcia, B., Shaw, D., Cooper, J.W., Karpinska, B., Quain, M.D., Makgopa, E.M. and Foyer, C.H. (2015). Redox markers for drought-induced nodule senescence, a process occurring after drought-induced senescence of the lowest leaves in soybean (*Glycine max*). *Annals of Botany* 116(4), 497-510.

Martin, R.J. and Jamieson, P.D. (1996). Effect of timing and intensity of drought on the growth and yield of field peas (*Pisum sativum* L.). *New Zealand Journal of Crop and Horticultural Science* 24(2), 167-174.

Martínez, J.P., Silva, H.F.L.J., Ledent, J.F. and Pinto, M. (2007). Effect of drought stress on the osmotic adjustment, cell wall elasticity and cell volume of six cultivars of common beans (*Phaseolus vulgaris* L.). *European Journal of Agronomy* 26(1), 30-38.

Maurya, D.K., Hasanain, M., Verma, S.K., Kumar, P., Singh, V.K., Singh, S. and Mishra, R. (2020). Seed priming and its effect on enhancing pulse productivity. *Food andScientific Reports* 1, 20-22.

Mishra, J.P., Praharaj, C.S. and Singh, K.K. (2012). Enhancing water use efficiency and production potential of chickpea and fieldpea through seed bed configurations and irrigation regimes in North Indian Plains. *Journal of Food Legumes* 25(4), 310-313.

Mohammadi, A., Habibi, D., Rohami, M. and Mafakheri, S. (2011) Effect of drought stress on antioxidant enzymes activity of some chickpea cultivars. *American-Eurasian Journal of Agricultural & Environmental Science* 11(6), 782-785.

Moradi, A., Ahmadi, A.H.M.A.D. and Hossain Zadeh, A. (2008). The effects of different timings and severity of drought stress on gas exchange parameters of mungbean. *Desert* 13(1), 59-66.

Nam, N.H., Chauhan, Y.S. and Johansen, C. (2001). Effect of timing of drought stress on growth and grain yield of extra-short-duration pigeonpea lines. *Journal of Agricultural Science* 136(2), 179-189.

Nayyar, H., Kaur, S., Singh, S. and Upadhyaya, H.D. (2006). Differential sensitivity of Desi (small-seeded) and Kabuli (large-seeded) chickpea genotypes to water stress during seed filling: effects on accumulation of seed reserves and yield. *Journal of the Science of Food and Agriculture* 86(13), 2076-2082.

Nuñez Barrios, A., Hoogenboom, G. and Nesmith, D.S. (2005) Drought stress and the distribution of vegetative and reproductive traits of a bean cultivar. *Scientia Agricola* 62(1), 18-22.

Ogbonnaya, C.I., Sarr, B., Brou, C., Diouf, O., Diop, N.N. and Roy Macauley, H. (2003). Selection of cowpea genotypes in hydroponics, pots and field for drought tolerance. *Crop Science* 43(3), 1114-1120.

Okçu, G., Kaya, M.D. and Atak, M. (2005). Effects of salt and drought stresses on germination and seedling growth of pea (*Pisum sativum* L.). *Turkish Journal of Agriculture and Forestry* 29(4), 237-242.

Pankhurst, C.E. and Sprent, J.I. (1976). Effects of temperature and oxygen tension on the nitrogenase and respiratory activities of turgid and water-stressed soybean and French bean root nodules. *Journal of Experimental Botany* 27(1), 1-9.

Pradhan, J., Katiyar, D. and Hemantaranjan, A. (2019). Drought mitigation strategies in pulses. *The Pharma Innovation Journal* 8(1), 567-576.

Praharaj, C.S., Singh, U., Singh, S.S. and Kumar, N. (2017). Microirrigation in rainfed pigeonpea-Upscaling productivity under Eastern Gangetic Plains with suitable land configuration, population management and supplementary fertigation at critical stages. *Current Science* 112(1), 95-107.

Praharaj, C.S., Singh, U., Singh, S.S., Singh, N.P. and Shivay, Y.S. (2016). Supplementary and life-saving irrigation for enhancing pulses production, productivity and water-use efficiency in India. *Indian Journal of Agronomy* 61, S249-S261.

Praharaj, C.S., Singh, U., Singh, S.S. and Kumar, N. (2015). Growing summer mungbean with less water. *Pulses Newsletter* 26(2): 4.

Purcell, L.C. and King, C.A. (1996). Drought and nitrogen source effects on nitrogen nutrition, seed growth, and yield in soybean. *Journal of Plant Nutrition* 19(6), 969-993.

Purushothaman, R., Zaman-Allah, M., Mallikarjuna, N., Pannirselvam, R., Krishnamurthy, L. and Gowda, C.L.L. (2013). Root anatomical traits and their possible contribution to drought tolerance in grain legumes. *Plant Production Science*16, 1-8.

Ramamoorthy, P., Lakshmanan, K., Upadhyaya, H.D., Vadez, V. and Varshney, R.K. (2016). Shoot traits and their relevance in terminal drought tolerance of chickpea (*Cicer arietinum* L.). *Field Crops Research* 197, 10-27.

Ramos, M.L.G., Gordon, A.J., Minchin, F.R., Sprent, J.I. and Parsons, R. (1999). Effect of water stress on nodule physiology and biochemistry of a drought tolerant cultivar of common bean (*Phaseolus vulgaris* L.). *Annals of Botany* 83(1), 57-63.

Ranawake, A.L., Dahanayaka, N., Amarasingha, U.G.S., Rodrigo, W.D.R J. and Rodrigo, U.T.D. (2011). Effect of water stress on growth and yield of mung bean (*Vigna radiata* L). *Tropical Agricultural Research and Extension* 14(4), 76-79.

Randhawa, N., Kaur, J., Singh, S. and Singh, I. (2014). Growth and yield in Chickpea (*Cicer arietinum* L.) genotypes in response to water stress. *African Journal of Agricultural Research* 9(11), 982-992.

Samarah, N.H., Mullen, R.E., Cianzio, S.R. and Scott, P. (2006). Dehydrin-Like Proteins in Soybean Seeds in Response to Drought Stress during Seed Filling. *Crop Science* 46(5), 2141-2150.

Sankaranarayanan, K., Praharaj, C.S., Nalayini, P., Bandyopadhyay, K.K. and Gopalakrishnan, N. (2010). Low cost drip as a precision irrigation tool in Bt cotton (*Gossypium hirsutum* L.) cultivation. *Indian Journal of Agronomy* 55(4), 312–318.

Scoffoni, C., Vuong, C., Diep, S., Cochard, H. and Sack, L. (2014). Leaf shrinkage with dehydration: coordination with hydraulic vulnerability and drought tolerance. *Plant Physiology* 164(4), 1772-1788.

Serraj, R. (2003). Effects of drought stress on legume symbiotic nitrogen fixation: physiological mechanisms. *Indian Journal of Experimental Botany* 41, 1136-1141.

Shamsi, K., Kobraee, S. And Haghparast, R. (2010). Drought stress mitigation using supplemental irrigation in rainfed chickpea (*Cicer arietinum* L.) varieties in Kermanshah, Iran.*African Journal of Biotechnology*, 9(27), 4197-4203.

Sharma, P., Abrol, V., Sankar, G.R.M. and Singh, B. (2009). Influence of tillage practices and mulching options on productivity, economics and soil physical properties of maize (*Zea mays*)–wheat (*Triticum aestivum*) system. *Indian Journal of Agricultural Sciences* 79(11), 865–870.

Sharp, R.E. and LeNoble, M.E. (2002). ABA, ethylene and the control of shoot and root growth under water stress. *Journal of Experimental Botany* 53(366), 33-37.

Sheoran, I.S. and Saini, H.S. (1996). Drought-induced male sterility in rice: changes in carbohydrate levels and enzyme activities associated with the inhibition of starch accumulation in pollen. *Sexual Plant Reproduction* 9(3), 161-169.

Srivastava R.N., Pathak, K., Aswathi, M.K. and Nema, R.K. (2018). A comprehensive study of haveli system of farming in central India. *International Journal of Agriculture Sciences* 10 (6), 5614-5617.

Subbarao, G.V., Johansen, C., Slinkard, A.E., Nageswara Rao, R.C., Saxena, N.P., Chauhan, Y.S. and Lawn, R.J. (1995). Strategies for improving drought resistance in grain legumes. *Critical Reviews in Plant Sciences* 14(6), 469-523.

Suresh, K. (2016). Morpho-Physiological Indices for Drought Tolerance in Pigeonpea [*Cajanus cajan* L. Millsp.] Genotypes under Rainfed and Irrigated Conditions. *International Journal of Agriculture Sciences* 8, 3367-3370.

Talukdar, D. (2013). Comparative morpho-physiological and biochemical responses of lentil and grass pea genotypes under water stress. *Journal of Natural Science, Biology and Medicine* 4(2), 396.

Tesfahunegn, G.B. and Wortmann, C.S. (2008). Tieridge tillage for high altitude pulse production in northern Ethiopia. *Agronomy Journal* 100(2), 447-453.

Turner, L.R., Holloway Phillips, M.M., Rawnsley, R.P., Donaghy, D.J. and Pembleton, K.G. (2012). The morphological and physiological responses of perennial ryegrass (*Lolium perenne* L.), cocksfoot (*Dactylis glomerata* L.) and tall fescue (*Festuca arundinacea* Schreb.; syn. Schedonorus phoenix Scop.) to variable water availability. *Grass and Forage Science* 67(4), 507-518.

Uddin, S.H.I.H.A.B., Parvin, S.H.A.H.N.A.J. and Awal, M.A. (2013). Morpho-physiological aspects of mungbean (*Vigna radiata* L.) in response to water stress. *International Journal of Agricultural Science and Research* 3(2), 137-148.

Vamsi Krishna, D., Prasada Rao, K. and Eugenia P. Lal. (2018). Physiological characterisation of green gram (*Vigna radiata* L.) genotypes for drought tolerance. *International Journal of Chemical Studies* 6(5), 589-593.

Vanaja, M., Maheswari, M., Sathish, P., Vagheera, P., Lakshmi, N.J., Kumar, G.V. and Sarkar, B. (2015). Genotypic variability in physiological, biomass and yield response to drought stress in pigeonpea. *Physiology and Molecular Biology of Plants* 21(4), 541-549.

Waggoner, P.E. (1966). Decreasing transpiration and the effect upon growth. WV. H. Pierre, D. Kirkliaii, J. Pesek, and R. Sliaw (eds.), *Plant Environment and Efficient Water Use*. Amnerican Society of Agronomists, Medison, Wisconsin. pp 49-72.

Webber, H.A., Madramootoo, C.A., Bourgault, M., Horst, M.G., Stulina, G. and Smith, D.L. (2006). Water use efficiency of common bean and green gram grown using alternate furrow and deficit irrigation. *Agricultural Water Management* 86(3), 259-268.

Yohan, Y., Umamahesh, Sudhakar, P., Reddy, D.M. and Sumathi, V. (2017). Screening of Blackgram genotypes for total drymatter and Yield components under imposition of Moisture stress. *Bulletin of Environment, Pharmacology and Life Sciences 6(3), 602-606.*

Zare, M. (2012). Evaluation of drought tolerance indices for the selection of Iranian barley (*Hordeum vulgare*) cultivars. *African Journal of Biotechnology* 11(93), 15975-15981.

CHAPTER 4

Production management technologies of pulse crops in acidic soils of NE region of India

K.S. Shashidhar[1]*, S. Bhuvaneswari[2], Samuel Jeberson[1], Hemant Kalita[3], L. Robert Singh[2] and A.K. Mishra[1]

1. *Central Agricultural University, Imphal*
2. *ICAR- RC for NEH Region, Manipur Centre, Lamphelpat, Imphal*
3. *Assam Agricultural University, Jorhat*

*Corresponding author email: shashiuas@gmail.com

Abstract

North East Region (NER) is the eastern-most part of India includes eight states, namely Arunachal Pradesh, Assam, Manipur, Meghalaya, Mizoram, Nagaland, Sikkim and Tripura. Major pulse crops cultivated in the region are urdbean, mungbean, rajmash, rice bean, pigeonpea and cowpea during kharif season and field pea, lentil, broad bean, chickpea and lathyrus during rabi season. Also, various locally important minor legume crops are grown both for human consumption as well as for feed/fodder purpose. As the topography is undulating, cultivated land area is limited, pulse crops are grown in small holdings adjacent to major staple crop of rice during kharif season and in rice fallows during rabi. In general, soil is rich in organic carbon content and acidic due to leaching of cations due to high rainfall. Soil acidity is known to reduce the nodulation in pulses necessitating soil amendment with lime. Pulses in this region are grown in total area of 2.41 lakh hectare with a production of 2.84 lakh tonnes and average productivity of 953 kg/ha. The region is 82% deficit in pulse production against its requirement of 8.74 lakh tonnes (2014-15). Hence, region specific production management technologies are highly important to improve the pulse production in the region. This chapter describes different management technologies available in various pulse crops recommended to improve the pulse production in NE region.

Keywords: Acid soils, Production, Pulses, Management, Nodulation, North eastern region, Technology

1. INTRODUCTION

North East Region (NER) of India comprising of 8 states, namely Arunachal Pradesh, Assam, Manipur, Meghalaya, Mizoram, Nagaland, Sikkim and Tripura, represent 8.3% geographical area and 4% populations of the country (Ngachan *et al.*, 2010). Out of the total 26.27 M ha geographical area, around 77% is hills and because of hilly topography only 12% net area (3.152 M ha) is under cultivation (Anup *et al.*, 2016). The majority of the farmers are small and marginal with lesser land holdings trapped in the traditional low input agricultural practices, making the production systems less viable (Barah, 2007).

The major pulses grown in the NER are urdbean, mungbean, rice bean, pigeonpea and cowpea in *kharif* season and field pea, broad bean, lentil, chickpea, frenchbean and lathyrus in *rabi* season (Anon, 2014). In addition, various lesser known pulse crops are also grown mostly in kitchen gardens for vegetable purpose (Table 1). In *jhum* areas with slash and burn practice, rice bean is a predominant pulse crop (Das *et al.*, 2016). In Sikkim, three different seed coloured urdbean are cultivated viz., green seeded '*Pahenlo dal*' (*Vigna mungo* ssp. *viridis*), brown and black seeded '*kalo dal*' (*Vigna mungo* ssp. *niger*).

Table 1. Diverse legume crops grown in NE region

Common Name	Scientific Name	Consumption	Benefits	Reference
Rice bean	*Vigna umbellata*	As a dal soaked overnight and boiled, as dal soups and sauces	Human, animal and soil health improvement	Dillon and Tanwar, 2018
Adzuki bean	*Vigna angularis*	Used as a red bean paste, sprouted or simply boiled	Weight loss, improved digestion and a lower risk of type 2 diabetes and heart disease	www.cabi.org
Broad bean	*Vicia faba*	Beans used in curries and also, fried, and then salted and/or spiced to produce a savory, crunchy snack,	help prevent birth defects, boost immunity, aid weight loss and lower cholesterol levels and blood pressure	
Winged bean	*Phosphocarpus tetragonlobus*	The leaves, flowers, roots, and bean pods can be eaten raw or cooked. Used in curries, stir-fries, and soups	Each of these parts contains vitamin A, vitamin C, calcium and iron	Yang and Tan, 2011
Tree bean	*Parkia roxburghii*	Vegetable is known as *nitta* sprout and it is added to curry. The flower is commonly eaten raw as a side dish in a local delicacy 'singju' or by mixing with traditionally fermented pickles	anti-diabetic, anti-microbial, antioxidants, anti-hypertensive and hypoglycemic effects	Chhikara *et al.*, 2018
Jack bean	*Canavalia ensiformis*	Industrial products such as isolates, starch, flakes, grits and flours can be produced from the bean	Rich in proteins and minerals	Akpapunam *et al.*, 1997
Bakthul bean	*Sesbania grandiflora*	both leaves and flowers are consumed	Helps in reducing reduce blood sugar, improve cholesterol levels and help maintain a healthy gut	www.cabi.org

Zombi pea	*Vigna vaxillata*	The young leaves, young pods, and seeds are also cooked and eaten as a vegetable	Used as a green manure and ground cover crop.	Somta *et al.*, 2019
Sword bean	*Canavalia gladiata*	used as a vegetable	Has medicinal properties	www.cabi.org
Pahenlodal	*Vigna Mungo vr. Viridis*	Consumed as dhal	-	Rahman and Karuppaiyan, 2017
Kalo dal	*Vigna mungo ssp.niger*	Consumed as dal	-	Rahman and Karuppaiyan, 2017
Lima/ Butter bean	*Phaseolus lunatus*	Consumed as cooked beans	-	
Velvet bean	*Dolichos pruriens*	Consumed as cooked bean	regulate blood sugar levels and lowers cholesterol, and insoluble fiber, which adds in the prevention of constipation, digestive disorders, irritable bowel syndrome, and diverticulitis.	Jayalakshmi *et al.*, 2015

2. PULSES PRODUCTION SCENARIO IN NE REGION

Area, production and productivity of different pulses grown in NE states are presented in the Table 2. The region is almost 82% deficit in pulse production against its requirements as per ICMR recommendation (Roy *et al.*, 2017). As the food pattern of people of NE region in general, is non-vegetarian, dependency on pulse as main protein source is comparatively low as in other parts of India.

Table 2. Area, production and productivity of different pulses in NE states (2014-15)

State	Pulses	Area (000 ha)	Production (000 tonnes)	Yield (kg/ha)
Assam	Urdbean	52.83	34.24	648
	Pea	29.95	27.67	924
	Lentil	29.05	22.52	775
	Mungbran	11.28	7.76	687
	Pigeonpea	6.05	5.74	949
	Chickpea	2.09	1.95	935
	Other Pulses	16.96	11.07	653
	Total Pulses	148.22	110.96	749
Meghalaya	**Total Pulses**	8.10	11.30	1405
Manipur	Urdbean	1.30	1.07	825
	Rice bean	1.84	1.41	765

	Pigeonpea	0.50	0.37	740
	Cowpea	0.95	1.07	825
	Mungbean	0.445	1407	765
	Peas	18.10	17.15	947
	Broad bean	6.00	5.19	865
	Lentil	0.56	0.463	826
	Chickpea	0.68	NA	NA
	Total Pulses	30.38	28.13	820
Tripura	**Total Pulses**	11.70	84.00	718
Sikkim	**Total Pulses**	6.30	8.50	925
Nagaland	Pigeonpea	2.97	2.66	890
	Mungbean	0.33	0.33	1000
	Naga Dal /Rice bean	4.71	5.23	1100
	Cow pea	0.48	0.70	1460
	Beans	2.32	3.08	1330
	Rajmash/ Kholar	14.84	18.59	1250
	Horsegram	0.44	0.46	1050
	Pea	7.09	7.64	1080
	Lentil	2.07	1.68	810
	Chickpea	0.74	0.62	840
	Urdbean	0.76	0.661	800
	Total Pulses	36.75	41.60	1055
Total of NE states		241.448*	284.495*	953*
All India (2018-19)		29.36 m ha	25.42 m t	865 kg/ha

(Source: Anon, 2014, Promotion of pulses in NE region, ICAR-IIPR, Kanpur. * Provisional)

The monocropping of rice is the principal cropping system in north eastern states. Legumes are grown as a sole crop, intercrop, catch crop, relay crop, cover crop or green manure crop under sequential cropping/mono-cropping in different agro-ecological sub-regions in NER (Ansari *et al.*, 2017). Sporadically some rice–based double cropping systems viz. rice–field pea/ lentil/ rapeseed mustard/ wheat/ fodder oat etc. also exist under rainfed conditions in the isolated areas. This practice is adopted particularly in the years receiving higher rainfall, which facilitate adequate residual soil moisture required for better establishment of succeeding crop after the harvest of rice. Crops like field pea, lentil, chickpea and lathyrus are commonly grown as succeeding crop by direct seeding on standing rice crop before its harvest (*Utera* or *Paira* cropping) in the areas with adequate irrigation facilities. The cluster demonstration of rice fallow pulses and oilseeds programme of KVKs, rice-utera cropping using lentil demonstrations sponsored by ICARDA through ICAR-RC for NEH region, Manipur centre, under Tribal Sub Plan (TSP) by AICRP (MULLaRP), CAU, Imphal has created awareness among farmers about zero tillage cultivation of pulses in rice fallows. The farmers seldom use fertilizers under this practice of *utera* cropping. Therefore, it is essential to develop improved agro-techniques for improving the productivity of rice-based cropping system.

Commonly practiced cereal cum pulse-based cropping system in the region includes:

- Intercropping of rice bean with maize, pigeonpea with sesame in the upland area during *kharif* season and intercropping of lentil with broad bean and mustard during winter season are commonly practiced.
- Also, maize/rice + soybean intercropping system in the hilly areas where rainfall ranges from 700 to 2000 mm spreading over 6-7 months is observed (Das *et al.*, 2019).
- Cultivation of blackgram, pigeonpea and soybean are taken up on bunds of rice fields in mid-lands and uplands of south and east Sikkim, border districts of Manipur and Nagaland.
- Mixed cropping of frenchbean along with maize observed in Jaintia hills, Ribhoi district of Meghalaya.
- Intercropping upland rice with soybean, frenchbean, or black/green gram in 4:2 ratio is recommended for enhancing the total system productivity per unit of land area (Das *et al.*, 2019).

Farming system approach is followed in the region for developing farm-household system built on the principle of productivity, profitability, stability and sustainability. Among various farming systems promoted, under agri-horti farming systems followed in hill slopes, two-third area (upper side) is covered under horticultural crops for which half-moon terraces and contour bunds are prepared on the hill slope and one-third area towards down side is used for the cultivation of cereals, pulses and oil seeds on the bench terraces. The interspaces in the contour are utilized for leguminous vegetables like bean, cowpea, guar, pea and cover crops like rice bean, blackgram and greengram (Sahoo *et al.*, 2015). The intercrops validated for mango based agri-horti systems in hilly tracts revealed that the pulse crops like cowpea, blackgram and greengram recorded highest system productivity in Manipur (Anon, 2019)

3. SOIL ACIDITY AND ITS EFFECT ON NUTRIENT AVAILABILITY

In pulses, soil pH determines the amount of nitrogen fixation, metabolic activity, nature of mineral nutrition, and nutrient availability which in turn influence the productivity. Majority of leguminous plants grow and develop better in neutral soils. Rhizosphere microbiocenoses are involved in the transformation of substances and actively affect soil composition and acidity (Zavalin *et al.*, 2019).

The soil type in the NER varies from clay to clay sandy loams which are acidic (4-6 pH) in nature. More than 80% cultivated area in the region is under acidic soils resulting in induced fertility stresses (Patiram, 2007). Phosphorus is the major nutrient required for protein synthesis in pulses. Soil acidity leads to aluminium (Al toxicity) which in turn causes conversion of phosphorus to unavailable form, restricting availability and uptake by pulses. Also, uptake of other nutrients like calcium, potassium, iron, sodium and boron is reduced by the root cells due to reduced cytoplasm membrane permeability. Soil acidity affects the mobility of heavy metals like manganese and their acquisition by plants' roots. The soil acidity also leads to poor absorption of magnesium. The ammonical and nitrate form influence the ionic balance in legumes which are mainly controlled by soil acidity and alkalinity. These ionic changes affect the plant assimilation capacity and thus dry matter accumulation is varied in legumes (Halen and Mark, 2018). Nitrate-based products are the least acidifying of the nitrogen fertilizers, while ammonium-based products have the greatest potential to acidify soil. Monoammonium phosphate (MAP), single superphosphate (SSP) and triple superphosphate (TSP) all add P to soil in the form of the $H_2PO_4^-$ ion, which can acidify soil with a pH greater than 7.2 but has no effect on soil pH in acidic soils. Application of DAP is restricted under acid soils due to increase in soil acidity. However, phosphate fertilizers like SSP and TSP which contains sulphur has beneficial effect on legumes. The form in which K is added to soil, either muriate of potash (KCl) or sulfate of potash (K_2SO_4), has no effect on soil acidification. The effectiveness of fertilizer sources on pulses is not well defined and no specific report on fertilizer sources on productivity of pulses is available.

4. SOIL ACIDITY MANAGEMENT FOR IMPROVED PRODUCTIVITY OF PULSES

Active nodulation is essential for growth and development of pulse crops. The soil conditions to which the rhizobia and host plant are exposed influence the success of the complex nodulation process. Most of the pulse crops and their associated rhizobia are sensitive to soil acidity. The nitrogen fixation varied in acid soils due to non-availability of molybdenum which is integral part of nitrogenase enzyme. Thus, correcting soil acidity improves the nodulation process which in turn increases the growth and vigour of plants.

For higher productivity, the acid soils in NE region should be ameliorated using lime. This improves nutrient availability in terms of their solubility and uptake into the plant system. Lime application in general, helps in increasing the availability of the secondary nutrients like calcium and magnesium. Recommended lime application ranges from 200-500 kg/ha in furrows once in two years depending upon the soil acidity and pulse crop to be grown. In urdbean 200 kg lime application hectare and seed treatment with sodium molybdate 4 g/kg seed helps to improve the nodulation and yield (Anon, 2014). Liming also helps in elevating the leghaemoglobin content in nodules and thus fixing more nitrogen in the soils.

Treating the pulse seeds with rhizobium and phosphorus solubilizing bacteria (*Pseudomonas striata, Bacillus polymixa, Aspergillus awamori*) helps in solubilizing phosphorus and its better acquisition. This in turn enhances the nodulation in pulses increasing the efficiency of nitrogen fixation. Our studies in urdbean, mungbean, lentil and field pea using the new strain of microbial consortia LNm 16* (Ludhiana Nutrient mobilizer 16) improved yield from 10-23% (Anon, 2017). Also, the combined application of calcium and magnesium increases the yield of the above-ground biomass in leguminous plants (Zavalin, 2019).

4.1 Variety selection

Pulse crops in general show variation for acidic soil adaption. Pulses like rajmash, pigeonpea and pea can be grown very well in soil with medium level of soil acidity (pH 5.5 to 6.0). In strongly acidic soils (pH up to 5.1) crop like cowpea, mungbean, urdbean, rice bean, rajmash etc. can be promoted. The recommended varieties in different pulses for the NER are given in the Table 3.

Table 3. List of recommended varieties of different pulse crops for NE region

Crop	Recommended varieties
Pigeonpea	UPAS 120, AL 201, VLA 1
Urdbean	IPU 2-43, Pant U 31, WBU 108, WBU 109, Uttara,
Mungbean	IPM 2-3, HUM 12, Meha, Samrat, IPM 2-14, IPM 205-7, IPM 2-14
Cowpea	V-385, V-282, V271, V269, Rogi, S-488
Rice bean	RCRB 1-6, RBL1, MNPL 1, PRR1, RBL 6, MNPL 2
Chickpea	*Desi*: IPC 97-67, Pusa 372, Pant G 186, Rajas, JG 14, JSC 55, JSC 56 *Kabuli*: Shubhra
Fieldpea	Aman, HUDP 15, TRCP 8, TRCP 9, IPFD 10-12, Pant P 74, Azad, Ambika (Vegetable pea)
Lentil	DPL 62, Tripura Lentil Sel. 1, HUL 57, Moitree, WBL 58 (Subrata), PL 7, PL 8, VL Masoor 125, VL Masoor 129
Lathyrus	Ratan (Bio L 212), Prateek, Mahateora
Rajmash	Utkarsh, Arun, Amber, Tripura Rajmash Sel. 1, Kailash, Arka Komal, Naga local, Meghalaya local, Manipuri, Tripura Selection-1, VL-63, PDR-3, PDR-20, PDR 14 (UDAY), HUR 15 (Malviya Rajmash), HUR 137, VL 63,

Source: Das *et al.*, 2016

4.2 Tillage management

Pulses generally do not need fine seed bed. The seedbed preparation is accomplished through primary tillage which consists of ploughing, harrowing, disking, planking etc. The region being high rainfall area requires tillage operations at optimum time failing which the rains hamper the tillage operation during *kharif* season. The objective of tillage is mainly to give anchorage and support to root and shoot growth. In *kharif* pulses, it is required to turn the soil using mould board plough followed by two cross harrowing and planking. This also helps in incorporation of the manures or crop residues, if any, for enriching the soil fertility. In hilly areas farmers still depend on the hand tools and manually perform tillage operations and grow pulses only for family consumption.

In *rabi* pulses, soil turning plough after *kharif* crops and harrowing followed by heavy planking is desired. However, the tillage operations in *rabi* season is mainly dependent on the timely harvest of *kharif* crops. A pre-sowing irrigation would ensure adequate moisture for good germination.

The soil surface crusting is the main problem in the region which affects the germination of the pulses. To tide over this problem, application of lime at the rate of 250 kg/ha along with FYM at 10 t/ha makes the soil loose and friable and would fetch an additional yield of about 15-20%. The encouraging results with the soil application of lime and seed treatment of sodium molybdate 2 g/kg of seeds has given better results in terms of nodulation, yield improvement in particularly *kharif* pulses like greengram and blackgram in our experiments under AICRP (MULLaRP) at Imphal and Shillongini centres (Anon, 2015).

The residue recycling through soil turning plough in pulse and cereal based cropping systems promoted the improvement of productivity, profitability and soil health besides improving the system productivity (Pooniya *et al.*, 2014). The deep ploughing helps in moisture conservation and root establishment and proliferation aiming at higher productivity over shallow cultivation (Vadi *et al.*, 2006). The heavy soils may require initial rains in the season for deep ploughing followed by 2-3 cultivations and harrowing after early shower. Tillage practices also depend upon climate and soil conditions.

Conventional tillage practice has beneficial effect in improving the productivity of lentil due to better aeration and root proliferation leading to extraction of more soil moisture and nutrients (Singh and Singh, 2008). In the recent past, to reduce energy use in tillage operations but still harness higher or equivalent yield, zero tillage practices under conservation agriculture system are being found promising in rice fallows (Das *et al.*, 2014) which is discussed in detail later in the chapter.

4.3 Sowing time

Sowing time has significant effect on crop phenophases, canopy development and insect-pest-weed dynamics influencing the crop productivity. This is one the major non-monetary inputs to be followed to reduce the cost of cultivation in pulses. The recommended optimum sowing time in various pulse crops in NER is given in the table 4. It is well established fact that delayed planting restricts vegetative phenophases and in turn reducing the pod bearing branches and forced maturity reduce the quality of the produce. Improper sowing time also decreases biological-nitrogen fixation and poor nodulation in legumes. Simultaneously, it has been reported to increase in incidence of pests, especially pod borer (*Helicoverpa armigera*) in chickpea (Ali *et al.*, 1998).

Off-season cultivation of peas, beans, cowpea for vegetable purpose fetches premium prices in the market though there will be reduction in yield (Pooniya *et al.*, 2015; Rahi *et al.*, 2013).

Summer pulses have wide potential in the whole north-eastern region. The early onset of pre-monsoon shower helps in better establishment of crops. However, the rainfall during harvesting stage may affect the summer crop. Sowing of summer greengram in first fortnight of March recorded higher yield as compared to last week of March (Patel *et al.*, 2003). Suitable time for summer blackgram is March (Jaiswal, 1995).

The climate resilient short duration crops like blackgram and greengram have vast sowing window. The varieties with 60-80 days crop duration are best fit in multiple cropping system. Thus, the crops can be sown from June-July months and can be extended up to September in *kharif* season. There are number of popular varieties of greengram like DGGS-4, Pusa Vishal, HUM 12, IPM 2-3 have the duration of 60-65 days with the yield potential of 1-1.2 t/ha and black gram varieties like Uttara, Pant U 31 (Anon, 2019). The crop like rajmash is grown during spring season in North eastern region. It is sown in the month of March and harvested during May-June months. This crop is also harvested for fresh pods which fetch good price in the market. This crop is also grown as winter crop in Assam and Tripura.

Table 4. Normal sowing time of various kharif and rabi and pulses in NE region

Season	Crop	Sowing time	
		From	Up to
Kharif / Rainy season	Urdbean	July II fortnight	September II fortnight
	Mungbean		
	Pigeonpea	May-June	July
Rabi/ Winter season	Lentil	October II fortnight	November II fortnight
	Pea		
	Chickpea		
	Grasspea		
Summer season	Urdbean	February II fortnight	March II fortnight
	Mungbean		
	Rajmash**		

**Rajmash is grown as winter season pulse in Assam and Tripura

4.4 Methods of sowing

Adoption of appropriate sowing method governs the success and productivity of crops in particular environmental, temporal and field variability regimes (Choudhary and Suri, 2014). The sowing method varies from crop to crop and season to season. This is the most important management practice for ensuring cultural operations, seed rate, plant establishment and effective utilization of resources. The sowing depth depends on type of crop/variety, season of cultivation, soil type and the seed size of the respective pulse crops. As a general rule sowing depth of 2.5 times the seed diameter of the respective crop is followed.

Since land holdings in the region are small, therefore farmers still follow broadcasting of pulse seeds in the seed beds followed by planking. The drilling of seeds in the furrow opened by plough is most commonly practiced. However, due to scientific interventions, the awareness is being created towards the use of seed drills and improvement of seed cum fertilizer drill/zero till planters to take up sowing of lentils, chickpea and field peas through KVKs.

In *kharif* pulses, raised bed/ridge-furrow planting technique has been found very effective in upland areas due to proper drainage of excess water from crop root zone under high rainfall and thus, 20-25% yield increased is observed over flatbed planting (Pramanik and Singh, 2000; Das *et al*., 2014). The yield improvement is due to adequate drainage and better aeration in the route zone. Raised and sunken bed technology has been standardized for crop intensification in lowlands for pulse intercropping with rice (Das *et al*., 2014). Cultivation of vegetables on raised beds in lowlands has been reported to enhance system productivity and income substantially over rice monocropping. Cabbage + pea (43.29 t/ha), cauliflower + methi (22.36 t/ha) and broccoli + pea (17.85 t/ha) recorded the highest system productivity in mid-altitude of Meghalaya (Das *et al*., 2019).

In *rabi* pulses, after harvesting of the *kharif* crops like rice/ maize the field should be immediately prepared for efficient utilization of residual soil moisture. The sowing can be done in rows behind plough or seed drill. North-south row orientation is generally beneficial over East-west orientation in terms of better light interception in crop rows.

4.5 Seed rate and planting geometry

The seed rate varies according to weather conditions and duration of crop growth. The plant population desired per unit of land area determines the seed rate (Prasad, 2012). Seed requirement varies for intercropping systems and sole cropping. It depends on the proportion of land to be occupied by the crop per unit area. In hill and mountain ecosystems, the seed requirements of pulses are higher due to varied moisture conditions leads to failure of germination. However, the higher rainfall, low temperature and bright sunshine are adequate enough to translate into higher crop yields (Choudhary, 2013). The seed rate and spacing recommended in pulse crops is given in the table 5.

The spacing of crops may vary based on the canopy architecture of the crop/variety for example, the tall variety of field pea needs spacing of 45 cm × 10 cm. Most of short duration pulse varieties need narrow spacing, while long duration varieties perform well under wider spacing. Sowing in first week of June recorded highest grain yield with both narrow and wider spacing in different varieties and sowing beyond that date reduced the pulse yield (Padhi, 1995). Thus the kharif crops requires more spacing than *rabi* and summer grown pulse crops as the availability of the resources and climatic factors helps in better establishment of crop leading to increase in branching and pod development which is most important to maintain yield levels in pulse crops (Prasad, 2012).

Table 5. Optimum spacing and seed rate recommended in pulses

Season	Crop	Seed rate (kg/ha)	Spacing (cm)
Kharif / Rainy season	Urdbean	15-20	30 × 10; 45 × 10
	Mungbean	15-20	30 × 10; 45 × 10
	Pigeonpea	12-15	60 × 20; 75 × 20
Rabi / Winter season	Lentil	30-40 50-60 (Late)	22.5 × 10
	Pea	75-80	30 × 10 (dwarf); 45 × 10 (tall)
	Chickpea	75-100	30 × 10
	Grasspea	50-60	30 × 10
Summer season	Urdbean	-	
	Mungbean	25-30	20 × 10; 30 × 10
	Rajmash	75-100	45 × 10

Source: Adopted from technical programmes of FLD under different AICRPs of ICAR-IIPR, Kanpur. 9711192196

4.6 Water management

In general, pulses require less water compared to cereals and oil seed crops. *Kharif* pulses in the region are grown as rainfed crops without any irrigation. However, due to heavy rainfall, proper drainage to remove excess moisture is required since pulses are very sensitive to water stagnation. The occasional dry spells in the *kharif* season affects pulse crops which can be managed through life saving irrigation using harvested rain water from farm ponds and *Jalkund.* Pulses cultivated during winter and spring/summer months require lifesaving irrigation. Experimental findings have shown that for winter pulses two to three irrigations (at germination, flowering and pod development stages) helps in sustaining the productivity (Anon, 2012). The use of furrow irrigation with raised-bed (FIRB) system improves the irrigation water-use efficiency where tillage is done on top of the beds (Lumpkin and Sayre, 2009). This system could further be explored and validated for pulse-based cropping systems to improve the productivity of rainfed areas. Application microbial consortia like of VAM fungi and other PSB culture has significance in improving the tolerance to water deficit in additions to releasing the complex form of phosphorus under rainfed pulses like field pea (Kumar *et al.*, 2014). Pre-*kharif*/summer greengram crop planted during the February-March needs 3 irrigations at critical stages at 10-15 days interval. Irrigation should be avoided during active flowering period otherwise flower drop and reversion to vegetative growth may occur. Good drainage facilities should be provided for higher yield.

4.7 Nutrient management

Pulses are efficient in fixing atmospheric nitrogen except rajmash. They seldom require any additional nitrogen. Phosphorus is often considered as prime limiting nutrient in pulses due to their high P requirements. Where P and/or pH levels are low, applying P near the seed in the root zone at the time of sowing is recommended. Potassium is another important macronutrient required for N fixation. Seed coating of lime and molybdenum improves the uptake of nutrient in pulses like greengram, blackgram, cowpea and horsegram (Padhi *et al.*, 2018).

Based on the AICRP agronomic experiments on nutrient management conducted in different pulse crops, the fertilizer dose recommendation is given in the table 6. On an average 15-20 kg nitrogen, 40-50 kg phosphorus and 15-20 kg potassium per ha should be applied for good yield of pulses. Rajmash requires higher dose of nitrogen compared to other pulses i.e. 100-120 kg/ha. Sulphur application of 20-25 kg/ha resulted in higher seed yield. Soil application of zinc sulphate 20-25 kg/ha or foliar application of 0.5% $ZnSO_4$ with 0.05% lime at pre-flowering/pod initiation stage have proved to be effective in controlling Zn- deficiency. Application of organic manures 5-10 t/ha in sandy soil increased the seed yield to a greater extent. The fertilizer should be applied either at the time of sowing or just before sowing in such a way that they are placed about 5-8 cm below the seed. Band placement of fertilizers is recommended over broadcasting for better results.

The foliar nutrition during pre-flowering and pod initiation has resulted in increase in yield of 5-10% in mungbean and urdbean at Imphal and Shillongini (Anon 2014). Thus, foliar application of 2% neem coated urea/DAP/19:19:19 has given encouraging results in greengram, blackgram, field pea and lentil crops in NE region.

Table 6. Nutrient requirement of different pulse crop in NE region

Season	Crop	Nutrient requirement (kg/ha)			
		Nitrogen	Phosphorus	Potassium	Sulphur
Kharif / Rainy season	Urdbean	18	46	20	20
	Mungbean	18	46	20	20
	Pigeonpea	25	50	25	20
Rabi / Winter season	Lentil	18	46	20	20
	Pea	18	45	20	20
	Chickpea	20	40	20	20
	Grasspea	20	40	20	20
Summer season	Urdbean	20	40	20	20
	Mungbean	20	40	20	20
	Rajmash	120	60	40	20

Source: Technical programmes of FLD programme of AICRP (MULLaRP) and chickpea of ICAR-IIPR Kanpur

4.8 Weed management

The most sensitive period of pulses for weed competition is initial 30-45 days of the crop. Unchecked weeds cause 20-90% yield loss in different pulse crops (Poonia *et al.*, 2015). Since the rainfall is higher with optimum growing conditions during *kharif* season, a number of weeds are existing in pulse-based cropping which compete with main crops for light, nutrient and moisture. Experiments were conducted under the AICRP on pigeonpea, MULLaRP and chickpea to identify the most effective strategy of efficient weed management. Farmers are reluctant to apply herbicide to control weeds in NER, though it is most effective, since mostly organic production system is preferred and also due to poor supply of selective herbicides for pulses in the market. *Eleucine indica, Digitaria marginata, Panicum psilopodium, Cynodon dactylon, lmperata cylindrica, Cyperus iria, Cyperus rotundus,* among grasses and *Bidden spilosa, Galinsoga parviflcra, Ambrossia sp., Ageratum conizoides, Eupatorium odoratum, Polygonum sp., Boerhaavia hispida, Mimosa pudica, Amranthus viridi. Commelina benghalensis, Euphorbia hirta, Portulaca oleracea, Argemone mexicana* among broad leaved weeds seen in *kharif* season and *Chenopodium album, Phlaris minor, Lathyrus spp,. Melilotus alba, Vicia sativa* are important weeds seen in *rabi* crops.

Use of herbicides for weed control has detrimental effect on the water bodies lying in the down streams of the valley region. It is essential to be cautious in adoption of chemical weed management options. Adoption of integrated weed management (IWM) where herbicide can be used at low rates by integration of mechanical methods helps in evolving eco-friendly options. In pulses, cultural practices like crop rotation, optimum dates of sowing, planting geometry, seed rates, methods of fertilizer application, etc. can be integrated with chemical or hand hoeing methods.

Two mechanical/hand weeding, one at 15-20 DAS and another at 35-40 DAS are recommended in greengram and blackgram (Shashidhar *et al.*, 2020). Further pre-emergent application of pendimethalin + imazethapyr (ready mix) 1.0 kg/ha with hand weeding at 20-30 DAS was equally effective in controlling weeds in other pulse crops. Some important post emergent herbicides like imazethapyr, quizolofop-ethyl, clodinafop-propargyl 8% + acifluorfen sodium 16.5% (ready mix) 125-187.5 g/ha at 15-20 DAS found to control weeds effectively in *kharif* pulse crops. Most of the post emergent herbicides are very effective in controlling grassy weeds.

4.9 Pest and disease management

The yield losses due to pests and diseases vary from 46 to 96% depending on the crop and variety of pulse crops (Pooniya *et al.*, 2015). The climatic conditions of NER favour diverse group of pests infesting to pulse crops (Sahay *et al.*, 1999). The pest problem also varies across altitude. The most important insect-pest of pigeonpea in NER is pod boring weevil (*Apion clavipes*). The other pests of pigeonpea are pod borer (*Heliothis armigera* and *Etiella zinckenellaj*), pod fly (*Agromyza obtusa*), blister beetles (*Mylabris phalerata, M. pustulataj*), aphids (*Myzus persicae*) and flea beetles (*Chaetocnema basalis*) found in almost all NE states. In dry areas, particularly in Tripura and Assam, the pests like termites (*Odontotermes* sp.) are serious pests in pigeonpea under upland/hilly conditions.

Integrated pest management (IPM) in pulses refers to application of an inter-connected set of principle and methods to minimize problems caused by insects, diseases, weeds and other agricultural pests. IPM includes deep summer ploughing, field sanitation, conservation of natural enemies like coccinellid beetles, syrphid beetles, hymenopteran parasitoid and predators, use of resistant/tolerant varieties, crop rotation with non-host crops in the field, intercropping practices like chickpea + linseed/mustard or chickpea + coriander encourages natural enemy population of pod borers. Entamophagous fungus like *Beaveria bassiana* helps in management of adult weevils. This can be sprayed at the rate of 10^9 spores/ml to control the pod boring weevil. Pea aphid fungus like *Eriniumneo aphid* can be used to control aphids (Azad Thakur and Firke, 2013). Use of bio-insecticide NPV 250-500 LE/ha controls pod borers. Use of neem seed kernel extract (5%) and pheromone trap is also helpful in controlling of pod borers (Pooniya *et al.*, 2015). Application of neem oil 0.3% or neem-seed kernel extract (NSKE) 5% can be used for effective control of white fly in blackgram. Sood *et al* (2013) from a study at Sikkim reported that pheromone-based-fruit fly management is one of the low-cost management option in beans (Sood *et al.*, 2013).

The north eastern region receive average annual rainfall of more than 200 cm and remain humid almost throughout the year which is major pre-disposing factor for a number of pathogens causing disease in pulses. As per the survey reports in NER for a number of years as reported by Sahay *et al* (1999) revealed that leaf spot, powdery mildew in greengram and blackgram, wilt and anthracnose in pigeonpea; leaf spot, anthracnose, rust, pod blight in frenchbean; leaf spot in cowpea; powdery mildew, leaf blight in peas; rust, powdery mildew and *Cercospora* leaf spot in rice bean, field peas are major diseases (Sahay *et al.*, 1999).

Integrated disease management (IDM) need to be followed to control the diseases in pulse crops in NER. In Arunachal Pradesh, straw mulching reduced the incidence of *Ascochyta blight* in field pea due to temperature variations and reduction in humidity suppressed the growth of pathogen (Das *et al.*, 2016). The foliar application of wettable sulphur 0.25% or eucalyptus leaf extract 10% in the initial stages of powdery mildew attack in pulses has been recorded effective in Sikkim (Babu *et al.*, 2015). The deep summer ploughing, field sanitation, adoption of resistant varieties, seed treatment with fungicides and fungal based spores, crop rotations with sorghum and tobacco, soil solarization and soil sterilization with formaldehyde, neem cake application 150 kg/ha helps to reduce root-rot (Pooniya *et al.*, 2015) in pulses. Arbuscular mycorrhizal fungi also induce disease tolerance in pulse crops (Kumar *et al.*, 2014). Seed treatment with carbendazim + thiram and bio-agent (*Trichoderma viride*) in combination with vitavax are also found effective in reducing the incidence of wilt in pulses.

5. POTENTIAL AREA EXPANSION UNDER PULSES IN NE REGION

NE region has about 4.5 M ha area under net cultivation including about 0.76 M ha area under *jhum* of which 3.5 M ha area is under rice and 0.23 M ha under maize. Out of 3.5 M ha area under rice, about 2.33 M ha rice fallow area can be brought under double cropping (Nagchan *et al.*, 2010). Thus, growing of pulses in non-traditional areas like rice fallow (pea and lentil after rice), *jhum* areas (frenchbean / rice bean / pigeonpea after maize), maize fallow areas in upland/terraced fields (frenchbean / blackgram / greengram / cowpea) (Table 7) can increase the area and production of pulses in NER. Also, area available under field bunds (blackgram/greengram especially on rice bunds) and field boundaries (rice bean / pigeonpea) can be also brought under pulses cultivation and area expansion.

Table 7. The potential area expansion under pulses in NER under different ecosystems

Crop	Ecosystem	State
Lentil	Rice fallow areas	Assam, Manipur, Tripura, Parts of Meghalaya (Garo Hills)
Grass pea (lathyrus)	Rice fallow areas	Assam, Tripura, Manipur
Pea	Rice fallow areas	All NE states
Mungbean	After maize	Assam, All NE states
Urdbean	After maize	Assam, All NE states
Frenchbean/Rajma	After maize in uplands/Jhum lands	Assam, All NE states
Rice bean	Uplands/jhum	All NE states

Selection of early varieties of pulses like lentil, field pea and lathyrus will enable the crop to escape from terminal moisture stress in rice-fallow cultivation (Erskine *et al.*, 2011) and provides an effective option to convert mono-cropped rice field into double cropped areas with rice-pulse crop sequence beside increasing the soil fertility and cropping intensity in the region (Gulab *et al.*, 2020).

Pulse in pre-rabi and summer season like short duration greengram and blackgram, frenchbean for vegetable purpose can be promoted in NER. Summer greengram is one of the multipurpose crop which helps in improving soil fertility and productivity of the system. Since most of the production systems, especially in hilly areas in the whole NER, are under organic cultivation, the organic pulse production may be encouraged with policy support for export from the region which in turn can fetch better price.

6. ZERO TILLAGE – AN INNOVATIVE APPROACH TO INCREASE AREA UNDER PULSES IN NE REGION

Zero tillage (ZT) is an innovative approach to increase area under pulses besides exploiting the residual moisture and nutrients in the rice fallows. This practice helps in timely planting of *rabi* pulses and saves time, energy besides reducing the cost of cultivation. The success depends on the effective management practices followed during cultivation. Transplanting rice seedling with a spacing of 20x20 cm or 25x25 cm, draining of excess water from the field at physiological maturity stage of rice to be followed to create favorable condition for sowing of pulses like field pea, lentil, chickpea and grass pea. These crops can be sown with ZT seed drill or manual sowing by maintaining 20-30 cm height of rice stubbles. The stubbles help in conserving the moisture and provide favourable condition for the growth and development of pulse crops under residual soil moisture in rice fallow. A common dose of 20 kg N, 60 kg P_2O_5 and 40 kg K_2O can be applied at the time of sowing under both minimum tillage and ZT practices. Two foliar spray of 2% DAP/urea, one at flowering and other at pod filling stages of crops, need to be applied for achieving better yield. Foliar nutrition of 19:19:19 NPK water soluble fertilizer @ 2% at pre-flowering and pod formation is recommended for better results under Assam and Manipur condition under AICRP experiments. When soil moisture content is very less and crop suffers from moisture stress, application of salicylic acid 75 ppm at pre-flowering and pod initiation stage improves the pod development and yield. The experiments conducted at CAU Imphal under AICRP (Dryland Agriculture) reported that using rice variety CAU R-1 in *kharif* followed by 150% seed rate (120 kg/ha) of field pea (variety Aman) and mulching with available paddy straw (4 Mg/ha) helps in increasing 15% system productivity and Rs. 6910/- additional net returns (Anon, 2019). Foliar application of 0.5% 19:19:19 NPK or 2% neem coated urea at branching stage and 15 days later and seed priming of field pea with sodium molybdate 0.5 g/kg seed would be helpful for increasing the productivity of the rice-field pea cropping sequence in foot hills of Manipur (Anon, 2020). Similarly, inclusion of pea, lentil and rapeseed with minimum tillage or ZT and mulching under rice fallow system recorded highest net return (Rs. 71.28×10^3, Rs. 72.65×10^3 and Rs. 29.26×10^3, respectively) (Baishya *et al.*, 2019)

7. POST-HARVEST TECHNOLOGY AND VALUE ADDITION

Harvesting is done either by collecting/picking pods separately or by harvesting whole plant at a time. However, this will be based on the availability of labour and extent of the area cultivated. The plants are cut as close as ground level and left in sun for drying for few days on threshing floor. Threshing should be done either manually (beating the pods with sticks) or by trampling under the feet of bullocks. The mechanical threshing has advantage in terms of reducing labour costs. The seeds to pod ratio is generally 40-60% in pulses. The threshed and cleaned produce should be sundried to bring down the moisture to 10-11% for safe storage. The humidity and rainfall are high in the region which makes pulse seeds storage very challenging.

Value addition is one of most important activities after harvest of pulses. Under AICRP (MULLaRP), CAU- Imphal has demonstrated and distributed IIPR Mini Dhal Mill to 4 village clusters under tribal area sub plan (TSP) that can be used for processing pulses into dal. Similar other value addition activities like making papad, badi, etc need to be promoted for popularizing pulses in NER.

8. CONCLUSION

Successful pulses cultivation to achieve maximum production and income depends on the available resources and their proper utilization by the farming community. North-eastern region is largely deficit in pulse production necessitating implementation of different production technologies recommended for acidic soils and high rainfall area. Area expansion in non-traditional agro-ecosystems like rice fallows, and inclusion of pulses with staple food crops like rice and maize, and horticultural crops helps in efficient utilization of available resources. Best management practices details given in this chapter would help in increasing area and production of pulses under diverse agro-ecologies of NER.

REFERENCES

Akpapunam, M.A., Sefa-Dedeh, S. (1997). Jack bean (*Canavalia ensiformis*): Nutrition related aspects and needed nutrition research. *Plant Food Human Nutrition* **50,** 93-99.

Ansari, M.A, Saraswat, P.K., Roy, S.S., Sharma, S.K., Punitha, P., Ansari, M.H., Prakash, N., Mishra, R.K., Lal, N. and Ramakrishna, Y. (2017). Significance and strategies of legume production for achieving nutritional security in North East Indian Himalayan Region. *Journal of Food Legumes* 30(3), 56-63,

Das, A., Layek, J., Babu, S., Krishnappa, R., ThoithoiDevi, M., Kumar, A., Patel, D.P., Ramkrushna, G.I., Yadav, G.S., Sarika, K., Tripathi, A.K., Ghosh, P.K. and Prakash, N. (2019) Intercropping for Climate Resilient Agriculture in NE Region of India. *Technical bulletin No 1 (Online). ICAR Research Complex for NE Region, Umiam – 793 103, Meghalaya.*

Das, A., Babu, S., Yadav, G.S., Ansari, M.A., Singh, R., Baishya, L.K., Rajkhowa, D.J. and Ngachan, S.V. (2016). Status and strategies for pulses production for food and nutritional security in north-eastern region of India. *Indian Journal of Agronomy* 61 (Special issue), 43-57.

Azad Thakur, N.S. and Firke, D.M. (2013). Insect pests of north eastern Himalayas and rationale for their management. https://www.krishisewa.com/disease-management/219-pulses-insect-pests.html, accessed on 18-09-2020.

Babu, S., Singh, R., Avasthe, R.K. and Yadav, G.S. (2015). Strategies for enhancing productivity of organic agriculture in North Eastern Region of India. (In) Souvenir and Conference Book of National Conference on Global Research Initiatives for Sustainable Agriculture and Allied Sciences, held at Gwalior, Madhya Pradesh, India, 12–13 December, 2015, pp. 55–58

Baishya, L.K., Rathore, S.S., Sarkar, D., Jamir, T. and Rajkhowa, D.J. (2019). Crop and varietal diversification for enhancing productivity and profitability of rice fallow system in eastern Himalayan region. *Indian Journal of Agricultural Sciences* 89(5), 800-805.

Barah, B.C. (2007). Strategies for Agricultural Development in the North-East India: Challenges and Emerging Opportunities. *Indian Journal of Agricultural Economics* 62(1), 13-31.

Chhikara, N., Devi, H.R., Jaglan, S., Sharma, P., Gupta, P. and Panghal, A. (2018). Bioactive compounds, food applications and health benefits of *Parkia speciosa* (stinky beans): a review. *Agriculture & Food Security* **7,** 46 (2018). https://doi.org/10.1186/s40066-018-0197-x

Choudhary, A.K. (2013). Technological and extension yield gaps in pulses in Mandi district of Himachal Pradesh. *Indian Journal of Soil Conservation* 41(1), 88-97.

Das, A., Patel, D.P., Ramkrushna, G.I., Munda, G.C., Ngachan, S.V., Kumar, M., Buragohain, J. and Naropongla. (2014). Crop diversification, crop and energy productivity under raised and sunken beds: results from a seven-year study in a high rainfall organic production system. *Biological Agriculture and Horticulture* 30(2), 73-87.

Dhillon, P.K. and Tanwar, B. (2018). Rice bean: A healthy and cost-effective alternative for crop and food diversity. *Food Security* 10**,** 525-535. https://doi.org/10.1007/s12571-018-0803-6

Yadav, G.S., Devi, A.G., Kandpal, B., Das, A., Barmanand, K.K. and Babu, S. (2020). Minimum till lentil (*Lens culinaris*): an efficient way for rice fallow utilization and income enhancement in subtropical Tripura. *Indian Journal of Agricultural Sciences* 90(1), 133-137.

Jaiswal, S., Naamala, J. and Dakora, F. (2018). Nature and mechanisms of aluminium toxicity, tolerance and amelioration in symbiotic legumes and rhizobia. *Biology and Fertility of Soils* 54, 309-318. https://doi.org/10.1007/s00374-018-1262-0.

Jayalaxmi, B., Vijayalakshmi, D., Usha, R., Revanna, M.L., Chandru, R. and Ramanjini Gowda, P.H. (2016). Effect of different processing methods on proximate, mineral and antinutrient content in lima bean (*Phaseolus lunatus*) seed. Legume Research 36(4), 543-549.

Kumar, A., Suri, V.K. and Choudhary, A.K. (2014). Influence of inorganic phosphorus, VAM fungi and irrigation regimes on crop productivity and phosphorus transformations in okra (*Abelmoschus esculentus* L.)-pea (*Pisum sativum* L) cropping system in an acid Alfisol. *Communications in Soil Science and Plant Analysis* 45(7), 953-967.

Lumpkin, A.T. and Sayre, K. (2009). Enhancing resource productivity and efficiency through conservation agriculture. (In) Proceedings of 4th World Congress on Conservation Agriculture Innovations for Improving Efficiency, Equity and Environment, 4-7 February 2009, New Delhi, pp 4–9.

Kumar, M., Rajkhowa, D.J., Rajesha, G., Meena, K.L., LireniKikon, E., Chowdhury, P., Lily Rangnamei, K., Zeliang, P.K. and Ezung, T.S. (2016). Production Technology of Pulses, *Krishi Vigyan Kendra (KVK) Longleng-798625*, Nagaland.

Kumar, N., Hazra, K.K., Nath, C.P., Praharaj, C.S. and Singh, U. (2018). Grain Legumes for Resource Conservation and Agricultural Sustainability in South Asia. Legumes for Soil Health and Sustainable Management (eds) R.S. Meena *et al*., Springer Nature Singapore Pte Ltd. 2018. 79-107 pages.

Ngachan, S.V., Anupam, M., Kadirvel, G., Das, A. and Saikia, K. (2010). Conservation of Natural Resources for Sustainable Hill Agriculture. ICAR Research Complex for NE Region, Umiam. pp. 428.

Padhi, A.K. (1995). Effect of sowing date and planting geometry on yield of redgram (*Cajanus cajan*) genotypes. *Indian Journal of Agronomy* 40(1), 72-76.

Patel, J.J., Mevada, K.D. and Chotaliya, R.L. (2003). Response of summer mungbean to date of sowing and level of fertilizers. *Indian Journal of Pulses Research* 16(2), 122-124.

Patiram. (2007) Management and Future Research Strategies for Enhancing Productivity of Crops on the Acid Soils. *Journal of Indian Society of Soil Science* 55(4), 411-420.

Patterson, C.A., Maskus, H. and Dupasquier, C. (2009). Pulse crops for health. AACC International Inc. *Cereals Foods World* 54(3), 108-112.

Pooniya, V., Choudhary, A.K., Sharma, S.N., Bana, R.S., Rana, D.S. and Rana, K.S. (2014). Mungbean (*Vigna radiata*) residue recycling and varietal diversification for enhanced system productivity and profitability in basmati rice (*Oryza sativa*)–wheat (*Triticum aestivum*)–mungbean cropping system. (In) Proceedings of National Symposium on Crop Diversification for Sustainable Livelihood and Environmental Security, held during 18-20 Nov., 2014 at PAU, Ludhiana, pp 6, 29-30.

Pooniya, V., Choudhary, A.K., Dass, A., Bana, R.S., Rana, K.S., Rana, D.S., Tyagi, V.K. and Puniya, M.M. (2015). Improved crop management practices for sustainable pulse production: An Indian perspective. *Indian Journal of Agricultural Sciences* 85(6), 747-758.

Praharaj, C.S. and Singh, N.P. (2019). Raising productivity of pulses in diverse agro-ecologies of North East Hill region of India - A case study. *Journal of Food Legumes* 32(2), 90-96.

Pramanik, S.C. and Singh, N.B. (2000) Boost pulse production through new planting techniques. *Indian Farming* 58(1), 4-6.

Prasad, R. (2012). Textbook of Field Crops Production-Food Grain Crops, Vol I, pp 248–319.

Premaradhya, N., Shashidhar, K.S., Jeberson, S., Krishnappa, R. and Singh, N. (2018). Effect and Profitability of Foliar Application of Thiourea on Growth & Yield Attributes of Lentil (*Lens culinaris* L.) under Manipur Conditions of North-East, India. *International Journal of Current Microbiology and Applied Sciences* 7(5), 1040-1050.

Padhi, P.P., Pattanayak, S.K. and Purushottam. (2018). Effect of lime coating and molybdenum seed treatment on productivity and nutrient uptake of different pulses grown in Alfisols. *International Journal of Chemical Studies* 6(1), 1503-1507.

Rahi, S., Thakur, S.K. and Choudhary, A.K. (2013). Off-season pea cultivation: An income enhancement venture in Mandi district of Himachal Pradesh. (In) Proceedings of National Seminar on Indian Agriculture: Present Situation, Challenges, Remedies and Road Map, held at CSK HPKV, Palampur during 4-5 Aug. 2012, CSK HPKV Publication, pp 47–8.

Rahman, H. and Karuppaiyan, R. (2017). Agrobiodiversity of Sikkim, in *Biodiversity of Sikkim: Exploring and Conserving a Global Hotspot* (eds) M.L. Arrawatia and S. Tambe., Gangtok: Information and Public Relations Department, Government of Sikkim. 403-426 pages.

Roy, A., Singh, N.U., Tripathi, A.K., Yumnam, A., Sinha, P.K., Kumar, B. and Das, S.K. (2017). Dynamics of Pulse Production in North-East Region of India-A State-wise Analysis. *Economic Affairs* 62(4), 655-662.

Sahay, G., Sarma, B.K., Gupta, H.S., Pathak, K.A. and Prasad, M.S. (1999). Biotic Stresses of Pulses in North Eastern Hill Region of India. *Indian Journal of Hill Farming* 12(1 & 2), 8-16.

Sahoo, M.R., Roy, S.S., Prakash, N. and Ngachan, S.V. (2015). Horticulture-based farming system for sustainability and higher profitability, in *A Comprehensive Manual on Integrated Farming System: An Approach towards Livelihood Security and Natural Resource Conservation* (eds) N. Prakash, S.S. Roy, M.A. Ansari and S.K. Sharma. Publication No. RCM (TM)-08. ICAR Research Complex for NEH Region, Manipur Centre, Lamphelpat, Imphal - 795 004 (India). 46-54 pages.

Shashidhar, K.S., Samuel Jeberson, M., Premaradhya, N., Singh, A.K. and Bhuvaneswari, S. (2020). Weed management effect in blackgram under acidic soils of Manipur. *Indian Journal of Weed Science* 52(2), 147-152.

Singh, V.P. and Singh, V.K. (2008). Effect of tillage practices and seed rates on the performance of large seeded lentil. *Journal of Food Legume* 21(1), 49-50.

Somta, P., Dachapak, S., Yimram, T., Srinives, P. and Poonchaivilaisak, S. (2019). Genetic diversity of zombi pea (*Vigna vexillata*) assessed by microsatellite markers. *Acta Horticulturae* 1241, 143-150. DOI: 10.17660/ActaHortic.2019.1241.21

Sood, P., Yadav, D.S., Thakur, S.K., Choudhary, A.K., Rahi, S. and Chauhan, K. (2013). Pheromone based fruit fly management for sustainability–A case study. (In) Proceedings of National Seminar on Indian Agriculture: Present Situation, Challenges, Remedies and Road Map, held at Chaudhary Sarvan Kumar Himachal Pradesh Krishi Viswavidyalaya, Palampur during 4-5 August, 2012, pp. 25–28.

Babu, S., Singh, R., Avaste, R.K. and Gopi, N. (2015). Phanlo dal cultivation under organic no till condition, technical bulletin, ICAR-RC for NEH region, Sikkim.

Vadi, H.D., Kachot, N.A., Polara, J.V., Sekh, M.A. and Kikani, V.L. (2006). Effect of tillage and mulching on yield and yield attributing characters of pigeon pea. *Advances in Plant Sciences* 19(2), 497-499.

Valentine, A., Benedito, V. and Kang, Y. (2010). Legume Nitrogen Fixation and Soil Abiotic Stress: From Physiology to Genomics and Beyond. *Annual Plant Review* 42, 207-248.

Yang, J. and Tan, H. (2011). Winged Bean Milk. International Conference on New Technology of Agricultural Engineering, Zibo. pp. 814–817. doi:10.1109/ICAE.2011.5943916.

Zavalin, A.A., Sokolov, O.A., Shmyreva, N.Ya. and Lukin. S.V. (2019). Legume reaction to soil acidity. *Amazonia Investiga* 8(23), 162-170.

CHAPTER 5

Prospects of grass pea in rice fallows of eastern India

Purabi Banerjee*, Bishal Mukherjee and Rajib Nath

Department of Agronomy, Bidhan Chandra Krishi Viswavidyalaya, Mohanpur-741252, Nadia, West Bengal, India

*Corresponding author email: itsmepurabi1@gmail.com

Abstract

Rice growing areas of eastern India consist of sufficient residual soil moisture after harvest, which can be utilized by cultivating low water requiring pulse crops. Grass pea (*Lathyrus sativus* L.), a multi-purpose and climate smart grain legume crop, offers a perfect choice for sustainable food production under this situation. It has wider adaptability to diverse soil types, including marginal lands. It also holds wonderful tolerance to several stresses like drought, water logging and salinity, even with minimum supply of external inputs. However, presence of a neurotoxin (β-ODAP) has created hindrance to its cultivation and consumption in the country. Adoption of improvised agro-techniques like *Rhizobhium* and PSB inoculation, nutrient seed priming, foliar fertilization and integrated pest management ramify upliftment of production. Strategies like development of grass pea varieties with low ODAP content coupled with multiple stress tolerance, effective seed detoxification methods, improved agronomic and breeding technologies to enhance yield require immediate attention. Thus, the agricultural science can provide a bright future to this promising crop for the sake of the economically weaker section of the society in Eastern India.

***Key words*:** Eastern India, Grass pea, Rice fallow, Relay crop, Stress tolerant, Utera

1. INTRODUCTION

The term 'rice fallows' applies to those lowland rainfed rice growing areas that are remained uncropped after rainy season rice crop (Ghosh *et al.*, 2016). These areas are mostly concentrated in Eastern India, accounting for about 84% (9.7 M ha) of total rice fallow areas (11.7 M ha) in the country (Ali *et al.*, 2014). Some major limitations such as lack of irrigation, fast depletion of the soil residual moisture and poor socio-economic condition of the farmers along with some regional constraints like free grazing of cattle have been observed under rice fallow situation (NAAS, 2013). Excessive soil moisture at the time of harvesting of monsoon rice in lowlands poses a potential threat to the sowing operations of successive crops. On the contrary, delayed harvesting of long-duration rice (150–160 days) restricts the sowing of winter crops in optimum time. This in turn aggravates seasonal moisture and temperature stresses in winter crops (Kumar *et al*., 2018a, 2018b) during later stages of their life cycle. Thus, combination of all these factors invariably affects the productivity of successive relay crops (Pande *et al*., 2012; Singh *et al*., 2016). After harvesting of rice crop in Eastern India, soil moisture reserve varies from 150 to 200 mm (Bandyopadhyay *et al*., 2016), whereas normal cumulative evapotranspiration (ET) is around 370 mm (Sarkar and Singh, 2007). Hostile soil physical condition due to puddling, soil compaction and acidic soil reaction further contribute to the poor performance of winter crops under rice fallows (Kumar *et al*., 2019).

In spite of several challenges in rice fallows, there is an extensive scope for growing relay pulse crops after rice with the dual objectives of crop diversification for sustainable production and enhancing productivity through area expansion of pulses to meet nutritional demand of human being (Singh *et al*., 2017). Inclusion of short duration legumes like lentil, chickpea, grass pea offered excellent opportunity of area expansion (Ghosh *et al*., 2014) and exploitation of residual soil moisture in rice fallow (Kar and Kumar, 2009). Characteristics of these crops such as deep root system, ability to utilize the moisture from deeper soil layers, low water requirement and crop establishment by surface seeding are the desirable characters for a moisture-limited environment of the rice fallows (Hazra *et al*., 2014; Kumar *et al*., 2016). Even this fact is equally applicable in case of coastal saline tracts of Eastern India (Sarkar *et al*., 2019).

Characterized by water scarcity, the dry areas are also taken into dispute by ever-increasing population, frequent droughts, unpredictable climatic phenomenon, land degradation and unstable economy. In this context, grass pea is an important component in production systems (Anonymous, 2010). Lower productivity, especially under *utera* system and presence of an anti-nutritional factor β-ODAP causing a neurotoxic disorder in monogastric animals including humans are the major problems associated with grass pea. Despite these bottlenecks, grass pea is still grown on a large area by the farmers. The reason behind this preference may be its hardy nature and least cost of cultivation (Hillocks and Maruthi, 2012), still offering a good quality feed for cattle. Grass pea renders an attractive choice for sustainable food production, owing to its intrinsic properties including limited water requirement than that of chickpea and lentil (Varma, 2019). Grass pea is inherently capable of withstanding temperature extremes. In addition to remarkable tolerance to drought (Talukdar, 2009) even at the later stages of crop growth, this crop has tolerance to water logging conditions (Girma and Korbu, 2012). It has a hardy and penetrating root system suited to a wide range of soil types including very poor soil and heavy clays (Parihar *et al*., 2013). Furthermore, late sowing cannot consequence in yield loss in grass pea unlike the other legumes. Its striking sturdiness along with intrinsic ability to biologically fix atmospheric nitrogen makes grass pea an attractive choice for adverse agricultural circumstances (Dixit *et al*., 2016).

2. CONSTRAINTS RELATED TO GRASS PEA CULTIVATION IN EASTERN INDIA

In the eastern part of the country, grass pea crop experiences a number of production constraints along with the common obstacles faced under rice fallow cultivation to reach the optimum yield potential. These are listed as follows:

(*i*) One of the major hurdle for grass pea cultivation is the presence of a neurotoxin known as β-N-Oxalyl-L-α,β-diaminopropionic acid (β-ODAP) in its seeds and plant system causing a neurological disease, lathyrism, if over consumed.

(*ii*) Grass pea is grown predominantly under rainfed conditions on marginal and sub-marginal lands which are usually characterized by poor soil fertility.

(*iii*) Inadequate plant population of grass pea under *utera* system.

(*iv*) Inadequate seed supply of improved and high yielding varieties.

(*v*) Delayed sowing and low seed rates.

(*vi*) Lack of good quality seed because of no systematic seed production of any of the released varieties of grass pea.

(*vii*) Though the crop establishes well with residual soil moisture, lack of winter rains from the very beginning to reproductive stage often leads to substantial loss of yield.

(*viii*) Poor soil fertility status coupled with inadequate or imbalanced use of fertilizers expose the crop to nutrient starvation.

(*ix*) Farmers rarely adopt novel practices like application of micronutrients, owing to their poor economic status. This compels the crop to suffer from the location specific nutrient deficiencies (P, Zn, S, B and Mo) in Eastern India.

(*x*) Unavailability of efficient strains of *Rhizobium* biofertilizerfor commercial use.

(*xi*) Weed infestation at the initial stage.

(*xii*) Lack of plant protection measures adopted by the farmers to encounter biotic stresses including diseases like powdery mildew (*Erysiphe pisi*), rust (*Uromyces fabae*) and downy mildew (*Peronospora* sp.) and insect pests like black aphids (*Aphis crassivora*), pod borer (*Helicoverpa armigera*) and thrips (*Caliothrips indicus*) causing considerable economic losses. Side by side, two species of nematodes (*M. incognita* and *R. reniformis*) are associated with this crop.

(*xiii*) Abiotic factors like soil moisture stress and water logging reduce the yield potential.

(*xiv*) Sometimes, animal grazing also poses a potential threat to grass pea in rice fallows.

(*xv*) Lack of awareness and knowledge about improved scientific production technologies of grass pea under rice fallows are very much predominant among farmers.

(*xvi*) Socio-economic constraints like poor purchasing capability of farmers, fragmented land holding, shortage of labour, unavailability of inputs, limited access to institutional credit and lack of market directly or indirectly discourage the farmers to go for grass pea cultivation after *kharif* rice.

(*xvii*) Irregular transfer of applicable technologies due to poor extension service.

3. ECONOMIC IMPORTANCE OF GRASS PEA

Grass pea (*Lathyrus sativus* L.) is an annual pulse crop belonging to the family *Fabaceae* and the tribe *Vicieae* (McCutchan, 2003; Gurung and Pang, 2010). It is commonly known by different names like grass pea, chickling pea, chickling vetch, dogtooth pea, grass pea vine, Indian pea, riga pea and wedge pea vine, *khesari*, *teora*, *lakhodi*, *lakh,* etc. (Chatterjee *et al*., 2019). Predominantly due to the high protein content (18-34%) in seeds of grass pea, it is valued as nutritious staple food and fodder (Kumar and Tripathy, 2009). Grass pea seeds are also rich in minerals (Table 1).

Table 1. Different components present in grass pea seeds

Components	Range (%)	Components	Range (mg/kg)
Starch	48.0-52.3	Calcium	0.07-0.12
Protein	25.6-28.4	Phosphorus	0.37-0.49
Acid detergent fibre	4.3-7.3	Lysine	18.4-20.4
Water	7.5-8.2	Threonine	10.2-11.5
Ash	2.9-4.6	Methionine	2.5-2.8
Fat	0.58-0.8	Cysteine	3.8-4.3

Source: Yerra and Kilary (2018)

Obviously, grass pea being a leguminous crop, fixes atmospheric N_2 and improves soil carbon and organic matter content. Thus, it collectively improves soil health and provides sustainable production systems. It is thought of as an '*insurance crop*' as it produces reliable yields when all other crops fail (Kalita and Chakrabarty, 2017). Grass pea is a remarkable drought-tolerant crop (Gusmao *et al.*, 2012) combined with outstanding ability to thrive with minimal external inputs (Nazrul and Shaheb, 2015). Therefore, nutrient rich grass pea crop is likely to play a key role in providing nutritional and livelihood security to the economically backward communities through alleviation of global malnutrition (Dixit *et al.*, 2016). Grass pea tender green leaves/plants, pods and seeds are consumed in different ways (Table 2).

Table 2. Different ways of consumption and uses of grass pea

Product	Usage
Leafy vegetable	Young leaves are plucked and used as green leafy vegetable; rolled and dried leaves are kept for off-season use
Pods	Green pods can be eaten directly as snacks; whole pods are cooked and eaten as vegetable
Dal	Dried grains are split to prepare *dal* and consumed with rice
Flour	Flour is used to prepare pancake like preparation; is used to adulterate high-priced legume flours made from chickpea or mungbean seeds
Feed	Ground splitted grain or flour are used as animal feed
Fodder	Grass pea plant at every stage of its growth can be used as forage

Source: Rizvi *et al.* (2016)

The seed protein of grass pea is composed of albumins (14%), globulins (66%), glutelins (15%), and prolamins (5%) (Lambeine *et al.*, 2019). It is rich in lysine unlike in most cereals, but deficient in sulphur containing essential amino acids like methionine and cysteine, (Mahler-Slasky and Kislev, 2010) and vitamin C, D and B_1 (Piergiovanni *et al.*, 2013). Moreover, the nutritional potential of grass pea is also limited by the combined action of several anti-nutritional compounds such as digestive enzyme inhibitors, BOAA, lectins, tannins, polyphenols and phytate (Enneking, 2011). However, the deficiency of essential sulphur containing amino acids, such as methionine that plays a pivotal role in the central nervous system, may be overcome with a balanced diet containing cereals (Lambein and Kuo, 2007). Larger proportion of the fatty acids of grass pea being polyunsaturated (58%) has made it more beneficial for human consumption (Grela *et al.*, 2010). Besides, it is the unique dietary source of the amino acid l-homoarginine which is of great importance in treatment of cardiovascular disease (Van Wyk *et al.*, 2016) and in overcoming the development of cancer tumour due to hypoxia at the tissue level (Jammulamadaka *et al.*, 2011). Interestingly, grass pea seeds are consisting of a striking amount of polyphenol and antioxidant that makes the crop potent enough for phytomedicinal and pharmacological use. The variety Ratan (Bio L-212) has performed best in this regard (Talukdar, 2012). Therefore, grass pea having nutraceutical properties, is an excellent example of a potential '*functional food*' (Singh and Rao, 2013; Llorent-Martínez *et al.* 2017).

4. ANTI-NUTRITIONAL FACTORS IN GRASS PEA

There is great potential to explore the dimensions of cultivation of grass pea in drier tracts which are becoming more droughts prone as a result of climate change. However, it has an ambivalent reputation due to which the production and the bright prospects of this valuable crop are being restrained. The seeds of the plant contain certain amount of a non-protein amino acid or more specifically, glutamate analogue neurotoxin named as β-N-oxalylamino-L-alanine (BOAA) or β-N-Oxalyl L-α, β-diaminopropionic acid (β-ODAP) that may cause neuro-lathyrism in humans and domestic animals (Khandare *et al.*, 2014; Patto and Rubiales, 2014; Gupta *et al.*, 2018). It is considered as a '*nutritional curse*' if consumed in excess (2/3rd of daily diet) for a prolonged period of 4-5 months continuously (Hillocks and Maruthi, 2012). Research indicates that β-ODAP forms chelates with bivalent cations such as zinc (Zn), manganese (Mn), and copper (Cu) which are essential for the activity of enzymes like superoxide dismutase (SOD) and peroxidase (POX). These enzymes have been found to protect plant body against the abrasive effects of reactive oxygen species (ROS) that play a key role in imparting oxidative stress (Getahun *et al.*, 2005). Neuro-lathyrismis a spastic para-paresis due to hyper-excitation of motor neurons resulting in spastic movements and crippling of the legs (lower limbs) in up to 6% of the population, mainly young male persons (Kumar *et al.*, 2011). Association between overconsumption of grass pea and neurolathyrism has come into light since pre-historic period. In fact, it has been described by famous Greek physician Hippocrates, as well as by ancient Indian documents. Still the conditions for disease development have been adebatable issue (Enneking, 2011). However, some cases of neurolathyrism disease were reported from various parts of India during 1960s, where it was consumed as staple food for survival, not merely as a pulse crop (Anonymous, 2010).

Despite this stigma, β-ODAP in grass pea has also been depicted as a multifunctional plant metabolite. One of the major physiological functions with possible therapeutic potentials is the activation of protein kinase C, which adds a new dimension to explore its potential in the treatment of Alzheimer's disease, hypoxia and long term strengthening of neurons essential for memory (Lambein *et al.*, 2019). Experimental evidences stipulate that β-ODAP being water soluble and heat sensitive amino acid, its large proportion is lost during traditional method of *dal* making. Steeping grass pea seeds in excess water leaches out upto 30% of the β-ODAP (Teshager, 2019). The enhancement of the bioavailability of nutrients contained in grass pea seeds may be achieved through two possible detoxification processes (Mahata *et al.*, 2018):

(a) *Genetic detoxification*: This method involves breeding and selection for low-ODAP (preferably < 0.1%) and dual purpose grass pea lines through classical hybridization and somaclonal variation.

(b) *Physico-chemical detoxification:* This involves educating the consumers, especially housewives in simple and effective steps of detoxification as follows:

(*i*) Boiling the seeds in water and draining away the first wash

(*ii*) Soaking the seeds overnight and draining away of excess water

(*iii*) Steeping the dehusked seed in water

(*iv*) Parboiling the seeds, resembling to parboiling of rice

(*v*) Roasting at about 150°C for 20 minutes

5. MODERN AGRO-TECHNIQUES

Profitable husbandry of a crop even like grass pea with wider adaptability, especially under rice-fallow conditions, requires major technological advancement for sustaining pulse production in the country. Grass pea is one of the vigourous legumes commonly grown in marginal lands which are generally considered unsuitable for cultivation of other legumes (Ali *et al.*, 2000). It is mostly grown on the residual soil moisture in rice-fallows as *utera* (relay) crop (Bhowmick *et al.*, 2014). In this context, lower productivity especially under *utera* system is the major problem associated with this crop.

5.1 Climatic and soil requirements

Grass pea tolerates a wide range of environmental conditions and climates but to achieve economic yields requires good growing conditions and management. Ideal growing conditions help to drop the levels of β-ODAP content of grass pea considerably as compared to grown under any kind of stress (Hillocks and Maruthi, 2012). Research efforts have found that grass pea crop generally can thrive in any type of soil whatever fertile or not, except highly acidic one.

5.2 Land preparation

Relay sowing does not require any land preparation. But one deep ploughing followed by cross harrowing and planking is required for line sowing of grass pea.

5.3 Sowing

Grass pea is generally sown in the month of October or November and harvested in February or March. Sowing date is largely determined by the soil type and soil moisture profile when the monsoon rains end. Seeding can be done using zero-till seed drill @ 40-50 kg/ha instead of *paira* cultivation about two weeks after harvesting of paddy crops as it gives better yields as compared to *utera* method of its cultivation. However, farmers generally broadcast grass pea seeds @ 80 kg/ha on standing rice crop field at suitable moisture condition as *paira* or *utera* crop in Eastern India.

5.4 Seed treatment

Seed treatment with fungicides should be done to encounter soil borne diseases like, *Fusarium* wilt, downy mildew and seed or seedling rots disease attack, ensuring higher yield. Generally after rice cultivation, the field lacks the native *Rhizobhium* along with mineral nutrients like P, Zn, Mo, B, etc. All of these are very much essential for proper development of pulse crops. Pelleting of seeds with *Rhizobhium* and phosphate solubilizing bacteria (PSB) culture improves the establishment, nodulation and grain yield of pulses under rice fallows. Seed treatment with Carbendazim (1 g a.i./kg seed) + Thiram (2 g a.i./kg seed) should be done a week before sowing followed by seed pelleting with *Rhizobhium* inoculum and PSB culture @15-20 g and 5-7 g/kg seed, respectively, just before sowing.

Table 3. Some improved varieties of grass pea

Variety	β – ODAP (%)	Average Yield (kg/ha)
Pusa 24 (P24)	0.20(duel)	1655
Bio L 202 (Ratan)	0.05	2530
Prateek (LS82046 × A60)	0.08	1560
Mahateora (Ratan × JRL2)	0.07	1550
Bio L 208 (Moti)	0.03	2900
Nirmal	0.15 (duel)	1500
Local cultivars	0.50-2.50	455-640

Source: Mahata *et al.* (2018)

5.5 Nutrient management

Mostly for sowing of grass pea after vacating the field by previous rice crop, a basal dose of N:P:K @ 20:60:40 kg/ha is applied at the time of sowing. Organic manures like phosphocompost or FYM or green manure @ 5 t/ha should be used as integrated nutrient management (INM) components. Generally, the soils of rice fallow regions are poor in some macro- and micro-nutrients like N, P, Mo, Zn, B, etc., but in case of relay cropping, the standing rice crop compels to completely exclude the basal application of fertilizers at the time of sowing of grass pea, drastically reducing its yield. In this unique context, seed priming can serve as a simple but cost-effective agronomic practice to improve the seed germination properties, to meet the nutrient demand of the crop in the early growth stages and eventually increasing the final yield. Application of various plant nutrients through seed priming *viz.*, sodium molybdate or ammonium molybdate @ 0.5 g/kg seed, KH_2PO_4 or $NaHPO_4$@ 20 g/kg seed have been reported earlier to increase the nodulation by 80-90% and grain yield up to 30%. Another productive intervention in this regard may be foliar spray of nutrient solutions at pre-flowering and pod initiation stages providing instant nutrition directly to the site of photosynthesis. Foliar feeding of 2% urea, DAP, KCL, 0.5% NPK (19:19:19) and micronutrients with special reference to B, Zn, Mo and Fe can be promoted to meet up the nitrogen demand after the active nodulation declines (Table 4).

Table 4. Effect of seed priming and foliar nutrition on seed yield of grass pea

Treatment	Seed yield (kg/ha)	
	1st Year	2nd Year
Seed priming		
No soaking	730.7	902.5
Water soaking	802.4	938.3
2% KH_2PO_4 soaking	910.0	1019.3
Sprouted seeds	938.2	1105.3
LSD (p=0.05)	52.5	53.3
Foliar nutrition		
No spray	687.7	902.5
Water spray	798.3	938.3
2% Urea spray	963.3	1116.7
2% DAP spray	912.5	1055.0
2% KCl spray	877.8	985.8
LSD (p = 0.05)	58.7	59.6

Source: Bhowmick *et al.* (2014)

5.6 Water management

Grass pea is grown across diverse regions that receive an average annual precipitation ranging from 300 to 1500 mm (Dixit *et al.*, 2016). The fast growing nature of this crop can effectively utilize residual soil moisture for their growth and development. In fact, the early ground cover may reduce the evaporative loss of soil moisture (Bandyopadhyay *et al.*, 2016). Thus, proper soil moisture is necessary for establishment and early growth (Table 5). Researchers have indicated that this crop can flourish well even under complete exclusion of irrigation in case of relay sowing. Nevertheless, one irrigation at pre-flowering period may result in remunerative production of grass pea. Use of conservation tillage (zero tillage with at least one-third rice residue retention) can also be promoted to enhance the yields under rainfed rice-fallow situations (Kumar *et al.*, 2018).

Table 5. Effect of different sowing methods and pre-emergence irrigation on yield of grass pea

Treatment	Seed yield (t/ha)	Stover yield (t/ha)
Broadcast sowing without irrigation	1.29b	3.76
Furrow sowing without irrigation	1.31b	3.62
Furrow sowing with furrow irrigation	1.69a	3.64
Broadcast sowing with primed seeds	1.64a	3.63
Furrow sowing with primed seeds	1.43b	3.63
Broadcast sowing with sprinkler irrigation at evening	1.67a	3.72

Broadcast sowing with sprinkler irrigation at evening and morning	1.75a	3.91
Furrow sowing with sprinkler irrigation at morning	1.66a	3.39
Broadcast sowing with post sowing flood irrigation	1.68a	3.60
CV (%)	5.72	10.52
Level of significance	0.01	NS

Source: Ahmed *et al.* (2019)

5.7 Weed management

Weeds drastically affect pulse production. The problem gets more complex as the habitat changes from waterlogged rice field to aerobic rainfed rice-fallows. Generally under rice-fallows, pulse crops are heavily infested with weeds of diverse nature. In order to get maximum yield benefits, the crops must be kept free from weeds. But grass pea can effectively smother weeds due to its fast spreading habit and bushy stature. Nevertheless, one hand weeding at 3 to 4 weeks after sowing proves an effective control of weeds like *Cyperus rotundus, Cynodon dactylon, Melilotus alba/indica, Chenopodium album, Vicia hirsuta,* etc. at initial stages of grass pea. In case of sole cropping of grass pea after rice, pendimethalin @ 1.5 kg a.i/ha can be applied immediately after sowing, followed by application of post-emergence herbicide quizalofop-p-ethyl (Targa super) @ 40-50 g a.i/ha at 15-20 DAS. Only post-emergence application of quizalofop-p-ethyl may be done under relay system.

5.8 Pest management

Grass pea being able to withstand several biotic stresses much better than the other pulses, should be better exploited through extensive cultivation of this crop. Integrated management of diseases and pests including use of healthy seeds, modification of cultural practices, and judicious use of chemical and bio-control agents may contribute substantially in stabilizing the yields. But still for a better crop husbandry, some chemical and biological control measures may be adopted. Spraying with 5% crude neem extract or 2% neem oil @ 3000 ppm or dimethoate 30 EC @ 1.7 ml/l and imidacloprid 17.8 SL @ 0.2 ml/l should be done for aphid management. Use of coccinellid beetles, their grubs and *Chrysoperla* may be done for bio-control of aphid. Foliar spray of wettable sulphur (Sulfex/Thiovit) @ 3 kg ha, or Carbendazin 50WP @ 1 g/l (0.05%), Benlate (0.05%) and Topsin-M (0.15%) are found to be efficient for management of powdery mildew. Spray with NSKE @ 50 g/l or eucalyptus leaf extract 10% at initiation of the disease may also be done.

6. CONCLUSION

Rice fallow, a rainfed lowland agro-ecology, is presently gaining particular attention for sustainable cropping intensification in the South Asia (Kumar *et al.*, 2019). Promotion of pulses in rice-fallows may enhance to meet up the nutritional demand of the country. In this context, grass pea is a perfect option with a combination of exceptional agronomic properties including tolerance to drought, flooding, salinity, a high nitrogen fixation capacity that increases the yield of subsequent crops, easy cultivation with minimal inputs, and adaptability to different climates and soils. As a legume, it is also a highly nutritious food and fodder crop. Grass pea holds a tremendous potential as a source of multiple stress-tolerant genes for general crop improvement (Hao *et al*., 2017). Considered as one of the most climate-resilient crop, grass pea survives in the field during severely dry years, thus saving lives of thousands of people during famines. The basis of the high tolerance to drought and salinity should be further investigated and the crop could be used as an elite germplasm resource for traits to adapt the world's most important crops to new climatic conditions

7. FUTURE PROSPECTS

Grass pea is one of the hardiest but most underutilized crops to acclimatize in fragile agro-ecosystems. Growing a legume crop like grass pea after *kharif* rice would have beneficial effects on soil fertility and soil health. Relay cropping of lathyrus, especially in medium-deep soils, could transform the mono-cropped rice fallow areas into double-cropping areas, increasing the productivity and sustainability of the agricultural system (Dixit *et al*., 2016). Besides, there is a great potential for the expansion in the utilization of grass pea in dry areas or zones which are becoming more drought prone, with increased salinity or increased tendency to suffer from biotic stresses (Almeida *et al*., 2015). An immediate attention is required by some areas of experimentations such as, systematic seed production of the released varieties of grass pea, development of low level β-ODAP genotypes with resistance against biotic stresses, accuracy of β-ODAP detoxification method, understanding of drought-tolerance mechanisms, breeding for increased forage and fodder production, refinement in production technologies used for *utera* cultivation, etc. Dissemination of these technologies are urgently needed to ensure the supply of cheaper source of protein and better returns to the farming community in adverse agro-climatic situations in India (Anonymous, 2010). Research on the use of grass pea for preparation of animal feed will be of great importance in order to stimulate the expansion of grass pea cultivation in sustainable and low-input agricultural systems (Chatterjee *et al*., 2019).

Progress in crop improvement of grass pea using traditional and molecular breeding tools has been slow. There is a tremendous need to develop a more comprehensive genetic map for grass pea with mapped positions of useful genes for marker-assisted breeding. Further, recent progress in protoplast fusion and anthers culture requires further advancement before these methodologies maybe used for grass pea improvement (Gurung and Pang, 2011). Relatively little research efforts have been devoted to improve this prospective crop. Sporadic efforts on its improvement through manipulation of genetic and agronomic interventions should be undertaken on large scale basis. To exploit the potential of rice fallow areas, location specific and economically viable technologies for better performance are required to standardize through proper understanding of the system ecology study. These efforts should have strong research support to deliver appropriate varieties and matching crop management technology suited to rice fallow cultivation.

REFERENCES

Ahmed, B., Khatun, K., Islam, M.R., Rahman, M.S., Hassan, M.Z. and Biplob, M.H.J. (2019) Growth and Yield Performance of Grass pea (*Lathyrus sativus* L.) under Different Methods of Sowing and Pre-Emergence Irrigation. *Advances in Biological Research* 13(3), 89-94.

Ali, M., Joshi, P.K., Pande, S., Asokan, M., Virmani, S.M., Kumar, R. and Kandpal, B.K. (2000) Legumes in the Indo-Gangetic Plain of India. In: C Johansen, JM Duxbury, SM Virmani, CLL Gowda, S Pande, and PK Joshi (Eds), Legumes in Rice and Wheat Cropping Systems of the Indo-Gangetic Plain-Constraints and Opportunities. International Crops Research Institute for the Semi-Arid Tropics, Cornell University. pp. 35-70.

Ali, M., Ghosh, P.K. and Hazra, K.K. (2014) Resource conservation technologies in rice fallow. In: Ghosh *et al.* (Eds), Resource Conservation Technology in Pulses. Scientific Publishers, Jodhpur. pp. 83-89.

Almeida, N.F., Rubiales, D. and Carlota, M. and Patto, V. (2015) Grass Pea. In: AMD Ron (Eds.), Grain Legumes, Handbook of Plant Breeding. 10: 251-262. DOI 10.1007/978-1-4939-2797-5_8

Anonymous. (2010) A Project Proposal for Financial Assistance under the NFSM-PULSES on Enhancing grass pea production for safe human food, animal feed, and sustainable rice-based production systems in India. Department of Agriculture and Cooperation (Crop Division), Ministry of Agriculture, Govt. of India.

Bandyopadhyay, P.K., Singh, K.C., Mondal, K., Nath, R., Ghosh, P.K., Kumar, N., Basu, P.S. and Singh, S.S. (2016) Effects of stubble length of rice in mitigating soil moisture stress and on yield of lentil (*Lens culinaris* Medik) in rice-lentil relay crop. *Agricultural Water Management* 173, 91–102.

Bhowmick, M.K., Dhara, M.C., Duary, B., Biswas, P.K. and Bhattacharyya, P. (2014) Improvement of lathyrus productivity through seed priming and foliar nutrition under rice-*utera* system. *Journal of Crop and Weed* 10(2), 277-280.

Chatterjee, C., Debnath, M., Karmakar, N. and Sadhukhan, R. (2019) Stability of grass pea (*Lathyrus sativus* L.) genotypes in different agroclimatic zone in eastern part of India with special reference to West Bengal. *Genetic Resources and Crop Evolution.* https://doi.org/10.1007/s10722-019-00809-2

Dixit, G.P., Parihar, A.K., Bohra, A. and Singh, N.P. (2016) Achievements and prospects of grass pea (*Lathyrus sativus* L.) improvement for sustainable food production. *The Crop Journal* 4, 407- 416.

Enneking, D. (2011) The nutritive value of grass pea (*Lathyrus sativus*) and allied species, their toxicity to animals and the role of malnutrition in neurolathyrism. *Food and Chemical Toxicology* 49, 694–709.

Getahun, H., Lambein, F., Vanhoorne, M. and Stuyft, P.V.D. (2005) Neurolathyrism risk depends on type of grass pea preparation and on mixing with cereals and antioxidants. *Tropical Medicine & International Health* 10, 169-178.

Ghosh, P.K., Hazra, K.K., Nath, C.P., Das, A. and Acharya, C.L. (2016) Scope, constraints and challenges of intensifying rice (*Oryza sativa*) fallows through pulses. In: PK

Ghosh, N Kumar, MS Venkatesh, KK Hazra and N Nadarajan (Eds). Scientific Publishers, Jodhpur, Rajasthan, India.

Girma, D. and Korbu, L. (2012) Genetic improvement of grass pea (*Lathyrus sativus*) in Ethiopia: an unfulfilled promise. *Plant Breeding* 131, 231–236.

Grela, E.R., Rybiński, W., Klebaniuk, R. and Matras, J. (2010) Morphological characteristics of some accessions of grass pea (*Lathyrus sativus* L.) grown in Europe and nutritional traits of their seeds. *Genetic Resource and Crop Evolution* 57, 693-701.

Gupta, P., Udupaa, S.M., Gupta, D.S. and Kumar, J. (2018) Population structure analysis and determination of neurotoxin content in a set of grass pea (*Lathyrus sativus* L.) accessions of Bangladesh origin. *The Crop Journal* 6, 435-442.

Gurung, A.M. and Pang, E.C.K. (2011) Lathyrus. In: C Kole (Eds.), Wild Crop Relatives: Genomic and Breeding Resources, Legume Crops and Forages. *Springer-Verlag Berlin Heidelberg*. DOI 10.1007/978-3-642-14387-8_6

Gusmao, M., Siddique, K.H.M., Flower, K., Nesbitt, H. and Veneklaas, E.J. (2012) Water deficit during the reproductive period of grass pea (*Lathyrus sativus* L.) reduced grain yield but maintained seed size. *Journal of Agronomy and Crop Science* 198, 430-441.

Hao, X., Yang, T., Liu, R., Hu, J., Yao, Y., Burlyaeva, M., Wang, Y., Ren, G., Zhang, H., Wang, D., Chang, J. and Zong, X. (2017) An RNA sequencing transcriptome analysis of grass pea (*Lathyrus sativus* L.) and development of SSR and KASP markers. *Frontiers in Plant Science* 8, 1873. DOI: https ://doi.org/10.3389/fpls.2017.01873

Hazra, K.K., Venkatesh, M.S., Ghosh, P.K., Ganeshamurthy, A.N., Kumar, N., Nadarajan, N. and Singh, A.B. (2014) Long-term effect of pulse crops inclusion on soil-plant nutrient dynamics in puddled rice (*Oryza sativa* L.)-wheat (*Triticum aestivum* L.) cropping system on an Inceptisol of Indo-Gangetic Plain zone of India. *Nutrient Cycling in Agroecosystems* 100(1), 95–110.

Hillocks, R.J. and Maruthi, M.N. (2012) Grass pea (*Lathyrus sativus*): Is there a case for further crop improvement?. *Euphytica* 186, 647–654.

Jammulamadaka, N., Burgula, S., Medisetty, R., Ilavazhagan, G., Rao, S. and Singh, S.S. (2011) β-N-Oxalyl-l-α, β-diaminopropionic acid regulates mitogen- activated protein kinase signalling by down-regulation of phosphatidylethanolamine-binding protein. *Journal of Neurochemistry* 118, 176-186.

Kalita, H. and Chakrabarty, R. (2017) Performance of grass pea (*Lathyrus sativus* L.) varieties under different seed rates in rice (*Oryza sativa*) – *utera* situations. *Journal of Crop and Weed* 13(1), 113-115.

Kar, G. and Kumar, A. (2009) Evaluation of post-rainy season crops with residual soil moisture and different tillage methods in rice fallow of eastern India. *Agricultural Water Management* 96(6), 931–938.

Khandare, A.L., Babu, J.J., Ankulu, M., Aparna, N., Shirfule, A. and Rao, G.S. (2014) Grass pea consumption & present scenario of neurolathyrism in Maharashtra state of India. *Indian Journal of Medical Research* 140, 96-101.

Kumar, G. and Tripathi, R. (2009) Influence of heat stress on genome of grass pea (*Lathyrus sativus* L.). *Journal of Environmental Biology* 30(3), 405-408.

Kumar, N., Hazra, K.K., Nadarajan, N. and Singh, S. (2016) Constraints and prospects of growing pulses in rice fallows of India. *Indian Farming* 66(6), 13–16.

Kumar, N., Yadav, A., Singh, S. and Yadav, S.L. (2018) Growing Pulses in Rice Fallow: Ensuring Nutritional Security in India. In : A Das, KP Mohapatra, SV Ngachan, AS Panwar, DJ Rajkhowa, GI Ramkrushna and J Layek (Eds.), Conservation Agriculture for Advancing Food Security in Changing Climate. Today & Tomorrow's Printers and Publishers, New Delhi - 110 002, India. pp. 107-122.

Kumar, R., Mishra, J.S. and Kumar, R. (2018a) Enhancing the productivity of rice fallows area of Eastern India through inclusion of pulses. *Indian Farming* 68(8), 7–10.

Kumar, R., Mishra, J.S., Rao, K.K., Kumar, R., Singh, S.K. and Bhatt, B.P. (2018b) Evaluation of crop establishment techniques in rice-fallows of Eastern Indo-Gangetic Plains. In: Proceeding of National Conference on Organic Waste management for Food and Environmental Security, 8–10February 2018, ICAR-IISS Bhopal, Madhya Pradesh, India.

Kumar, R., Mishra, J.S., Rao, K.K., Bhatt, B.P., Hazra, K.K., Hans, H. and Mondal, S. (2019) Sustainable intensification of rice fallows of Eastern India with suitable winter crop and appropriate crop establishment technique. *Environmental Science and Pollution Research*. https://doi.org/10.1007/s11356-019-06063-4

Kumar, S., Bejiga, G., Ahmed, S., Nakkoul, H. and Sarker, A. (2011) Genetic improvement of grass pea for low neurotoxin (β-ODAP) content. *Food and Chemical Toxicology* 49, 589–600.

Lambein, F., Kuo, Y.H., Kusama-Eguchi, K. and Ikegami, F. (2007) β-N-oxalyll-α,β-diaminopropionic acid, a multifunctional plant metaboliteof toxic reputation. *Arkivoc Part* ix, 45–52.

Lambein, F., Travella, S., Kuo, Y.H., Montagu, M.V. and Heijde, M. (2019) Grass pea (*Lathyrus sativus* L.): orphan crop, nutraceutical or just plain food? *Planta* 250, 821–838.

Llorent-Martínez, E.J., Zengin, G., Fernández-de Córdova, M.L., Bender, O., Atalay, A., Ceylan, R., Mollica, A., Mocan, A., Uysal, S., Guler, G.O. and Aktumsek, A. (2017) Traditionally used *Lathyrus* species: phytochemical composition, antioxidant activity, enzyme inhibitory properties, cytotoxic effects, and in silico studies of *L. czeczottianus* and *L. nissolia*. *Frontiers in Pharmacology* 8, 83. https :// doi. org/10.3389/fphar .2017.00083

Mahata, D., Chhetri, B., Ghosh, A. and Chongdar, S. (2018) Improved technology of Grasspea (*Lathyrus sativus*) cultivation. *Journal of Pharmacognosy and Phytochemistry* 7(5), 1141-1143.

Mahler-Slasky,Y. and Kislevm, M.E. (2010) Lathyrus consumption in late Bronze and iron age sites in Israel: an Aegean affinity. *Journal of Archaeological Science* 37, 2477-2485.

McCutchan, J.S. (2003) Review: A brief history of grasspea and its use in crop improvement. *Lathyrus Lathyrism Newsletter* 3, 18-23.

NAAS. (2013) Improving productivity of rice fallows (ed), Policy Paper No. 64. National Academy of Agricultural Sciences, NewDelhi.16pp.

Nazrul, M.I. and Shaheb, M.S. (2015) Adaptation of grass pea as sole and relay cropping systems with transplanted aman rice in Sylhet region of Bangladesh. *American Journal of Biology and Life Sciences* 3(6), 254-259.

Pande, S., Sharma, M., Ghosh, R., Rao, S.K., Sharma, R.N. and Jha, A.K. (2012) Opportunities for chickpea production in rainfed rice fallows of India-Baseline Survey Report (ed), Grain Legumes Program Report No.1. ICRISAT, Patancheru-502 324, Andhra Pradesh, India. 56 pp.

Parihar, A.K., Dixit, G.P. and Singh, D. (2013) Multivariate analysis of various agronomic traits in grass pea (*Lathyrus* spp.) germplasm. *Indian Journal of Agricultural Sciences* 83, 570–575.

Patto, M.C.V. and Rubiales, D. (2014) Lathyrus diversity: available resources with relevance to crop improvement – *L. sativus* and *L. cicera* as case studies. *Annals of Botany* 113, 895–908.

Piergiovanni, A.R., Bisignano, V., Polignano, G. and Lioi, L. (2013) Dark and Bright Facets of the Nutritional Value of Grass Pea (*Lathyrus sativus* L.) Seeds. In: H Satou and R Nakamura (Eds.), Legumes: types, nutritional composition and health benefits. Nova Science Publishers, New York. pp. 179-192.

Rizvi, A.H., Sarker, A. and Dogra, A. (2016) Enhancing grasspea (*Lathyrus sativus* L.) production in problematic soils of South Asia for nutritional security. *Indian Journal of Genetics and Plant Breeding* 76(4), 588-592.

Sarkar, S., Samui, I., Brahmachari, K., Ray, K., Ghosh, A. and Nanda, M.K. (2019) Management Practices for Utera Pulses in Rice-fallow System under Coastal Saline Zone of West Bengal. *Journal of the Indian Society of Coastal Agricultural Research* 37(2), 98-103.

Sarkar, S. and Singh, S.R. (2007). Interactive effect of tillage depth and mulch on soil temperature, productivity and water use pattern of rainfed barley (*Hordium vulgare* L.). *Soil & Tillage Research* 92, 79–86.

Singh, N.P., Praharaj, C.S. and Sandhu, J.S. (2016) Utilizing untapped potential of rice fallow of East and North-east India through pulse production. *Indian Journal of Genetics and Plant Breeding* 76(4), 388-398.

Singh, S.S., Kumar, N., Praharaj, C.S. and Singh, N.P. (2017) Agro-technologies for Pulses in Rice Fallows. In: NP Singh (Eds.), Technical bulletin. ICAR-Indian Institute of Pulses Research, Kanpur 208 024 (U.P.), India. pp. 1-10

Talukdar, D. (2009) Recent progress on genetic analysis of novel mutants and aneuploid research in grass pea (*Lathyrus sativus* L.). *African Journal of Agricultural Research* 4, 1549–1559.

Talukdar, D. (2012) Total flavonoids, phenolics, tannins and antioxidant properties in seeds of lentil and grass pea. *International Journal of Phytomedicine* 4(4), 537-542.

Teshager, A.A. (2019) Role of grass pea (*Lathyrus sativus* L.) in nutrition security and risk of neurolathyrism. *Global Scientific Journals* 7(5), 382-390.

Singh, R.N., Praharaj, C.S., Kumar, R., Singh, S.S., Kumar, N. and Singh, U. (2017) Influence of rice (*Oryza sativa*) habit groups and moisture conservation practices on soil physical and microbial properties in rice + lathyrus relay cropping system under rice fallows in Eastern Plateau of India. *Indian Journal of Agricultural Sciences* 87(12), 1633–9.

Van-Wyk, S.G., Kunert, K.J., Cullis, C.A., Pillay, P., Makgopa, M.E., Schlüter, U. and Vorster, B.J. (2016) Review: the future of cystatin engineering. *Plant Science* 246, 119–127.

Varma, P. (2019) Production, Markets and Trade: A Detailed Analysis of Factors Affecting Pulses Production in India. In: Final report. Centre for Management in Agriculture Indian Institute of Management, Ahmedabad-380015. Supported by Ministry of Agriculture and Farmers Welfare, Government of India.

Yerra, S. and Kilari, E.K. (2018) β-ODAP As An Anti-Nutritional Component In *Lathyrus* Species - A Review. *Journal of Global Trends in Pharmaceutical Sciences* 9(2), 5466- 5479.

CHAPTER 6

Rainfed greengram production techniques in Tamil Nadu under climate change

S Marimuthu[1] and U. Surendran[2]

1. *National Pulses Research Centre, Vamban, Pudukkottai, Tamil Nadu*
2. *Centre for Water Resources Development and Management, Calicut, Kerala, India*

*Corresponding author email: agrimuthu76@rediffmail.com

Abstract

Pulses are the important source of vegetable protein in the world. In Tamil Nadu, greengram is cultivated in marginal land under rainfed conditions and only 15% of the area under pulses has assured irrigation. Although these crops are nutrient rich, they are cultivated under nutrient starved low soil fertility environments, and the average yield of greengram in Tamil Nadu is as low as 574 kg/ha. It is relatively climate-hardy crop and efficient producers of protein. Sustainable efforts are needed to increase the yield of greengram under rainfed through developing resource conservation technologies. These crops contribute sustainability by mitigating and adapting climate change, reducing poverty and hunger, improving health by providing nutrition, and help to promote economic stability. Adaptability to fit into various cropping systems by introducing greengram into farm production can be a key to increase resilience to climate change. Thus, greengram is good for people, good for soil and good for the planet.

Keywords: Drought, Greengram, Rainfed, Moisture conservation, Resource conservation, Tillage

1. INTRODUCTION

Pulses constitute the major source of protein for majority of population in India that is predominantly vegetarian in dietary habits. India contributes 25-28% of the total global pulse production and it is the largest producer and consumer of pulses in the world. Pulses are known to have higher protein content and other essential nutrients and considered as important component in nutrition for poor in developing countries (Varshney *et al.*, 2013). Pulses are not only important sources of proteins, but also offer vitamins and minerals, popularly known as '*Poor Men Meat*' and '*Rich Men Vegetable*' (Reddy *et al.*, 2013). In total pulses production, chickpea stands first with 48%, followed by redgram with 17%, blackgram with 10% and greengram with 7%. Remaining quantity is contributed by other pulses. India is currently importing approximately 3 to 4.5 Mt of pulses every year. The projected requirement of pulses for 2030 is 32 Mt, which requires a growth rate of 4.2%.

Majority of pulses cultivated area falls under rainfed, resource poor and harsh environments where it frequently prone to drought and other abiotic stress condition, resulting in low productivity. Greengram (*Vigna radiata* (L.) R. Wilczek) also known as mungbean, is an excellent source of protein (24.5%), 1.15% fat, 62.6% carbohydrate with high quality of lysine (460 mg/g) and tryptophan (60 mg/g). Whole, dehusked and split grains are extensively used for human consumption. The germination of the grains increases their vitamin C level. However, in some, riboflavin as well as niacin content is also increased. The activity of many enzymes such as amylase, protease, phytase and lipase also increase during germination. Heat treatment increases the palatability of food and bioavailability of nutrients. Greengram crop is generally cultivated on marginal lands having low soil fertility and under rainfed conditions. India is the major producer of greengram in the world, it is cultivated in 4.25 M ha with production of 2.41 Mt and the productivity of 567 kg/ha. In Tamil Nadu, it is cultivating in an area of 1.73 lakh ha with production of 0.993 lakh tonnes and productivity of about 574 kg/ha (Table 1). A short duration legume crop, greengram is an important component of tropical farming system and offer great opportunity for intensification and diversification.

Impacts of climate change mainly the temperatures are likely to increase from 1.0 to 2.0°C by 2025 AD and consequently there may be shifts in the rainfall pattern and in its temporal and spatial distribution. Increase in greenhouse gas emissions will also have an impact on agriculture and water resources. Short-term or long-term fluctuations in weather parameters associated with climate variability, climate change can influence crop yields. Fall in water table and increased use of energy to pump water makes the practice of irrigation more expensive. The impact of climate change is going to vary with crops and it is expected that pulses and oilseeds will be benefited while rice and wheat will have a negative impact. This chapter discussed about the various drought mitigation techniques and ways for increasing greengram productivity in Tamil Nadu.

Table 1. Area, production and productivity of greengram in Tamil Nadu from 1981 to 2019.

Year	Area (000 ha)	Production (000 tonnes)	Productivity (kg/ha)
1981-85	91.4	23.3	254
1986-90	114.8	41.8	364
1991-95	124.2	52.7	427
1996-20	133.8	52.8	392
2001-05	130.05	49.05	378
2006-10	155.8	57.3	454
2011-15	196.5	119.7	580
2016-17	167.6	50.4	300
2017-18	184.5	86.1	487
2018-19	173.2	99.3	574

Source: Annual report of AICRP on MULLaRP, 2019.

2. CONSTRAINTS IN GREENGRAM CULTIVATION IN TAMIL NADU

Greengram is cultivated under rainfed dryland situation in Tamil Nadu. In general, the common features of this region are poverty and food insecurity, acute fodder shortage and poor livestock productivity. There exists short growing season, limited choice of the crops and varieties, existence of marginal soils with poor soil fertility, low soil biological activity, existence of sodic soils, improper land management systems, lower farm mechanization levels and low socio-economic condition of the farmers.

2.1 Climatic constraints

Based on rainfall distribution, irrigation pattern, soil characteristics, and other physical, economic and social characteristics, Tamil Nadu is classified into seven agro-climatic zones: northeast, northwest, west, south, Cauvery delta zone, high rainfall zone and high altitude zone. The two monsoons *viz*., southwest and northeast monsoon are dominant source of precipitation providing almost 80 percent of the total annual rainfall. Northeast monsoon is associated with cyclonic storms causing wide spread damage to property. Inter annual rainfall variability causes significant damage to agriculture. The annual co-efficient of variation is around 30 percent. Despite of inter annual variability, intra-seasonal variability also causes frequent crop failures and loss to major agricultural resources. In general, rainfed greengram suffers from the climatic constraints such as wide range of annual rainfall (600–2000 mm), higher co-efficient of variation in seasonal rainfall, occurrence of long dry spells within a season and higher potential evapotranspiration. The aberrant weather includes erratic/uncertain rainfall, drought, short length of growing period, high temperature, wind erosion, etc. which affect the greengram production in the state.

2.2 Soil related constraints

Continued decline in soil organic matter (SOM) is a challenge in rainfed areas due to poor biomass production and inadequate integration of crop and livestock farming. Organic manures require a decomposition period of at least 40 days while inorganic fertilizers are readily available. Research has to be carried out to find ways and means to shorten the duration for the preparation of the organic fertilizers. Nutritional disorders occur in rainfed crop production includes deficiencies of nitrogen and phosphorus. Potassium deficiency is common in coarse textured soils. There is a widespread occurrence of zinc deficiency while iron deficiency is common in calcareous soils. Decline in soil biodiversity, an indicator of soil health, makes the soil vulnerable to soil degradation. Soil compaction and sealing occur due to use of heavy machinery or over grazing. Soil compaction affects soil structure which intends root growth and hydrological activity. Salinity and alkalinity is an issue in rainfed agriculture due to excess salt accumulation and high exchangeable sodium percentage in soil. Efficient nutrient management has a greater relevance under rainfed conditions where every drop of water makes a difference in getting higher response (Jayaraman, 2011).

2.3 Production related constraints

There has been a high degree of risk in greengram production in Tamil Nadu. More than 90% of the greengram area is presently under rainfed in the 7 rainfed ecozones of Tamil Nadu (Fig 1). The characteristics of these rainfed ecozones are given in Table 2. Looking to the situation it is important to analyze the constraints in greengram production in Tamil Nadu. The main constraints are:

- Cultivation of traditional/local varieties
- Mostly grown as relay cropping
- Limited areas under irrigated condition
- Improper agronomic management
- Biotic and abiotic stresses

Rainfed agriculture is characterized with erratic rainfall leading to frequent drought events, degraded lands, resource poor farmers, small and marginal holdings and lowering of groundwater table. The productivity of many rainfed crops in farmers' field is still less than one tonne per hectare, while the achievable yield is quite high. The main reason for such lower productivity is always attributed to characteristics of rainfall. The primary concern in rainfed is 'buffering' the moisture availability from the rains received due to monsoon. Low productivity of greengram is mainly due to rainfall variability and also droughts during critical period of crop growth and poor management practices (Marimuthu *et al*., 2020).

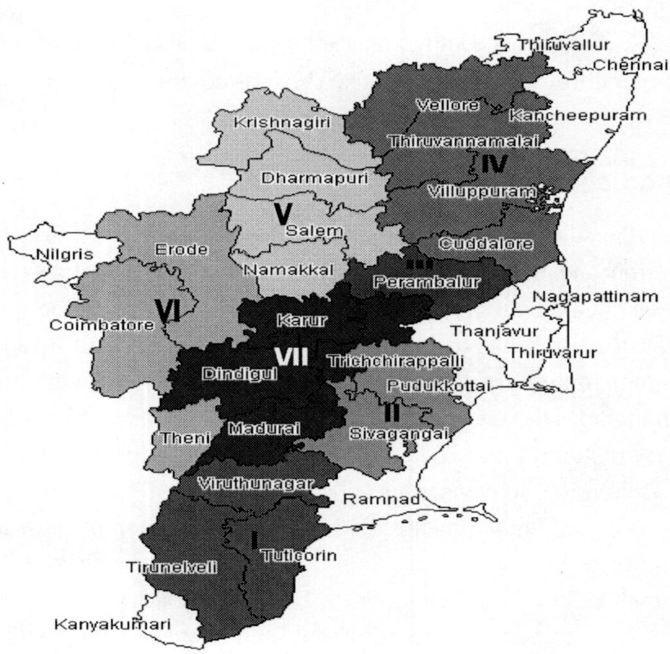

Fig. 1. Map of rainfed ecozones of Tamil Nadu

Table 2. Characteristics of rainfed ecozones of Tamil Nadu

Zone	District	Rainfed Area (Lakh ha)	Rainfall	LGP (days)	Soil	Specific Information
I	Virudhunagar, Tirunelveli, Thoothukudi	2.62 (57.7 %)*	425 mm (NEM)	115	Black soil	Unimodel rainfall, single crop
II	Sivagangai, Pudukkottai	0.916 (36.8 %)	300 mm (SWM) 400 mm (NEM)	165	Laterite soil	Bimodel rainfall, double crops
III	Perambalur	1.64 (74.3 %)	350 mm (SWM) 450 mm (NEM)	170	Black soil	Bimodel rainfall, double crops
IV	Vilupuram, Cuddalore, Vellore, Thiruvannamalai	5.46 (53.9 %)	390 mm (SWM) 500 mm (NEM)	190	Red and Laterite soil	Bimodel rainfall, assured double crops
V	Krishnagiri, Dharmapuri, Salem, Part of Namakkal	4.99 (73.9 %)	350 mm (SWM) 270 mm (NEM)	160	Red soil	Bimodel rainfall, double crops (NEM season risky)
VI	Erode, Coimbatore, Theni, Parts of Namakkal and Karur	4.91 (58.3 %)	325 mm (NEM)	100	Red and Black soil	Unimodel rainfall, single crop (risky due to intermittent dry spell)
VII	Karur, Trichy, Dindigul, Madurai	3.35 (60.4 %)	360 mm (NEM)	95	Red and Black soil	Unimodel rainfall, assured single crop

*( ) Percentage to total cultivable area

3. STRATEGIES FOR INCREASING GREENGRAM PRODUCTION IN TAMIL NADU

Greengram is an annual herb with the height ranging from 30 to 120 cm, leaves are alternate, trifoliate and dark green in colour, stipules ovate, long petiole, leaflet ovate, deep rooted and branched plants. Flowers are in auxiliary racemes, peduncle upto 13 cm in length with clusters of 10-12 flowers, corolla yellow in colour sometimes curved, 5-10 cm long. Number of seeds per pod ranges from 8-14. The seeds are oblong, green or olive green in colour, sometimes yellow and brown. Seed weight generally varies from 30 to 40 g/1000 seeds. Soil moisture is the major factor influencing germination and plant performance. Greengram seed has the remarkable capacity for absorption of water and germination, and is relatively earlier compared with other major pulses. Almost two-thirds of the total dry matter accumulation takes place during one-third of the total life span of greengram. The NAR is maximum before flowering and gradually declines afterwards due to dropping of lower leaves. The RGR is maximum in the initial stage of growth but continue to decline with advancement in growth. Duration of vegetative phase increases with increase in photoperiod from 12 to 16 hours. Major yield components are pods per plant, seeds per pod and test weight. Soil moisture deficit hastens flowering and reduces flowering and pod filling period leading to reduced pod number and yield. However, seed size may increase following reduced seeds per pod. Greengram is a tropical pulse crop largely grown under semiarid and sub-tropical environment. It is one of the drought resistant plant but susceptible to frost, water logging and salinity. A hot humid with sufficient soil moisture is ideal. It is grown from sea level to 2000 m above mean sea level. Thought, it can be grown in temperature regime of 20 to 40 °C, optimum mean temperature is 28 to 32 °C. An annual rainfall of 600 to 750 mm with uniform distribution is sufficient for the crop growth and development. Rains during flowering stage are having detrimental effect on the crop. It is the short day plant and there foremost cultivars flower in 12-13 hours photoperiod. Flowering gets delayed with extended photoperiod but hastened with a rise in temperature. Greengram is cultivated in various seasons in different parts of Tamil Nadu. The main growing seasons are June - July (Adipattam), September - October-(Puratasipattam), December – January (Margazhi-Thai Pattam), January - February (Rice fallows) and February - March (Summer). The recommended advanced agro-techniques for different agro-ecological conditions may be followed to achieve higher yield of greengram in different season in the state. The recommended practices are discussed below in details.

3.1 Suitable soil and tillage

Soil is one of the most important non-renewable basic resources on the earth's surface. The major soil groups of Tamil Nadu are Red soils (62%), Black soils (12%), Laterite soils (3%) and Coastal soils (7%). Greengram is cultivated on a wide range of soils from sandy loams to black cotton soils. Ideal soils are well drained loam or sandy loam. Optimum pH range is 6.5 to 7.5. It is grown on a variety of soils ranging from red lateritic to clay loam in Tamil Nadu (Table 3). Saline alkali and acidic soils are not suitable for greengram cultivation. Soil should be well drained because temporary waterlogging may cause the considerable damage to the crop.

Table 3. Agro-climatic region with rainfall and soil types

S. No	Agro Climatic Zones	Districts Covered	Soil Type	Altitude (m)	Annual rainfall (mm)	Annual PET (mm)	Method of land configuration
1	North Eastern Zone	Kancheepuram, Tiruvallur, Cuddalore, Vellore, Villupuram and Tirunvannamalai	i. Red Sandy Loam ii. Clay Loam iii. Saline coastal Alluvium	100-200	1105	1700	Ridges and furrow/ Compartmental bunding
2	North Western Zone	Dharmapuri, Krishnagiri, salem and Namakkal (Part)	i. Non Calcareous Red, ii. Non Calcareous Brown, iii. Calcareous Black	200-600	875	1727	Boarded furrow
3	Western Zone	Erode, Coimbatore, Tiruppur, Theni, Karur (part), Namakkal (part), Dindigul, Perambalur and Ariyalur (part)	i. Red Loamy ii. Black	200-600	715	1622	Compartmental bunding / Broad bed furrow
4	Cauvery Delta Zone	Thanjavur, Nagapattinam, Tiruvarur, Trichy and parts of - Karur, Ariyalur, Pudukkottai and Cuddalore	i. Red Loamy ii. Alluvium	100-200	984	1932	Compartmental bunding
5	Southern Zone	Madurai, Sivagangai, Ramanathapuram, Virudhunagar, Tirunelveli and Thoothukudi	i. Coastal Alluvium ii. Black iii. Red Sandy soil iv. Deep red soil	100-600	857	1825	Beds and channel/ ridges & furrow
6	High Rainfall Zone	Kanyakumari	i. Saline Coastal ii. Alluvium iii. Deep Red Loam	100-2000	1420	1816	Ring basin/Tied ridges
7	Hilly Zone	The Nilgiris and Kodaikanal (Dindigul)	Lateritic	2000	2124	1213	Contour bunding & Bench terrace

Source: http://tnhorticulture.tn.gov.in/horti/agro-climatic-zones

The main purpose of tillage is to develop desired soil structure for a seedbed, which allows rapid infiltration and good retention of rainfall, control weeds and remove unwanted crop plants, manages crop residues and its decomposition. The depth of tillage depends on soil type, crop and time of tillage. Deep tillage of 25–30 cm is beneficial for deep heavy clay soils to improve permeability and to close cracks formed while drying (Chandrasekaran *et al.*, 2010). In soils with hard pans, deep tillage once in 2–3 years with chisel plough up to 35–45 cm depth at 60–120 cm interval will increase effective depth for rooting and moisture storage. Medium deep tillage of 15–20 cm depth is generally sufficient for most soils and crops. Shallow tillage up to 10 cm should be followed in light textured soils, shallow soils and in soils highly susceptible to erosion. Summer tillage or off-season tillage done with pre-season rainfall causes more conservation of moisture and also enables early and timely sowing. It is particularly useful for pre-monsoon sowing. For moisture conservation, ploughing across slope or along contour is very effective. Plough furrows check the velocity of runoff, promote more infiltration when water stagnates in the depressions caused by plough furrows and improves soil moisture storage.

3.2 Suitable varieties

In dry farming regions, traditional local crop varieties still dominate. The preference for these local varieties is based on their pronounced drought tolerance. But they are usually longer in duration susceptible to moisture stress at maturity. They have low yield potential even under favourable rainfall. They do not respond significantly to improved management practices. The criteria needs to be adopted for selection of varieties include drought tolerance, short or medium duration, high yield potential, response to nutrient supply, high water use efficiency, moderate resistance to pest and MYMV and other diseases. Greengram varieties suitable for rainfed zones of Tamil Nadu are given in Table 4.

Table 4. Greengram varieties suitable for pulse growing zones of Tamil Nadu

Name of the Varieties	Parentage	Duration (days)	Grain yield (kg/ha)		Special features
			Rainfed	Irrigated	
CO 1	Pureline selection from Coimbatore local (culture No.68)	135	750	---	Long duration and suitable for rainfed
ADT 1	Pureline selection from Aduthurai local	80	500 (rice fallow)	---	Suitable for rice fallow
JGG 1 (Rajendran)	Selection from Jayankondam local	65	625	---	Tolerant to drought
CO 2	Pureline selection from PLS 365/3	65	600	1200	Suitable for irrigated conditions
CO 3	Pureline selection from PLS 362/2	65-75	600	1250	Suitable for rainfed condition, tolerant to root rot and yellow mosaic virus.
KM 1	S.8 × PS 16	65-70	708	---	Tolerant to YMV and pod borer
KM 2	No.122 × S.9	6--65	767	---	Tolerant to YMV and pod borer
CO 4	Gamma irradiated mutant of CO 1 (20kR)	85	700-900	1250-1500	High biomass, suitable for rainfed condition

ADT 2	ADT 1/AB 33	70	850 (rice fallow)	---	Resistant to powdery mildew, suitable for rice fallow
ADT 3	H 70 - 16 / Rajendran/G.65	70	500 (rice fallow)	---	Resistant to YMV and stem fly.
Paiyur1	Pureline selection form DPT 703	85-90	700	---	Higher grain yield,low incidence of YMV, suited for rainfed tracts of Dharmapuri, Madurai, Ramnad, Tirunelveli and Periyar Districts.
Vamban 1	Hybrid derivative of S.8 × PIMS 3 (Culture: NPRG 1)	65-70	800	---	High yield, tolerant to YMV
CO 5	KM 2 × MG 50-10 (G)	75	900	---	High biomass, resistant to tip blight, tolerant to powdery mildew and root rot. Suitable for rainfed cropping.
K1	CO 4 x ML 65	70-75	670	---	Tolerant to drought. Suitable for pure and cotton based intercropping systems.
CO 6	WGG 37 × CO 5	62-67	850	1300	Resistant to yellow mosaic virus, suited to all greengram growing tracts of Tamil Nadu.
VBN (Gg) 2	VGG 4 × MH 309 (Culture: VGG 9)	65-70	750	900	Shiny grains, lobed leaves and resistant to YMV
VRM (Gg)1	Selection from K851	60-70	1100	-	Location specific variety suited to Vellore and Tiruvannamalai districts
COGG 912 (CO 7)	MGG 336 × COGG 902	62	978	-	High yielder, short duration, high protein content (25.2%). Seeds are medium bold with good cooking quality
VBN (Gg) 3	K1 × Vellore local (Culture: VGG 04-001)	65 - 70	775	878	Resistant to MYMV, powdery mildew.
CO 8	COGG 923 X VC 6040A	55 - 60	-	845	Short duration, determinate plant type with synchronized maturity, Resistant to yellow mosaic disease, stem necrosis and moderately resistant to root rot.
VBN 4	PDM 139 x BB 2664 (Culture : VGG 10-008)	65-70	940	1250	Multi bloom, non shattering type, moderate resistance to MYMV and powdery mildew diseases and resistance to blackgram leaf crinkle virus disease

(Source: www.arsvamban@ac.in, 2019)

3.3 Seed hardening

Seed hardening is done to induce drought tolerance in emerging seedlings. It is the process of soaking seeds in chemical solution and drying to induce tolerance to drought. Seed hardening enables seedlings to survive early moisture stress. During seed hardening, seeds are subjected to partial hydration followed by dehydration before sowing. Seeds are soaked for specified time in chemical solutions of prescribed concentration. Soaked seeds are then dried in shade back to original moisture content. During soaking, seeds imbibe water and germination process is started but not completed. The hardened seeds are thus in a ready state for germination. When sown in moist soils, seeds germinate immediately. Such early germination helps in seedling emergence before surface soil dries up. The seed hardening ensures early germination by 2–3 days compared to untreated seeds and induces better root development, which enables utilization of more soil moisture (Solaimalai and Subburamu, 2004). The seed hardening is considered as low cost technology and is the most important requirement for pre-monsoon sowing. For success in seed hardening, attention must be paid in selection of right chemical, its concentration, time of soaking, volume of solution and drying under shade to original moisture content. Greengram seeds can be hardened in 100 ppm $MnSO_4$ for 3 hours in a seed to solution ratio of 1:0.3 and dry back the seeds to original seed moisture content (8 - 9%) under shade (Crop Production Guide, 2020). In another process, 4 kg of wood ash is collected and powdered in which 30% *Acacia* gum is added and mixed thoroughly so that wood ash-gum paste is obtained (Srimathi *et al.*, 2007).

3.4 Seed treatment

Seed treatment is done for many purposes such as protection against pests and diseases, inoculation of bio-fertilizers and inducing drought tolerance. Seed treatment with insecticides and fungicides is a low cost technology for protection against pests and diseases. In drylands, spraying of chemicals for pest control is difficult due to scarcity of water. Greengram seeds should be treated with carbendazim or thiram 2 g/kg of seed 24 hours before sowing or with talc formulation of *Trichoderma viride* 4 g/kg of seed. Biofertilizer treatment with 60 g/kg of rhizobial culture CRM 6 and 60 g/kg of phosphobacteria 60 g/kg of PGPR developed at TNAU using rice kanji as binder should be done at 3-24 hours before sowing. If the seed treatment is not carried out, apply 2 kg of phosphobacteria and 2 kg of PGPR with 25 kg of FYM and 25 kg of soil mixture per hectare before sowing. Pre-sowing treatments on storability of greengram seed explained that designer seeds formulated with seed hardening, treated with polymer at 3 ml/kg and applied with carbendazim 2 g/kg + imidacloprid 1 ml/kg of seed maintained the vigour and viability of seed upto nine months without pathogen and pest infestation while storing at polythene bag container (Ananthi *et al.*, 2007).

3.5 Seed rate and sowing time

Seed should be pure, free from diseases and physical damage with 90-95% viability. Seed rate of greengram is 15-20 kg/ha in rainy season and 20-25 kg/ha in summer greengram. Optimum plant population is 33-40 plants/m^2 recommended for higher yield. The optimum time of sowing is first fortnight of July and first fortnight of August for effective utilization of the rain water. Sowing time also decided on the basis of length of growing period. Length of growing period is defined as a period in which the available soil moisture is enough to meet the evapotranspiration requirement of dryland crops and hence the productivity is assured (Table 5). The district-wise growing period of Tamil Nadu is given in Table 6.

Table 5. Crop status under different length of growing period

Length of growing period	Crop status
Less than 5 weeks	Crop failures will occur
Less than 14 weeks (98 days)	Permit the dryland crop to attain its potential productivity
14 weeks	A single dryland crop can be cultivated
14 to 20 weeks	Suitable intercropping system can be cultivated
More than 20 weeks	Long duration crop / double crops can be cultivated

Table 6. District wise time of soil moisture availability for growing of rainfed crops.

District	Station	Growing period (Met. Standard week)	No. of growing weeks	District	Station	Growing period (Met. Standard week)	No. of Growing weeks
Ramanathapuram	Kamudi	41-47	7	Vellore	Ambur	33-46	14
	Mudukulattur	41-49	9		Arakkonam	29-51	13
	Pamban	42-52,1	12		Gudiyattam	33-47	15
	Paramakudi	40-48	9		Sholingnur	31-49	19
	Ramanathapuram	41-52	12		Tiruppattur	31-45	15
	Theethanathanam	41-51	11		Vaniyambadi	32-45	14
	Tiruvadanai	41-50	10		Vellore	30-50	21
	Vattaram	41-51	11		Walajapet	30-50	21
Virudhunagar	Arupukottai	39-48	10	Tiruvannamalai	Arani	30-50	21
	Sattur	41-48	8		Chengam	31-49	19
	Sivakasi	41-48	8		Cheyyar	30-50	21
	Srivilliputtur	41-49	9		Polur	30-50	21
	Virudunagar	38-48	11		Tiruvannamalai	31-50	20
	Watrap	39-50	12		Vandavasi	29-51	23
Tuticorin	Arasadi	43-49	7	Dharmapuri	Denkanikota	32-46	15
	Kayattur	41-49	9		Dharmapuri	32-46	15
	Kovilpatti	41-49	9		Harur	33-47	15
	Kulasekarapatnam	42-52	11		Hosur	33-45	13
	Morekulam	42-51	10		Krishnagiri	33-45	13
	Ottappidaram	41-48	8		Palacode	32-46	15
	Sattankulam	42-50	9		Pennagaram	33-45	13

	Srivaikuntam	42-50	9		Rayakottai	33-46	14
	Tiruchchendur	42-52	11		Thalli	31-44	14
	Tuticorin	43-50	8		Uttangarai	31-46	16
Tirunelveli	Ambasamudram	42-52,1	12	Perambalur	Ariyalur	35-50	16
	Ayikudi	42-51	10		Perambalur	35-50	16
	Kadaiyam	42-52,1	12	Coimbatore	Annur	38-47	10
	Kadaiyanallur	42-51	10		Avanashi	38-47	10
	Kirnurnam	43-51	9		Coimbatore	41-47	7
	Nanguneri	42-51	10		Mettupalayam	39-50	12
	Palayamkottai	42-50	9		Palladam	41-47	7
	Radhapuram	42-49	8		Periyanaiakampalayam	38-49	12
	Sankarankovil	41-49	9		Pollachi	24-31, 41-47	8, 7
	Shencottah	41-51	11		Sulur	41-46	6
	Sivagiri	41-52	12		Tiruppur	38-47	10
	Tenkasi	41-52	12		Udumalaipet	41-48	8
	Tirunelveli	42-51	10	Erode	Bhavani	34-47	14
Sivaganga	Manamadurai	39-48	10		Dharapuram	40-47	8
	Sivaganga	35-48	14		Erode	34-47	14
	Tirupattur	33-48	16		Gopichettipalayam	35-47	13
Madurai	Cholavandan	36-48	13		Kangayam	38-47	10
	Madurai	34-48	15		Kodumudi	38-44	7
	Melur	33-49	17		Perundurai	35-47	13
	Nattam	33-49	17		Sathyamangalam	35-47	13
	Peraiyur	36-49	14		Talavadi	34-45	12
	Tirumangalam	34-49	16	Namakkal	Tiruchengodu	34-45	12
Pudukkottai	Adanakottai	37-49	13	Karur	Aravakkurichi	39-46	8
	Alangudi	36-50	15		Karur	39-45	7

	Annavasal	36-47	12	Dindigul	Nilakottai	36-47	12
	Ilupur	36-48	13		Palani	40-49	10
	Karambakkudi	38-50	13	Theni	Periakulam	38-49	12
	Kilanilai	38-49	12		Uttamapalayam	40-48	9
	Marungapuri	35-49	15	Madurai	Usilampatti	36-49	14
	Ponnamaravati	34-48	15	Tiruchchirapalli	Manaparai	36-48	13
	Pudukkottai	35-49	15	Dindigul	Chattrapatti	38-50	13
	Tirumayam	36-48	13		Dindigul	36-49	14
	Udyalipatti	37-47	11		Vedasunthur	38-48	11
	Viralimalai	38-48	11	Theni	Bodinayakanur	39-48	10

Source: http://agritech.tnau.ac.in/agriculture/agri_index.html

Plant population play a major role to decides the grain yield. For maintaining optimum plant population proper method of sowing is needed with the plant geometry of 30 x 10cm. Seed drill can be used for sowing of greengram in red and lateritic soils for effective utilization of soil moisture. A tractor drawn precision pulses seed drill is developed by TNAU, Coimbatore. The special feature of this seed drill is saving of 19.1 and 97.0 percent in cost and time of operation respectively and 40 percent saving in seed rate.

Fig. 2. Tractor drawn precision pulses seeder

3.6 Cropping system

Cropping system, in simple terms, is the intensification of crops in space and time dimensions. A chosen system has to be efficient in resource use and ensure sustained production preferably with no adverse effect on the bio-ecosystem. Not only should it meet the varied needs of the grower but also the demands of the dependent institutions. Crop rotation is also an important tool for enhancing productivity of pulses. Maize/ groundnut-rice - greengram /blackgram cultivation of short duration redgram in fallows are some of the crop rotations which will be helpful in increasing pulses production. In general pulses are short duration in nature so timely following of agronomic practices results in better yield. Rice-rice-greengram in Tamil Nadu is important sequential cropping system. Important intercropping systems are redgram + sorghum / groundnut / greengram / blackgram in western zone of Tamil Nadu. Ninety percent of pulses are grown under rainfed conditions as intercrops or in sequence cropping system all over the country. Pulses have shown a positive response to elevated CO_2. The main cropping systems under inter cropping where pulses are cultivated are mixed inter cropping (homesteads), row inter cropping (maize + greengram), relay intercropping (greengram followed by maize). The prominent sequential cropping systems in which pulses are included are double cropping and triple cropping. *Paira/utera* cultivation is actually practiced to use the residual moisture left by the preceding crop (Palaniappan and Sivaraman, 1996). In the southern zone comprising of Tamil Nadu, Andhra Pradesh and Karnataka, blackgram and greengram performed better in 1:2 and 2 (paired): 3 row intercropping system with cotton. North and north-western region pearlmillet / cotton + greengram, maize + cowpea / greengram / redgram, pearlmillet / sorghum / cotton + greengram and sorghum + blackgram / greengram are performing better than other systems (Latha, 2011). There is scope for intercropping with spring planted sugarcane with greengram / blackgram.

3.7 Nutrient management

Rainfed soils are both hungry and thirsty, but due to inherent risk of the un-irrigated rainfed crops, the farmers are always reluctant to invest in basal dose of fertilizers. Top dressing of fertilizer is done generally after establishing good crop stand. The fertilizer use in dryland crops might vary between 5 and 40 kg/ha ($N+P_2O_5+K_2O$). Soils are low in N and P, and Zn is the most limiting factor among micronutrient. The response for Ca, Mg and S has also been recorded. The reasons for poor soil fertility are slow weathering of minerals, low organic matter content, reduced microbial activity, erosion, very low addition of manures and fertilizers, soil salinity and alkalinity, and reduced mobility of nutrients and nutrient fixation. Pulses do not show much response to higher dose of nitrogen, whereas they respond significantly to applied phosphorus. Application of 15-20 kg N, 40-60 kg P_2O_5 and 20 kg K_2O and 20 kg S/ha as basal placement is recommended for normal conditions. Application of zinc, iron, molybdenum and boron also increased yield (Kuttimani and Velayutham, 2011). With the introduction of high yielding varieties, it is essential to give balanced fertilizer to the greengram. Farm yard manure and vermi-compost improve soil physical condition and thus enhances water retention in the soil for longer period along with increased availability of essential nutrients. Application of NPK at 12.5:25:12.5 kg/ha along with composted coir pith (CCP) at 12.5 t/ha increased the yield attributes, dry matter and grain yield of greengram (53%) than nitrogen and phosphorus alone. Nitrogen content was significantly increased in P applied plot along with CCP. Application of K at 12.5 kg/ha significantly increased the P and K content in grain (Saraswathy *et al.*, 2004). The integration of 50% RDF of nitrogen by fertilizers and 50% by vermicompost with pulse wonder @ 5 kg/ha found to be best practice for greengram grown under rainfed in vertisol (Balraj *et al.*, 2018). Applications of 125% NP along with foliar sprays of 2% DAP and 1% SOP twice recorded higher N, K and S uptake by greengram and higher soil available N (Sathyamoorthi *et al.*, 2007). Canopy bottom light quantum was the lowest and differential light quantum was the highest with 125% recommended NP along with foliar spraying of DAP and SOP (Sathyamoorthi *et al.*, 2008).

Phosphorus, sulphur and nutrient solubilising bacteria, fungi, mycorrhiza and poly-culture are other ways of increasing soil fertility of soil. Soil Health Cards can also be used for optimizing application of fertilizers. Green manures/green leaf manure (GLM) can also be used for improving soil health. In Tamil Nadu, Kolinji and Aavarai (*Cassia auriculata*) are used as green leaf manure in rainfed areas. Application of biochar 5 t/ha is ideal to overcome the problem of soil encrustation in red lateritic soil and enhanced greengram seed yield to a tune of 32% over control. Moreover, basal application of redgram biochar 5 t/ha with phosphobacteria 2 kg/ha has been found to be best soil conditioner for blackgram under rainfed *Alfisol* and it also enhanced the soil moisture and carbon under changing climate (Marimuthu and Kannan, 2019). The reasons for low yield of greengram are the slow rate of dry matter accumulation during the pre-flowering phase, poor pod setting, onset of leaf senescence during the period of pod development and low partitioning efficiency of assimilates to grain. Moreover, greengram being cultivated under rainfed/summer season in soil moisture deficit conditions, the application of recommended fertilizers at right time and right quantity may not be efficient to harvest a good crop. Under such situations, foliar application of nutrients may be an option to meet out the nutrients demand of crop to some extent. The maximum grain yield (928 kg/ha) was recorded by foliar application of

2% DAP +100 ppm salicylic acid + 0.05% sodium molybdate spray at vegetative stage (30 DAS) and flowering stage (45 DAS) followed by 2% urea + 100 ppm salicylic acid + 0.05% sodium molybdate enhanced the yield of greengram. The yield increase over no spraying or control was of 30.3% (Kuttimani and Velayutham, 2011). Number of pods per plant is an important yield component in greengram and it was found that individual or combined use of urea, DAP, salicylic acid and sodium molybdate increased number of pods per plant which may be due to efficient translocation of photosynthates from source to sink (Rajendran, 1984). All the foliar spray treatments (TNAU Pulse wonder 5 kg/ha, 1% Urea + 2% DAP + 1% KCl, 1% polyfeed and water spray), recorded significantly higher in terms of number of seeds/pod, number of pods/plant and grain yield except water spray. Interaction effect study revealed that planting geometry of 25 x 10 cm with foliar application of 1% Urea + 2% DAP + 1% KCl gave highest greengram yield followed by 30 x 10 cm with 1% polyfeed (Elamathi and Manoharan, 2019). The significantly higher plant height (55.4 cm), pods/plant (24 pods), 100 grain weight (3.5 g) and grain yield (854 kg/ha) were obtained with 2% neem coated urea spray at flower initiation which was at par with Pulse wonder 5 kg/ha spray (Umamageshwari, 2019).

3.8 Soil moisture management

Being grown on marginal lands, greengram is largely considered as a drought tolerant crop (grow with a limited soil moisture). However, like any other plants, it responds to decrease in available soil moisture by reducing its growth and hence productivity. It is evident from the experiment that 30% decrease in water supply relative to water optimum for crop growth results in nearly 20% decrease in seed weight per plant if the soil moisture stress imposed around a vegetative stage. The decrease in total plant dry weight and harvest index were the main reasons for reduced seed yield due to drought stress in mungbean (Sadasivan *et al.*, 1988). Soil moisture stress during vegetative stage is more detrimental. Some of the genotypes of greengram *viz*., COGG 1332, VGG 15029, VGG 16069, VGG 17003, VGG 17004, VGG 17009 and VGG 17019 were found to be tolerant to drought and high temperature stress. These genotypes showed higher proline content, chlorophyll stability index, catalase activity and relative water content, whereas the oxidants such as hydrogen peroxide, superoxide radical, melondialdehyde content were decreased in tolerant genotypes. Foliar spray of osmolites such as proline (50 mM) during stress followed by KCl (2.0%) were found to be effective in improving growth and yield parameters in greengram (Amutha, 2019). Soil moisture conservation can be done through different practices are discussed below.

3.8.1 Land configuration

Land configuration is key to conserve moisture in soil profile. The basic principle are: (*i*) shaping the land surface manually or with implements in such a way as to reduce the velocity of runoff, (*ii*) to allow more time for rainfall to stand on soil surface, and (*iii*) to facilitate more infiltration of rainfall into soil layers. During summer, land should be ploughed in a crossed way to remove the weeds. It facilitates to hold summer rains in soil particles or raise the green manure crop incorporated within the field at the time of flowering by using rotavator. With this operation land is kept ready for sowing. Choice of any particular method under a given situation is influenced by rainfall characters, soil type, crops, sowing methods and slope of land. The common methods used for conserving soil moisture are basin listening, bunding, furrows irrigated raised bed (FIRB) and broad bed and furrows (BBF).

3.8.1.1 *Basin listing*

Formation of small depressions (basins) of 10–15 cm depth and 10–15 cm width at regular intervals using an implement called basin lister. The small basins collect rainfall and improve its storage. It is usually done before sowing. It is suitable for all soil types and crops.

3.8.1.2 *Bunding*

The bunds check the free flow of runoff water, impound the rainwater in the inter-bund space, increase its infiltration and improve soil moisture storage. Leveling of inter-bund space is essential to ensure uniform spread of water and avoid water stagnation in patches. It is of three types *viz*., countour bunding, graded bunding and compartmental bunding. In case of contour bunding, bunds of 1 m basal width, 0.5 m top width and 0.5 m height are formed along the contour. The distance between two contour bunds depends on slope. It is suitable for deep red soils with slope less than 1%. It is not suitable for heavy black soils with low infiltration where bunds tend to develop cracks on drying. Contour bunds are permanent structures and require technical assistance and heavy investment. Under graded/field bunding, bunds of 30-45 cm basal width, and 15-20 cm height are formed across slope at suitable intervals of 20-30 m depending on slope. The inter-bund area is leveled and cropped. It is suitable for medium deep to deep red soils with slopes up to 1%. It is also not suitable for black soils due to susceptibility to cracking and breaching. Bunds can be maintained for 2-3 seasons with reshaping as and when required.

3.8.1.3 *Compartmental bunding*

Small bunds of 15 cm width and 15 cm height are formed in both directions (along and across slope) to divide the field into small basins or compartments of 40 sq. m. size (8 × 5 m). It is suitable for red soils and black soils with a slope of 0.5-1%. The bunds can be formed before sowing or immediately after sowing with local wooden plough. It is highly suitable for broadcast sown crops. It registered higher grain yield of greengram, rain water use efficiency, relative water content and post harvest soil available NPK. In *alfisol*, 13% increase in grain yield of greengram was recorded under this system over flat bed (Marimuthu *et al*., 2020). In addition to compartmental bunding, application of Pusa hydrogel @ 5.0 kg/ha + mulching with crop residue @ 5 t/ha + PPFM 1% spray or Pusa hydrogel @ 5.0 kg/ha + mulching with crop residue @ 5 t/ha + 1% KCl spray at moisture stress period recorded higher yield of rainfed greengram (Marimuthu, 2019).

3.8.1.4 *Furrow irrigated raised bed (FIRB)*

Ridges of 30-45 cm width and 15-20 cm height are formed across slope. The furrows guide runoff water safely when rainfall intensity is high and avoid water stagnation. They collect and store water when rainfall intensity is less. It is suitable for medium- deep to deep black soils and deep red soils. It is not suitable for shallow red soils, shallow black soils and sandy/ gravelly soils. It is also not suitable for broadcast sown crops and for crops sown at closer row spacing less than 30 cm. Since furrows are formed usually before sowing, sowing by dibbling or planting alone is possible.

Tied ridging (Figure 3) is a modification of the above system of ridges and furrows where in the ridges are connected or tied by a small bund at 2-3 m interval along the furrows. Random tied ridging is another modification where discontinuous furrows of 20-25 cm width, 45-60 cm length and 15 cm depth are formed between clumps or hills of crops at the time of weeding. Yet another modification of ridges and furrows method is the practice of sowing in lines on flat beds and formation of furrows between crop rows at 25-30 DAS. This enables sowing behind plough or through seed drill (Chandrasekaran *et al.*, 2010)

Fig. 3. Tied ridges

3.8.1.5 *Broad Bed Furrow (BBF)*

The beds of 1.5 m width, 15 cm height and convenient length are formed, separated by furrows of 30 cm width and 15 cm depth. Crops are sown on the beds at required intervals. It is suitable for heavy black soils and deep red soils. The furrows have a gradient of 0.6%. In rainfed areas, incorporation with soil conditioners like coir pith, land is configured into broad beds and furrows, it would help to conserve the soil moisture. BBF method recorded 13% higher yield compared to the other two methods *viz.*, flat bed sowing followed by ridge making and flat bed sowing (Subbulakshmi, 2019).

3.8.2 Measures to reduce evaporation loss

Evaporation happens to maintain soil thermal regime and is governed by soil moisture content, vegetative cover on surface, soil type, temperature gradient between soil and atmosphere and atmospheric water demand. Higher soil moisture content, especially a wet surface soil increases evaporation rate. As the surface soil dries up, continuity of capillary pores is disrupted and moisture movement upwards from deeper layers is reduced. Evaporation loss in a cropped field is more in the early growth stage when canopy cover is less, especially in widely spaced crops and slow growing species. Black soils tend to absorb more heat and may evaporate more water. When cracks are formed during drying, evaporation takes place from the sides of the cracks also. With high temperature, low humidity and dry winds, atmospheric water demand increases the rate of evaporation. Shallow or surface tillage or dust mulch can be done to disrupt the continuity of capillary pores is broken and the rise of water through capillary movement is obstructed. Shallow tillage after summer showers is beneficial in this regard. This process is called dust mulching. Inter tillage between crop rows during early dry spells has a similar effect. Guntaka (Blade harrow)/Danti/hand hoe are the implements used for dust mulching. Mulching is another way to conserve the soil moisture. The organic wastes used for mulching include crop stubbles, straw, coir pith, groundnut shell, husk etc. These wastes at 5–10 t/ha are spread on the soil surface to a thickness of 5-10 cm. Some of the anti-evaporating chemicals like hexadecanol can also be used as to check the evaporation loss. When sprayed and mixed with soil surface, it reduces evaporation by 43%. The treated surface layer dries up fast and creates a diffusion barrier for upward movement of water vapour. It is resistant to microbial activity and degradation. It remains in soil for more than a year and also increased the soil aggregate stability.

3.8.3 Measures to reduce transpiration loss

Though transpiration is necessary and unavoidable evil, excessive transpiration has to be controlled especially when soil moisture stress develops during critical stages of crop growth. The rate of transpiration is governed by soil moisture potential, atmospheric water demand and plant canopy characters such as leaf area, leaf orientation, stomatal resistance, etc. Transpiration loss can be reduced by the use of antitranspirants and by some cultural methods also. When reflectant type antitranspirant (Kaolin and lime solution) sprayed on leaf surface, it increases the leaf albedo or leaf reflectance of sunlight, as a result heating is reduced, leaf temperature inside decreased and thus need for transpiration is also reduced. Spraying of 3-6% concentration kaolin reduces leaf temperature by 3-4 °C and transpiration by 22-28%. These are less expensive, non phytotoxic and do not interfere with photosynthesis, since stomatal closure does not take place (Somasundaram and Arokiaraj, 2000).

3.9 Weed management

Weeds are major problem in rainy and post-rainy season greengram compared with summer irrigated crop. The predominant weeds in greengram are sedges viz., *Cyperus rotundus*, broad-leaved weeds viz., *Amaranthus viridis, Boerhavia diffusa, Commelina benghalensis, Convolvulus arvensis*, *Corchorus fascicularis, Corchorus olitorius*, *Desmodium triflorum*, *Digera arvensis*, *Euphorbia hirta*, *Euphorbia microphylla*, *Phyllanthus niruri*, *Phyllanthus maderaspatensis*, *Trianthema portulacastrum* and *Tribulus terrestris*. Most weeds have a high transpiration coefficient *i.e.*, amount of water transpired to produce unit quantity of dry matter. Early weed control prevents unwanted transpiration loss of conserved soil moisture through weeds. First 30 DAS is the critical period for weed competition. Since row seeding with a row spacing of 30 × 10 cm is recommended, there is scope for inter-cultivation using local blade harrows. A single hand weeding around 40 DAS can remove weeds within the row. If the row spacing or soil condition does not permit inter-cultivation and hand weeding, use of herbicides is the only option. Application of pre-emergence (PE) oxyfluorfen 0.10 kg/ha + early post-emergence (POE) imazethapyr 50 g/ha at 20 DAS or PE pendimethalin 0.75 kg/ha + POE imazethapyr 50 g/ha at 20 DAS provided a broad spectrum of weed control by significantly reducing weed density and dry weight which resulted in higher weed control efficiency (90.2%) and lower weed index. PE application of oxyfluorfen 0.10 kg/ha + POE application of imazethapyr 50 g/ha at 20 DAS or application of pendimethalin 0.75 kg/ha + imazethapyr 50 g/ha at 20 DAS recorded significant improvement in plant height, yield attributes like number of pods/plant, number of seeds/pod, test weight (g) and produced higher grain yield of rainfed greengram (Sobhana *et al.*, 2018).

4. CONCLUSION

In short, to increase area and production of greengram in rainfed regions of Tamil Nadu, crop specific and region specific approaches should be adopted in the overall framework of systems approach. The major thrust areas to be addressed are replacement of cereal crops in the prevailing rice-rice cropping systems with high yielding varieties of pulses, inclusion of short duration varieties of greengram as catch crop, development of multiple disease and pest resistant varieties, adoption of soil moisture conservation techniques and co-ordination of research, extension and farmers to encourage farmer's participatory research. The increase in productivity of any system comes through adoption of improved crop production technologies with commitment.

5. FUTURE THRUST

Greengram is primarily a rainy season crop, but with the development of short duration and photo-thermo-insensitive varieties, it can also be grown in spring and summer season. Since there is little scope for horizontal growth in acreage, the only alternative left is to achieve vertical growth either through genetically improvement or through improved management practices. Attention for sowing on bed and ridges and use of mulches, deficit irrigation etc are helpful for water stress management in pulses. Adoption of integrated drought mitigation management practices like seed hardening, land configuration, use of hydrogel and nutrient management practices needs to be implemented at larger scale. Phosphorus needs critical attention for increasing its use efficiency in pulses production systems. Susceptibility of the pulses to aberrant weather conditions, especially waterlogging and adverse soils making them highly unstable in performance, which should also be taken care before cultivation of pulses. More emphasis needs to be given for research on physiological and genetic attributes like high harvest index, reduced flower dropping, develop determinate growth habit and develop fertilizers responsive genotypes in pulses crop. Promotion and adoption of short duration, disease and insect-pest resistant high yielding greengram cultivars with added ability to adjust in the climate resilient intensive cropping system is need of the hour for higher productivity and also to meet out the protein requirement for growing population of the country.

REFERENCES

All India Coordinated Research Project on MULLaRP. (2019). Project coordinator report (Mungbean and Urdbean), Group meet held at Lam, Guntur on May 19-21, 2019. ICAR-Indian Institute of Pulses Research, Kanpur -208 024, India.

Amutha, R. (2019). Management of drought by osmolytes in greengram under deficit irrigation.In: Report on 37th Crop scientist meet on pulses-2019 held on 9.04.2019 & 10.04.2019 at Tamil Nadu Agricultural University, Coimbatore.

Ananthi, M., Sasthri, G., Srimathi, P. and Malarkodi, K. (2007). Evaluation of storage potential of pre-sowing seed treatment in greengram *Journal of Pharmacognosy and Phytochemistry* 6(3), 502-505.

Chandrasekaran, B., Annadurai, K. and Somasundaram, E. (2010). A text book of Agronomy. New Age International (P) Ltd., Lucknow, India. pp. 835.

Crop Production Guide. (2020). Directorate of Agriculture, Chepauk, Chennai and Tamil Nadu Agricultural University, Coimbatore. pp.163.

Elamathi, S. and Manoharan, S. (2019). Studies on planting geometry and foliar spray application for yield maximization in green gram under dryland vertisols condition. In: Report on 37thCrop scientist meet on pulses-2019 held on 9.04.2019 & 10.04.2019 at Tamil Nadu Agricultural University, Coimbatore.

Geethalakshmi, V. and Ganesan, D. (2008). Impact of climate change on agriculture over Tamil Nadu. Climate Change and Agriculture over India. 80-93 pages.

Jayaraman, S. (2011). Dryland agriculture: prospects and retrospects. In. Productivity enhancement of in dryland through diversified farming options. Centre of Advanced Faculty Training in Agronomy, Tamil Nadu Agricultural University, Coimbatore. pp. 1-9.

Kuttimani, R. and Velayutham, A. (2011). Foliar Application of Nutrients and Growth Regulators on Yield and Economics of Greengram. *Madras Agricultural Journal* 98(4-6), 141-143.

Latha, K.R. (2011). Enhancing productivity of pulses and pulses based cropping system in rainfed areas. In: Productivity enhancement of in drylands through diversified farming options. Centre of Advanced Faculty Training in Agronomy, Tamil Nadu Agricultural University, Coimbatore. pp. 254-262.

Marimuthu, S., Kannan, P., Babu Rajendra Prasad, V. and Manivannan, N. (2020). Enhancing mungbean production under resilient climatic condition in Tamil Nadu. In: GP Dixit, PS Basu, Ammrit Lamichancey, CS Praharaj, Narendra Kumar, Abhishek Bohra, Naimuddin, PK Katiyar, AK Parihar, Meenal Rathore and Jitendra Kumar (eds) Abstracts: International conference on Pulses as the Climate Smart Crops: Challenges and Opportunities, ICAR- Indian Institute of Pulses Research, Kanpur. pp. 380.

Marimuthu, S. and Kannan, P. (2019). Biochar and phosphobacteria on carbon buildup, phosphorous availability and blackgram yield in rainfed Alfisol. In: Report on 37thCrop scientist meet on pulses-2019 held on 9.04.2019 & 10.04.2019 at Tamil Nadu Agricultural University, Coimbatore.

McCua, P. and Shetty, K. (2002). A Biochemical analysis of mungbean response to microbial polysaccharides and potential phenflac enhancing effects for nutraceutical applications. *Food Biotechnology* 16 (1), 57-64.

Palaniappan, S.P. and Sivaraman, K. (1996). Cropping Systems in the Tropics - Principles and Management. New Age International (P) Limited, New Delhi. pp. 211.

Raina, S.K., Govindasamy, V., Kumar, M., Singh, A.K., Rane, J. and Minhas, P.S. (2016). Genetic variation in physiological responses of mungbeans (*Vigna radiate* (L.)

Wilczek) to drought. *Acta Physiologiae Plantarum* 38, 263. doi: 10.1007/ s11738-016-2280-x

Rajendran, R. (1984). Studies on the response of greengram Co4 to foliar nutrition of urea and DAP under irrigated condition. M.Sc. (Ag.) Thesis. Tamil Nadu Agricultural University, Coimbatore.

Reddy, A.A., Bantilan, M.C.S. and Mohan, G. (2013). Pulses production scenario: policy and technological options. Policy brief no. 26, ICRISAT May 2013. pp. 1–8.

Sadasivan, R., Natrajaratnam, N., Dabu, R., Muralidharan, V. and Rangasmay, S.R. (1988). Response of mungbean cultivars to soil moisture stress at different growth phases. Mungbean Proceeding of the Second International Symposium. AVRCD. pp. 260–262.

Saraswathy, R., Krishnasamy, R. and Singaram, P. (2004). Nutrient management for rainfed greengram. *Madras Agricultural Journal* 91(4-6), 230-233.

Sathyamoorthi, K., Mohamed Amanullah, M., Vaiyapuri, K. and Somasundaram, E. (2008). Growth and yield of greengram (*Vigna radiata* (L.) Wilczek) as influenced by increased plant density and nutrient management. *International Journal of Agricultural Sciences* 4(2), 499-505.

Sathyamoorthi, K., Mohamed Amanullah, M., Vaiyapuri, K. and Somasundaram, E. (2007). Influence of Increased Plant Density and Fertilizer Application on the Nutrient Uptake and Yield of Greengram (*Vigna radiata* (L.) Wilczek). *Research Journal of Agriculture and Biological Sciences* 3(6), 886-895.

Singh, N.P. and Praharaj, C.S. (2020). Achieving self- sufficiency in pulses in India: Lessons learnt. In: GP Dixit, PS Basu, Ammrit Lamichancey, CS Praharaj, Narendra Kumar, Abhishek Bohra, Naimuddin, PK Katiyar, AK Parihar, Meenal Rathore and Jitendra Kumar (eds), Abstracts: International conference on Pulses as the Climate Smart Crops: Challenges and Opportunities, ICAR- Indian Institute of Pulses Research, Kanpur. pp.1-6.

Sobhana. E, Velayutham, A. and Sujithra, P. (2018). Evolving Viable Weed Management Practices for Rainfed Greengram. *Madras Agricultural Journal* 105(1-3), 13-15.

Solaimalai, A. and Subburamu, K. (2004). Seed hardening for field crops - A review. *Agricultural Reviews* 25(2), 129-140.

Somasundaram, E. and Arokiaraj, A. (2000). Text Book on Principles of Agronomy, Crystal Offset Printers, Tiruchirappalli.

Srimathi, P., Kavitha, S. and Renugadevi, J. (2007). Influence of seed hardening and pelleting on seed yield and quality in greengram (*Vigna radiata* (l.) Hepper) cv. Co 6. *Indian Journal of Agricultural Research* 41(2), 122-126.

Subbulakshmi, S. (2019). Integrated drought mitigation technologies for blackgram.In: report on 37[th]Crop scientist meet on pulses-2019 held on 9.04.2019 & 10.04.2019 at Tamil Nadu Agricultural University, Coimbatore.

Umamageshwari, C. (2019). Land configuration and foliar spray of nutrients for yield maximization of greengram. In: report on 37th Crop scientist meet on pulses-2019 held on 9.04.2019 & 10.04.2019 at Tamil Nadu Agricultural University, Coimbatore.

Varshney, R.K., Roorkiwal, M. and Nguyen, H.T. (2013). Legume genomics: from genomic resources to molecular breeding. *Plant Genome* 6(3), 1-7.

CHAPTER 7

Indeterminate pigeonpea: An ideal leguminous crop for rainfed cropping systems

V. Maruthi[1]*, K.S. Reddy[2], P.K. Pankaj[3], K. Srinivas[4], S.M. Vidyasekhar[5] and G. Sri Krishna[6]

Central Research Institute for Dryland Agriculture, Hyderabad-59, Telangana, India

*Corresponding author email: v.maruthi@icar.gov.in

Abstract

Pigeonpea is the most consumed pulse in India. This chapter highlights the advantages of indeterminate nature of pigeonpea as the best fit in rainfed agro-ecosystems. Indeterminate nature of pigeonpea gives an opportunity to fetch excellent yield in rainfed regions of India through improved resource use efficiency, best fit in the cropping systems such as strip intercropping, drought resilience in medium and long-duration genotypes and ratoon habit extending support to livestock feeding as well as livelihoods. Being perennial, these varieties are deep-rooted making them resilient to drought. Strip intercropping of sorghum with pigeonpea in 4:4 ratio has yielded 68.9% higher yield as compared to 2:1 intercropping system as additive series. In addition, as a value added product, pigeonpea milling byproduct (chunni) with low fibre, high crude protein of about 15-20% and high energy can be used as concentrate to feed the animals and protein supplement for human consumption. The pigeonpea ratoons can bridge the gap of fodder demand during lean period.

Keywords: Monsoon, Intercropping, Legume, Fodder, Drought, Ratoon

1. INTRODUCTION

Pigeonpea (*Cajanus cajan* (L.) Millspaugh) is an important drought tolerant legume crop, which can be grown in areas receiving annual rainfall even less than 500 mm. Though, India accounts for nearly 90% of the world's pigeonpea acreage (3.73 M ha) and production (3.07 Mt), but the productivity (707 kg/ha) is lower than the world average. Pigeonpea is one of the most consumed pulses in the form of *dal* in India comprising 28.18% of total pulses consumption (11.0 kg per capita per year) in both rural and urban areas. To fulfill the demand for pigeonpea in the country, it is being imported from Asian and African countries. In addition to contributing directly to food security, pigeonpea also plays an important role in providing valuable byproducts for animal feeding and thus indirectly contributing to fodder security. There is also considerable potential to use crop byproducts (crop residues) left after harvesting the seeds, as sources of dry fodder for livestock. In addition, these byproducts do not compete with human food, and contribute to decreasing cereals and soybean levels in the diets of livestock in intensive livestock production systems.

Sorghum and pigeonpea intercropping are one of the established recommendations in semi-arid tropical (SAT) rainfed regions for the stability and productivity of the cropping systems (Subba Reddy *et al.*, 2001). In dryland areas, frequent droughts are the chronic problem with frequencies of dry days increasing due to climate change, it is essential to look into those types of pigeonpea which can survive well under the odd weather. The indeterminate type of pigeonpea has the unique characteristic of hibernating during long dry spells and triggering the vegetative growth with rainfall or favourable soil moisture. Since, pigeonpea has a tap root system and the indeterminate plant type has been bred and evolved for long duration standing, their roots are deep enough to sustain the active plant life. So, this chapter has been designed to understand the prospects and constraints of indeterminate types of pigeonpea under rainfed situation.

2. DEFICIT AND EXCESS RAINFALL CONDITIONS

Inter and intra-seasonal rainfall variation, fluctuating intensities of drought, pest and disease complexes restrict the pigeonpea crop performance in rainfed regions. However, various maturity duration pigeonpea genotypes accomplish the completion of lifecycle differently. Plasticity of the pigeonpea crop instills the characteristic of drought resistance as the indeterminate crop type grows along with overlapped vegetative, flowering and pod formation stages (Robertson *et al.*, 2001). In order to transform into reproductive phase from vegetative stage, physiological parameters like the economic threshold status of leaf area and shoot biomass have to be established besides relative water content and osmotic adjustment. Further, the soil moisture stress affects the formation of root nodules restricting the symbiotic nitrogen fixation (Kumar *et al.*, 2011) while the trait of deep root system imparts drought resistance in pigeonpea to achieve better yield attributes and best yields (Choudhary *et al.*, 2011). Determinate type of pigeonpea varieties is mostly preferred in the irrigated or assured rainfall regions of India. These varieties are more sensitive towards the soil moisture deficit conditions (Mula *et al.*, 2016), which reduce the vegetative growth (otherwise are fast growing), and the plants are not able to conserve enough energy during this phase to be viable to reproduce. Further, the plant loses its capacity to recoup from the deficit moisture shock. If grown as an intercrop, there is a chance of competing for resources (soil, water, light, nutrients, etc.) with the main crop due to synchrony in the cropping period.

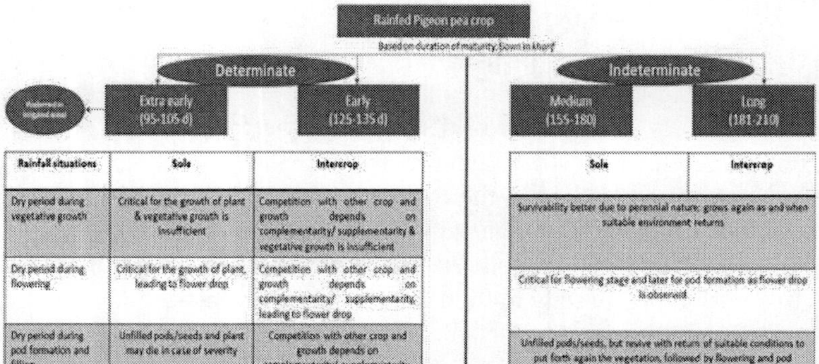

Fig. 1. Pigeonpea types and their characteristics concerning different maturity period

The indeterminate types of pigeonpea are a crop of choice both as sole as well as intercrop in the rainfed regions of India (Fig. 1). Being perennial, these varieties are deep rooted to sustain crop for longer duration of time as well as a coping mechanism to bring resilience to drought. Depending on the occurrence of dry periods during different phenophases of pigeonpea crop, the indeterminate varieties may become dormant to withstand the adversity. They again begin their respective growth at the stage where dry period incidence has begun. If the dry period is coming at the vegetative stage, the plant will begin its journey from vegetative stage once the suitable soil moisture returns. However, if the dry period is initiated at the flowering stage, there will be incidence of flower drop. If the dry period is coinciding with pod formation and filling stage, the plant will produce either unfilled pods or shrivelled seeds. This unique coping mechanism in indeterminate varieties makes them more suited for chronic drought prone areas.

3. CROPPING SYSTEM WITH PIGEONPEA IN RAINFED REGIONS

Studies at ICRISAT showed that the pigeonpea plant could accumulate only 16% of the total drymatter yield in sorghum+ pigeonpea intercropping system, of its sole crop while it picked up later by 53% of the sole crop dry matter after the harvest of the sorghum main crop. According to Natarajan and Willey (1980), low resource use efficiency of the pigeonpea crop in the intercropping system could be improved by manipulating geometry of the component crops as it was shown in 2:1 additive series of sorghum+pigeonpea (Table 1). In order to understand the suitable strategy for intercropping of the indeterminate variety of pigeonpea with sorghum crop under extended monsoon situations in rainfed regions and to deal with the reduced resource use efficiency of 2:1 additive series intercropping, an experiment was conducted at ICAR-CRIDA Research Farm, Hyderabad with sorghum and pigeonpea in 4:4 strips (replacement series) and compared with 2:1 ratio (additive series). The extended monsoon with the rainfall of 315 mm was experienced during 2013 as compared to the rainfall of 51 mm during 2014. The plasticity of pigeonpea crop is well exploited in case of strip system with extended monsoon rainfall situation. This was done by transforming 2:1 additive series system into 4:4 strip system with replacement series which enhanced yields by 22% and returns by Rs. 14,910/- per ha due to indeterminate nature of pigeonpea and generation of more number of branches (Maruthi *et al.,* 2017). This phenology was exhibited due to increased opportunity for pigeonpea growth even during or after sorghum as main crop was harvested. Hence, indeterminate pigeonpea could be the matching crop for most of the prevalent cropping systems, which vary spatially and temporally. Boosted production efficiency, heightened water productivity are some advantages recorded in strip intercropped pigeonpea (Maruthi *et al.*, 2017) besides the indirect benefits of resource conservation and environmental protection. Although both the systems received same amounts of rainfall, pigeonpea in 4:4 strip system had luxuriant vegetative growth (Fig 2), capitalizing on the extra rain events occurred, and extra aerial space accessible (after the harvest of tall growing main crop of sorghum) for the crop to expand while pigeonpea in 2:1 additive series system failed to capitalize on the extra rain events due to space congestion unable to expand its growth (Table 1). Besides this, relay or sequence cropping of horsegram for fodder security in place of harvested sorghum is a possibility in rainfed lands. Further, the strip system addresses the cereal and pulse needs of the family of five for 81-87 days and 415-565 days, respectively. During the scarce rainfall year, both the systems performed at par. Hence plasticity or indeterminate nature of the pigeonpea crop could address food/fodder security through crop diversification and intensification.

Table 1. Sole and intercropping yields of indeterminate pigeonpea in different farming situations and competition ratios

Intercropping Systems	Farming Situation	Competition	Pigeonpea Yields (q ha^{-1})	Reference
Sorghum + pigeonpea (4:4)	Alfisols; Rainfed areas of SAT; High rainfall extended monsoon years	Replacement series	Sole crop: 16.1 Intercrop: 17.4	Maruthi *et al.*, 2017
Sorghum + pigeonpea (2:1)	Alfisols; Rainfed areas of SAT; High rainfall extended monsoon years	Additive series	Sole crop: 16.1 Intercrop: 10.4	
Pigeonpea + maize (2:1)	Sandy Loams; Rainfed areas of Bihar; High rainfall years	Additive series	Sole crop:15.1 Intercrop: 16.3	Pandey *et al.*, 2013
Pigeonpea + finger millet (2:4)	Sandy Loams; Rainfed areas of Odisha	Replacement series	Sole crop:8.2 Intercrop: 4.9	Padhi *et al.*, 2010
Pigeonpea + finger millet (2:8)	Sandy Loams; Rainfed areas of Odisha	Additive series	Sole crop:8.2 Intercrop: 2.9	

Fig. 2. Strip intercropping of pigeonpea+ sorghum (4:4)

4. ROOT SYSTEM FOR DROUGHT MITIGATION

A deep and extensive root system attributes pigeonpea plants to withstand several types of water-limiting stresses under rainfed conditions. Medium and long duration pigeonpea varieties have the advantage of deeper root system as a presupposition for sustaining growth and avoiding terminal water deficits (Chauhan *et al.*, 1992). A compromise in the root system happened to be seen in the early and extra early varieties (prolific but shallow root system), as a part of the reduction in its crop duration to fit into the intensive cropping systems, which make them more vulnerable to soil moisture deficit stress. Further, the medium and long duration types have recorded more primary roots as compared to the early maturity types, a trait, which features drought tolerance (Upadhyaya *et al.*, 2012; Mula *et al.*, 2016). Under deficit moisture stress, root depth and biomass are positively correlated with grain yield of pigeonpea (Choudhary *et al.*, 2011).

5. VARIETIES SUITABLE FOR RAINFED CROPPING SYSTEMS

Pigeonpea varieties popular among farmers in semi-arid tropics (SAT) are mostly of medium (maturing in 150-200 days) and long-duration (maturing in >200 days) types, which are basically indeterminate type and the yields are higher than determinate genotypes (extra early and early maturing varieties). These varieties produce flowers throughout the growing season (longer period) under the suitable moisture regimes unlike the determinate genotypes. Some popular indeterminate pigeonpea varieties grown in India are Hy-3C, BSMR-736, BSMR-853 (in Karnataka and Telangana), LRG-52 (in Andhra Pradesh), PRG-158, PRG 176 (in Telangana), ICPH-2740, ICPH-2741, ICPL-87119 (in Maharashtra, Andhra Pradesh, Telangana and Karnataka), IPA-203 (in Uttarakhand, Jharkhand, Bihar and Uttar Pradesh), BDN-711 (in Maharashtra); ICPH-2671 (in Madhya Pradesh), Rajeev Lochan (in Chhattisgarh), Pusa 2002 (in central India); CORG 8 (in Tamil Nadu) (Tiwari *et al.*, 2017).

6. WATER AND FERTILIZER MANAGEMENT

Mostly the medium duration and long duration pigeonpea genotypes suffer from terminal drought or early drought in SAT regions of peninsular India. Crop requires supplemental irrigation if the harvested rainwater during May month available, for timely sowing and better germination. Studies in CRIDA explained the advantage of farm ponds when the crops required early season soil moisture support which increases the seed yields of pigeonpea by 20-100%. If soil moisture is not supplemented, the crop season gets jeopardized with the delayed sowing rains and sowing season. An experiment was conducted by authors with pigeonpea as a test crop in semi-arid-tropic (SAT) region to evaluate the water harvesting in on-farm reservoir (OFR). An OFR with capacity of 600 m^3 and a catchment of 1.5 ha size farm land was constructed to harvest rain water. Micro-irrigation system like sprinkler and raingun were used for effective utilization of harvested water in OFR along with organic mulch @ 5 t/ha. The results revealed that the crop of pigeonpea was sown with pre-sowing irrigation of 30 mm due to lack of soil moisture during sowing with the OFR water stored during the pre-monsoon rainfall. The pigeonpea crop also faced terminal drought in the same year and the crop was again irrigated with OFR stored water by using micro-irrigation system (sprinkler & raingun). These two irrigations during dry spells resulted in increased yield of pigeonpea (1.4 t/ha) by 3.0 times over the rainfed (0.45 t/ha) with organic mulching. Therefore, on-farm rainwater harvesting is an efficient climate resilient practice to be adopted in SAT for cultivation of pigeonpea.

Sprinkler irrigation with OFR at critical stage	Organic mulching (*Glyricidia* leaves)

Fig. 3. Dry spell management through OFR and organic mulching

The response of pigeonpea to applied fertilizer nutrients is limited by soil moisture availability and rainfall pattern and therefore, response varied from location to location and year to year. Imbalanced nutrients application is one of the major abiotic constraints limiting pigeonpea productivity. In general, farmers either do not use fertilizer or apply sub-optimal dose in pigeonpea under rainfed areas. Due to the inherent capability of fixing of atmospheric nitrogen in association with rhizobium, pigeonpea requires only 20-30 kg N/ha as starter dose to support initial plant growth. However, in case of poor nodulation due to soil moisture stress, foliar application of 2% urea was found beneficial. Phosphorus is required for proper root growth and thus, positive response to phosphorus upto 60 kg/ha was reported by many researchers (Ahlawat and Saraf, 1998; Deshbhratar *et al*., 2010). In general, the response to potassium is poor in pigeonpea, but in the area where soil is deficient, potassium response is reported up to 40 kg K_2O/ha (Deshmukh *et al*., 2018). Soil application of micronutrients like $ZnSO_4$@ 25 kg/ha and boron @ 20 kg/ha to pigeonpea increased the seed yields depending upon the micronutrient status of soils. The foliar application of nutrients like 2% urea/DAP/KCl is also recommended under soil moisture stress conditions (Selvi *et al*., 2009). The integrated nutrient management practices recommended in pigeonpea by different researchers are given in Table 2. Further, with the high decomposition rates of litter fall of pigeonpea (1.9-2% nitrogen content), available soil nitrogen and soil health improved under this crop, particularly with the indeterminate genotypes, which is beneficial to sustain the smallholder agriculture in rainfed areas.

Table 2. Integrated nutrient management in pigeonpea

Integrated nutrient management	**Reference**
50% RDF + 2.5 t/ha vermicompost or phosphocompost	Sharma *et al*., 2010; Pal *et al*., 2016
100% RDF + 50% N through vermicompost + 5 kg Zn/ha	Kumawat *et al*., 2013
50% RDF + Rhizobium and PSB inoculation	Reddy *et al*., 2011
FYM @ 5 t/ha + 100% RDF + Biofertilizers	Sharma *et al*., 2012
50% RDF + FYM 5.0 t/ha + Biofertilizers (Rhizobium + PSB+PGPR)	Mittolia *et al*., 2015; Pandey *et al*., 2015

7. RATOONABILITY

Ratooning as they say is cutting plants at branching level for re-growth in pigeonpea is one option to exploit the indeterminate nature of pigeonpea either for grain (in early and medium maturing genotypes) or for fodder (in late maturing genotypes). The factors responsible for success of a ratoon crop may be the soil moisture availability, ratoonability of the plant and crop maturity duration. Sharma *et al.* (1978) concluded that the potential re-growth in indeterminate pigeonpea is high which in medium duration types could be exploited for 50% of the yield of main crop harvest and for green biomass in case of late maturing pigeonpea genotypes. Pandey (2000) also reported similar yield in pigeonpea ratoons. Pigeonpea's regenerative nature is much more remunerative in these days of climate change to capitalize on the extended South-West monsoon situation. The forage so available is much more valuable for the winter and early summer seasons, if long duration indeterminate genotypes are cultivated.

8. PEST AND DISEASE DYNAMICS

Pigeonpea has the risk of economic damage by about 12 insect-pests although the crop has around 300 insect pests. Among insect pests, *Helicoverpa armigera* and *Maruca* Spp. are predominant, while sterility mosaic virus and wilt are perilous among diseases. In general, these pigeonpea pests are managed through cultural, biological and chemical methods. Pests on indeterminate pigeonpea are being managed by growing trap crops like Marigold, cowpea, coriander, chickpea, sorghum etc for encouraging natural enemies. However in this case, sorghum as a crop for natural enemies needs to be freshly evaluated in view of the climate change and variations in temperatures over seasons etc. Also neem based formulations are applied as anti-feedants against lepidopteran pod borers (Lal *et al.*, 1996). In case of short duration pigeonpea, the short time span available with the plant to recoup from the pest incidence causing total crop failure is a major disadvantage. Further, the relationship between the sweetness of seed and the pest incidence in short duration pigeonpea is yet to be established.

Fusarium wilt (*Fusarium udum* Butler) and sterility mosaic disease (virus) are two major diseases that all the types of pigeonpea suffer from but differently based on its soil endemic nature, intensity of infestation and agronomic practices etc., while *Phytophthora* blight is a major problem in short and extra short duration pigeonpea (Reddy, *et al.*, 1996).

9. POST-HARVEST PROCESSING AND VALUE ADDITION

Reducing phytate and tannin levels improve the quality of pigeonpea seed for consumption through heat processing. Dehulling is one another technique used for pigeonpea processing as recovery of *dal* in this process is around 70-80%. The byproducts of dehulling process, hulls, shriveled seeds, broken seeds and endosperms form '*chunni*' mostly contain about 15-20% crude protein, 3.5-4.5% ash and 55-65% total digestible nutrient (TDN). It is valued as a concentrate feed due to its comparatively low fibre and more in energy and protein content as compared to roughage. Pigeonpea *chunni* has been reported to be safely included at $\geq$10 per cent level in pig and other animal rations and thereby compensate for the shortage of cereals and cereal byproducts for animal feeding and to utilize locally available feed resources more efficiently (Barman *et al.*, 2012). Apart from this, *chunni* byproducts are also used to fill feed gaps during periods of acute shortage of other feed resources and used as adjuncts to the natural pastures and planted forages. Pepsin hydrolysed protein concentrate (PC) of pigeonpea milling byproducts exhibited antioxidant activity which on heating resulted in the formation of a gel which could be exploited as a protein based carrier as a functional food material for human consumption and would be safe compared to synthetic materials (Tapal and Tiku, 2013). As a product of ratooning of indeterminate pigeonpea varieties, forage is of high value with 24% crude protein, 36% crude fibre and significant amount of minerals, could be used as a part of ration for livestock.

10. CONCLUSION

Pigeonpea is the most important rainy season crop in rainfed regions of the country. The indeterminate varieties of medium and long-duration pigeonpea have the potential to perform better than other crops in rainfed regions where rainy season crops encounter soil moisture deficit stress or drought at different crop growth stages. The indeterminate pigeonpea is also a suitable candidate of intercropping with cereals (sorghum, pearl millets, etc) and oilseed (groundnut, soybean). The leaf litter and root biomass added by pigeonpea crop improves the soil health. In true sense, we can say that indeterminate pigeonpea is a climate resilient crop and it supports the livelihood and nutritional security of resource poor farmers of rainfed regions.

REFERENCES

Ahlawat, I. and Saraf, C. (1981). Response of pigeon pea (*Cajanus cajan* (L.) Millsp.) to plant density and phosphorus fertilizer under dryland conditions. *The Journal of Agricultural Science* 97(1), 119-124.

Barman, K., Das, A., Tamuli, M.K., Banik, S. and Pankaj, P.K. (2012). Effect of replacing maize with steamed flaked rice byproducts (SFRB) on the performance of growing crossbred (Hampshire x Niang Megha) pigs. *Indian Journal of Animal Nutrition* 29(2), 201-203.

Chauhan, Y.S., Saxena, N.P. and Johansen, C. (1992). Abiotic factors limiting chickpea and pigeonpea production. In J.N. Sachan (ed.) Proceedings of the National Symposium on New Frontiers in Pulses Research and Development. Directorate of Pulses Research, Kanpur, India.pp. 111–123.

Choudhary, A.K., Sultana, R., Pratap, A., Nadarajan, N. and Jha, U.C. (2011). Breeding for abiotic stresses in pigeonpea. *Journal of Food Legumes* 24, 165–174.

Deshbhratar, P.B., Singh, P.K., Jambhulkar, A.P. and Ramteke, D.S. (2010). Effect of sulphur and phosphorus on yield, quality and nutrient status of pigeonpea (*Cajanus cajan*). *Journal of Environmental Biology* 31(6), 933-937.

Deshmukh, M.S., Zade, S.P. and Ajabe, M.A. (2018). Effect of Potassium Management on Nutrient Uptake (N,P,K) in Pigeon Pea under Vertisols. *Journal of Agriculture Research and Technology* 43(2), 301-304.

Kumar, R.R., Karjol, K. and Naik, G.R. (2011). Variation of sensitivity to drought stress in pigeon pea (*Cajanus cajan* (L) Millsp) cultivars during seed germination and early seedling growth. *World Journal of Science and Technology* 1(1), 11-18.

Kumawat, N., Singh, R.P., Kumar, R. and Hari, O. (2013). Effect of integrated nutrient management on the performance of sole and intercropped pigeonpea (*Cajanus cajan*) under rainfed conditions. *Indian Journal of Agronomy* 58(3), 309-315.

Lal, S.S., Yadava, C.P. and Ahmad, R. (1996). Insect pests of short duration pigeonpea and their management: current status andscope. Pp. 69-72. In Prospects for growing extra short duration pigeonpea in rotation with winter crops: proceedings of the IARI/ICRISAT Workshop and Monitoring Tour. 16-18 Oct 1995, New Delhi, India. Laxman Singh, Chauhan Y.S., Johansen, C., and Singh S P (eds.). New Delhi, and Patancheru, Hyderabad, India.

Maruthi, V., Reddy, K.S. and Pankaj, P.K. (2017). Strip cropping system as a climate adaptation strategy in semi-arid Alfisols of South-Central India. *Indian Journal of Agricultural Sciences* 87(9), 1238-45.

Mittolia, V.K., Singh, R.K. and Verma, S. (2015). Performance of integrated nutrient management in pigeonpea (*Cajanus cajan*). *Journal of Community Mobilization and Sustainable Development* 10(1), 5-8.

Mula, M.G., Patil, S.B., Aden, J., Rathore, A., Vemula, A.K. and Kumar, R.V. (2016). Screening of pigeonpea geotypes for drought stress at early vegetative phase in Alfisol and Vertisol. *Green Farming* 7(3), 507-511.

Natarajan, M. and Willey, R.W. (1980). Sorghum-pigeonpea intercropping and the effects of plant population density 1. Growth and Yield. *Journal of Agricultural Science (Cambridge)* 95, 51-58.

Padhi, A.K., Panigrahi, R.K. and Jena, B.K. (2010). Effect of planting geometry and duration of intercrops on performance of pigeonpea- fingermillet intercropping systems. *Indian Journal of Agricultural Research* 44(1), 43-47.

Pal, A.K., Singh, R.S., Shukla, U.N. and Singh, S. (2016). Growth and production potential of pigeonpea (*Cajanus cajan* L.) as influenced by intercropping and integrated nutrient management. *Journal of Applied and Natural Science* 8(1), 179-183.

Pandey, I.B., Pandey, R.K. and Kumar, R. (2015). Integrated nutrient management for enhancing productivity and profitability of long-duration pigeonpea (*Cajanus cajan*) under rainfed condition. *Indian Journal of Agronomy* 60(3), 436-442.

Pandey, I.B., Singh, S.K. and Tiwari, S. (2013). Integrated nutrient management for sustaining the productivity of pigeonpea (*Cajanus cajan*) based intercropping systems under rainfed condition. *Indian Journal of Agronomy* 58(2), 192-197.

Pandey, N. (2000). Assessment of yield performance of long duration pigeonpea hybrids in ratoon crops. *Legume Research- An International Journal* 23(3), 159-162.

Ravindranath, N.N.V.S., Satyanarayana, N.V., Prasad, P. and MadhavRao, M.V. (1985). Foliar application of potassium on the growth and yield components of pigeonpea (*Cajanus cajan* (L.) Millsp.). *Proceeding of Indian Academy of Sciences (Plant Sci.)* 94, 671–676.

Reddy, A.S.R., Babu, J.S., Reddy, M.C. and Khan, M.M. (2011). Integrated nutrient management in pigeonpea (*Cajanus cajan*). *International Journal of Applied Biology and Pharmaceutical Technology* 2(2), 467-470.

Reddy, M.V., Varaprasad, K.S., Raju, T.N. and Lenné, J.M. (1996). Disease and nematode problems of extra short duration pigeonpea and their management. Pp 73-79. In Prospects for growing extra short duration pigeonpea in rotation with winter crops: proceedings of the IARI/ICRISAT Workshop and Monitoring Tour. 16-18 Oct 1995, New Delhi, India (Laxman Singh, Chauhan Y.S., Johansen, C., and Singh S P eds.) New Delhi, and Patancheru, Hyderabad, India.

Robertson, M.J., Silim, S., Chauhan, Y.S. and Ranganathan, R. (2001). Predicting growth and development of pigeonpea: Biomass accumulation and partitioning. *Field Crops Research* 70(2), 89-100.

Selvi, R.V., Srinivasan, S., Ramasamy, M. and Marimuthu, R. (2018). Agronomic management of pigeon pea under drought conditions. *Legume Research-An International Journal* 32(2), 139-141.

Sharma, A., Rathod, P.S. and Chavan, M. (2010). Integrated nutrient management in pigeonpea (*Cajanus cajana*) based intercropping systems under rainfed conditions. *Karnataka Journal of Agricultural Sciences* 23(4), 584-589.

Sharma, A., Rathod, P.S., Dharmaraj P.S. and Chavan, M. (2012). Response of Pigeonpea to biofertilizers in pigeonpea based intercropping systems under rainfed conditions. *Karnataka Journal of Agricultural Sciences* 25(3), 322-325.

Sharma, D., Saxena, K.B. and Green, J.M. (1978). Potential of ratooning in pigeonpea. *Field Crops Research* 1, 165-172.

Subba Reddy, G., Maruthi, V. and Vanaja, M. (2001). Effect of soil depth on productivity of sorghum (*Sorghum bicolor*) and pigeonpea (*Cajanus cajan*) in sole and intercropping systems. *Indian Journal of Agricultural Sciences* 71(8), 510-515.

Tapal, A. and Tiku, P.K. (2013). Utilization of pigeonpea milling waste by-product as a protein source. *International Journal of Agriculture and Food Science Technology* 4(8), 741-742.

Tiwari, A.K., Shivhare, A.K. and Kumar, V. (2017). Pigeonpea production technologies. Directorate of Pulses Development, Department of Agriculture Cooperation & Farmers Welfare, Vindhyachal Bhavan, Bhopal, India. pp -12.

Upadhyaya, H.D., Kashiwagi, J., Varshney, R.K., Gaur, P.M., Saxena, K.B., Krishnamurthy, L., Gowda, C.L.L., Pundir, R.P.S., Chaturvedi, S.K., Basu, P.S. and Singh, I.P. (2012). Phenotyping chickpeas and pigeonpeas for adaptation to drought. *Frontiers in Physiology*. https://doi.org/10.3389/fphys.2012.00179.

CHAPTER 8

PGPR for abiotic stress management in pulses

A. Velmurugan[1*], T.P. Swarnam[1] and A.K. Singh[2]

1. *ICAR-Central Island Agricultural Research Institute, Port Blair-744101*

2. *ICAR-Indian Institute of Pulses Research, Kanpur – 208 024*

*Corresponding author email: vels_21@yahoo.com

Abstract

Food production system is greatly challenged with inconsistent rainfall, rise in temperature, and salinization of farmlands more than before. These challenges in the form of abiotic stress (drought, salinity, alkalinity, and heat) affect the normal functioning of plants which impair crop performance. This affects food security, and yield reduction ranging from 5-50% has been reported for pulse crops. Though plants are inherently equipped with stress tolerance ability, heightened stress level negatively affect plant growth, nutrient and water relations, photosynthesis, assimilate partitioning, N-fixation and reduction in crop yield. In majority of the pulse crops such as chickpea, pigeonpea, greengram, blackgram, lentil and beans, studies revealed supplementary production of phytohormones by PGPR inoculation that induce root growth, thus allowing better water absorption, production of deaminase enzyme responsible for the disintegration of ACC molecules, nitrogen fixation, and solubilization of P under abiotic stress conditions. Considering the importance of PGPR in alleviating abiotic stresses in plants, this article has reviewed the different aspects and mechanisms of rhizobacteria that play key role in promoting growth and yield of major pulse crops. It is concluded that PGPR can be used as a cost effective and ecologically sustainable tool for abiotic stress management in pulses aimed at enhancing crop productivity and total production.

***Keywords*:** Adaptation, Climate change, Crop production, Drought, Rhizobacteria, Salinity

1. INTRODUCTION

Climate change and burgeoning human population are the two major factors affecting agricultural production. They are interrelated and one influences the other with the consequent impact on natural resource degradation and heightened challenge to food production. Salt accumulation, water logging, erosion, pollution are some of the major soil degradation concerns directly related to food production for human population. On the other hand, climate change can result in an unregulated supply of rain water which may occur either in excess or in prolonged short supply. Extreme rainfall gives rise to erosion and flooding, while drought results in loss or poor performance of crops. At the same time, in a bid to boost crop productivity and yield and to combat drought-induced stress, farmers resort to the use of irrigation which adds further salt to the soil. All the above factors and processes ultimately create abiotic stress on crop plants which turned out to be a major challenge to sustainable food production.

Abiotic stresses are stress conditions to plants arising from the plant growing environment. They are the non-living part of the ecosystem whose effect is felt by the living component of the system that might jeopardize the well-being of living things. Water, nutrients, salts, temperature, and pH are among the basic abiotic components of the agricultural ecosystem having direct influence on plants (Enebe and Babalola, 2018). Different plant species vary in their vulnerability and adaptation to abiotic stresses that also affect several soil dwelling microbes linked with crop production. Some of these soil dwelling microbes particularly in the rhizosphere region have evolved several mechanisms to protect themselves to survive in these adverse conditions and also trigger consequential changes in plants leading to crop adaptation to abiotic stresses. Further, in an effort to survive and prevail at the rhizosphere, some microbes possessing the cellulase enzyme capable of dissolving the cellulose cell wall of plant roots gain entrance into the apoplast of plants, the cell wall interior as well as the vascular bundle where they live and undergo normal metabolic activities (Khan *et al.*, 2016). They also provide the plant with useful metabolites that are very essential to overcome abiotic stress.

Researchers are engaged in finding a viable strategy for managing these stress inducers so as to enable plants to increase their survival and performance (Alori *et al.*, 2017). However, more needs to be done on the integrative aspect of this enhancement strategy of protecting and empowering plants to resist and grow better under drought, salinity, or alkaline conditions. Suitable adjustment in soil management and adequate nutrient management with respect to improving the fertility of the soil have been suggested in solving the problem of climate change challenges on crops (St Clair and Lynch, 2010) together with the use of rhizobacteria to boost crop production (Babalola, 2010). But the rapidly growing demand to increase food production together with diversion of agricultural land to non-production purposes has been pushing agriculture to marginal and degraded land. Further, the condition is aggravated by climate change that challenges the production system with drought and salt stress. This is more pertinent to pulses which are very important constituent of food crops in India. Pulses are the major source of protein in vegetarian diet that grow quickly, maintain soil fertility and generate good return to farmers. India is the largest producer (25%), consumer (27%) and importer of pulses in the world (DES, 2019). Pulses are grown in 29.03 M ha area with the total production of 23.40 Mt and average productivity of 806 kg/ha (2018-19). The productivity is low due to salinity and water stress among other factors. This constitutes a compelling reason for evolving a suitable way to successfully manage pulses under these emerging conditions. In this context, understanding the pulse (plant) - microbial interactions and microbial aided plant adaptation mechanism is of great help for abiotic stress management.

Several studies have proved the usefulness of plant growth-promoting rhizobacteria (bio-inoculants) in combating these abiotic induced changes (Alori *et al.*, 2017; Staudinger *et al.*, 2016). PGPR induce tolerance to abiotic stress in crops/pulses by production of phytohormones, volatile compounds, exo-polysaccharides, ACC deaminase and osmolyte, and triggering antioxidant activities (Wang *et al.*, 2003; Glick, 2012). Therefore, isolation, characterization and evaluation of PGPR favouring plant performance under stress conditions followed by its large scale application will constitute a viable strategy for enhancing pulse productivity. With this background, this chapter aims to understand the influence of plant growth-promoting rhizobacteria in plant tolerance to abiotic stress, and attempts to highlight some of the measures involving PGPR to support plant exposed to abiotic stress to overcome. Further, an attempt has been made to compile and analyze literatures focused on studying the effect of PGPR on different pulse crops and based on that suggest ways to improve the pulse production by harnessing the benefit of PGPR in alleviating abiotic stresses.

2. SIGNIFICANCE OF ABIOTIC STRESS

Abiotic stresses such as drought, salinity, and high or low temperatures are among the major cause of poor crop growth and reduced yield. At the global level, it is reported that high temperature and precipitation occurrence since 1980 have lowered yield of foodgrains by 3-6% (Borlaug Institute for South Asia, 2015). With the projected climate change and land degradation, production performances in the coming decades are expected to be much worse than before with the consequent impact on world peace (Lal, 2015). This is more pertinent to pulses and millets that are mostly grown in marginal lands under rainfed conditions. Without adaptation measures or use of improved technologies, productivity of pulses may decrease upto 6-12% with lots of regional level and seasonal variations. For example, the yield of chickpea is project to decrease upto 50% due to salinity or water stress (Table 1). Similar is the case in respect of lentil and greengram (Sarma and Saikia, 2014; Kumar *et al.*, 2016; Rana *et al.*, 2016). This would jeopardize the efforts to achieve food security and sustainable development.

Although abiotic stresses are found to affect food production, the severity and duration may vary at different locations and seasons. Accordingly crops are able to tolerate certain environmental stresses through their innate defense mechanism. The degree of tolerance and adaptability to abiotic stresses may differ from species and varieties, and even seasons. Higher temperature and moisture stress would aggravate the salinity stress and in many cases drought and salinity may occur simultaneously. Crop in their early stages do not show any visible symptoms but their physiology can undergo significant changes with the progress of intensity of stress (Cramer *et al.*, 2011). Changing rainfall patterns and reduced precipitation is causing the frequent onset of drought around the world (Lobell *et al.*, 2011). Plant subjected to drought interfered with growth, nutrient and water relations, photosynthesis, assimilate portioning and reduction in crop yield (Farroq *et al.*, 2009). Reduced absorption of photosynthetically active radiation, radiation use efficiency and decreased harvest index are the major yield-reducing factors under low soil moisture (Earl and Davis, 2003). In chickpea, terminal drought stress reduced seed yield which ranged from 26 to 61% (Kumar *et al.*, 2016) and in pigeonpea where seed yield decreased by 40-55% when imposed drought at flowering stage (Nam *et al.*, 2001).

Heat stress greatly affects the morphological, biochemical and physiological changes in pulse crops particularly in rainfed areas (Wahid *et al.*, 2007). It also reported to cause severe damage to membrane integrity, damaged protein functions and significantly reduced yield and quality of crops (Smertenko *et al.*, 1997). The rising temperature or long dry spell may cause a shift of growing periods and distribution of the agricultural crops (Porter,

2005). Similarly, salinity is also a major worldwide problem which can develop naturally due to geographical conditions. However, due to human intervention, landscape hydrology has changed and salinization of waterways and land has accelerated, affecting agricultural productivity.

Yield reduction due to abiotic stresses in major cereals and pulses would be significant but the yield loss would mostly be felt in the small holder farms. The estimated yield losses in major pulses due to abiotic stresses are summarized in Table 1. However, these studies didn't project any clear trend or degree of vulnerability of different species to abiotic stresses. There are lots of uncertainty and region level variations based on the resource management and adaptive capacity of the farmers. Despite this, all these studies have highlighted yield loss in all the major pulse crops across various agro-climatic regions.

Table 1. A glimpse of recorded yield loss in major pulses due to abiotic stress

Crops	Abiotic stress	Yield loss (%)	Reference
Chickpea	Moisture stress (Terminal)	30-60	Kumar *et al.* (2016)
	Salinity (ESP >10)	22-50	Rana *et al.* (2016)
Lentil	Moisture stress (Terminal)	6-54	Kumar *et al.* (2016)
	Salinity (ESP >15)	Upto 50	
Faba bean	Moisture stress (Terminal)	Upto 70	Rana *et al.* (2016)
Field pea	Moisture stress (Terminal)	20-54	Kumar *et al.* (2016)
Greengram	Moisture stress	30-45	Saravanakumar *et al.* (2011) Sarma and Saikia (2014)

Chickpea, pigeonpea, blackgram and greengram are the major pulse crops grown in India with the total pulse production of around 23 Mt. Global yields of both chickpea (968 kg/ha) and pigeonpea (762 kg/ha) have been relatively stagnant (FAOSTAT, 2013) for the last five decades in spite of using various conventional and molecular breeding approaches and extensive use of inorganic fertilizers and pesticides that created environmental and health concerns. Similar is the case in India with respect to the productivity of almost all pulse crops. In greengram, salt stress along with other pests (stem-and pod-borer) and yellow mosaic disease caused 80-100% yield loss particularly during rainy season (Sehrawat *et al.*, 2015). In India, especially in Indo-Gangetic Plains, area under pulse crop has declined gradually primarily due to salinity stress and high uncertainty associated with rainfall. For example the area under chickpea cultivation has reduced considerably due to soil salinity and increased use of brackish water for irrigation (Gowda *et al.*, 2009). This necessitates developing viable alternative strategy to alleviate the abiotic stress on pulse crops supported by agronomic practices and genetically well adapted cultivars among other measures.

3. CROP RESPONSE TO ABIOTIC STRESS

Crops are able to tolerate certain level of stresses either biotic or abiotic through their innate defense mechanism. With reference to pulse crops, the degree of tolerance and adaptability to abiotic stresses may differ from species and varieties. In certain cases the tolerance is influenced by other climatic and edaphic factors. Abiotic stress on plant occurs mainly due to salinity, drought and heat that trigger a sequence of physiological changes in plant system termed as plant response (Fig. 1). Understanding plant response assumes significance not only to identify stress tolerant cultivars but also to appreciate the effect of PGPR in alleviating abiotic stress on pulses.

3.1 Drought / moisture stress

Plant mitigating drought stress can be broadly grouped into two mechanism *viz*., avoidance and tolerance. Under avoidance, plant are able to escape water stress through increased root system, reduce stomatal number and conductance, decreased leaf area, increased leaf thickness and leaf rolling to reduce evaporation (Goufo *et al*., 2017). Whereas, tolerance mechanism help to maintains tissue hydrostatic pressure by cellular and biochemical modifications, mainly through osmotic pressure (Blum, 2017). Water uptake by the plants is primarily determined by soil water availability and atmospheric vapour deficit, creating turgor pressure within the cells. Regulations of this turgor pressure through physiological adjustment are important aspect to get drought tolerance. Studies have shown decrease in photosynthetic efficiency and transpiration rate under drought condition (Zandalinas *et al*., 2016). This may be due to stress induced stomatal closure and other non-stomatal limitations such as decreased leaf expansion, leaf senescence and inappropriate functioning of stomatal regulation (Rahnama *et al*., 2010). Recent studies reported that decreased in net photosynthetic activity associated with stomatal closure resulted in increased WUE (net CO_2 assimilation rate/transpiration) (Ruggiero *et al*., 2017). In pulses, drought stress during the vegetative stage greatly reduces plant growth and developmental that leads to weak reproductive phase (Manikavelu *et al*., 2006). These studies indicated that understanding the physiological processes and crop responses at different growing stages are prerequisite for mitigation of drought induced stress through management practices or alter the cropping pattern of the regions.

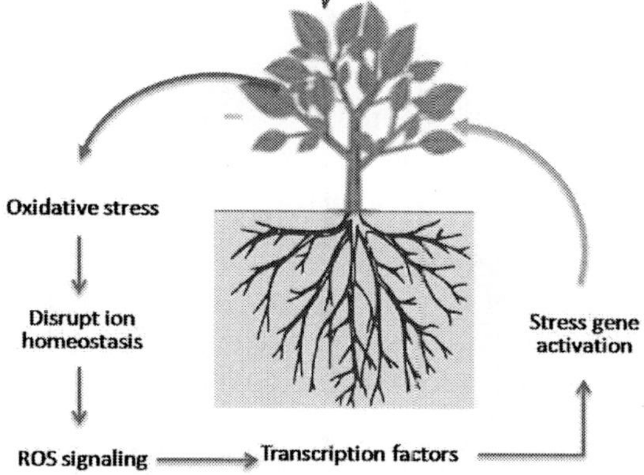

Fig. 1. Crop responses to abiotic stresses

3.2 Heat stress

With the increasing global temperature, heat stress is also recognized as a major abiotic stress which restricts crop yield and losses. This is more pertinent to pulses grown under rainfed conditions in tropical region. Heat stress often coincides with drought in field condition especially in tropical and sub-tropical environments (Simoes-Araujo *et al.*, 2003). In pulses, it was reported that heat stress causes substantial yield reductions in common bean (*Phaseolus vulgaris* L.) and lentil (Rainey and Griffiths, 2005). However, the effect of high temperature on crops varies with the timing, duration, and sternness of the heat stress.

3.3 Salinity stress

Arable lands are getting transformed into saline soil due to anthropogenic acceleration which may go globally upto 50% land loss by 2050 (Hasanuzzaman *et al.*, 2013). Consequently agriculture is forced to be taken up in these soils exposing crops to salinity. Salinity stress can be referred to the accumulation of soluble salts like Cl, SO_4, HCO_3, Na^+, Ca^{2+}, and Mg^{2+}. These ions chiefly come from the irrigation, poor soil drainage and sea water intrusion in coastal areas (Velmurugan *et al.*, 2014). Globally, in terms of environmental stresses, saline stress is considered the most severe stress that affects not only the soil and plant growth but also the living organisms. Salinity reduces plant growth chiefly in three ways, (i) osmotically induced water stress, (ii) specific ion toxicity due to high concentration of sodium and chloride ion and (iii) nutrient imbalances which hampers the uptake of water by plants (Greenway and Munns, 1980). Ultimately photosynthetic rate decreases not only because of reduced leaf area and lesser gas exchange but also due to feedback inhibition of unused photosynthates, after exposure to salinity (Gayathri and Smith, 2017). Pulses are more sensitive to saline soil as compare to cereals and oilseeds. This might be due to the fact that, pulses accumulate excess salts that quickens the anthocyanin pigmention in leaves and stems which ultimately reduce germination and seedling establishment (Kumar *et al.*, 2016).

4. ABIOTIC STRESS MANAGEMENT

With the better knowledge and understanding of plant responses to drought, heat and salinity stresses associated their morphological and physiological responses, it may be possible to mitigate the stress through agronomic, chemical and biological methods.

4.1 Agronomic methods

Better crop management practices like tillage, planting time, planting geometry, quality seed, optimum seed rate, method of sowing; mulching, nutrient, weed and irrigation provide greater opportunities in mitigating those stresses and increasing yields. In many agro-climatic regions farmers heed little attention for pulse cultivation. In dryland area, deep ploughing is beneficial for pulses which led to better root proliferation and higher productivity over shallow condition (Hidangmayum *et al.*, 2018). Dry farming technique may be practiced in pulse cultivation to conserve soil water reservoir and limiting water consumption.

4.2 Chemical methods

External (exogenous) introduction of beneficial chemical substances as supplement has been used to improve resilience, yield, and tolerance of plants to the toxicity of these stress imposed conditions, such as the application of caffeic acid (Klein *et al.*, 2015), sodium polyacrylate (Hong *et al.*, 2016), jasmonic acid (Khan *et al.*, 2016), genistein, chitosan, humic acid (Kasmani *et al.*, 2013), gibberellic acid and other hormones (Shaddad *et al.*, 2013), and molybdenum.

4.3 Plant-microbes interaction

Plant-associated microbes can be classified into beneficial, deleterious and neutral groups on the basis of their effects on plant growth. The review of studies on the beneficial effects of plant-microbes interaction revealed several possibilities for application to improve the plant health and crop production even under stress conditions. In other words, plant–microbe interactions have been successfully utilized to improve plant growth and development for the production of food, fibre, biofuels and key metabolites. The summary of applications resulting from microbe–plant (shoot and root) interactions is given in Fig 2.

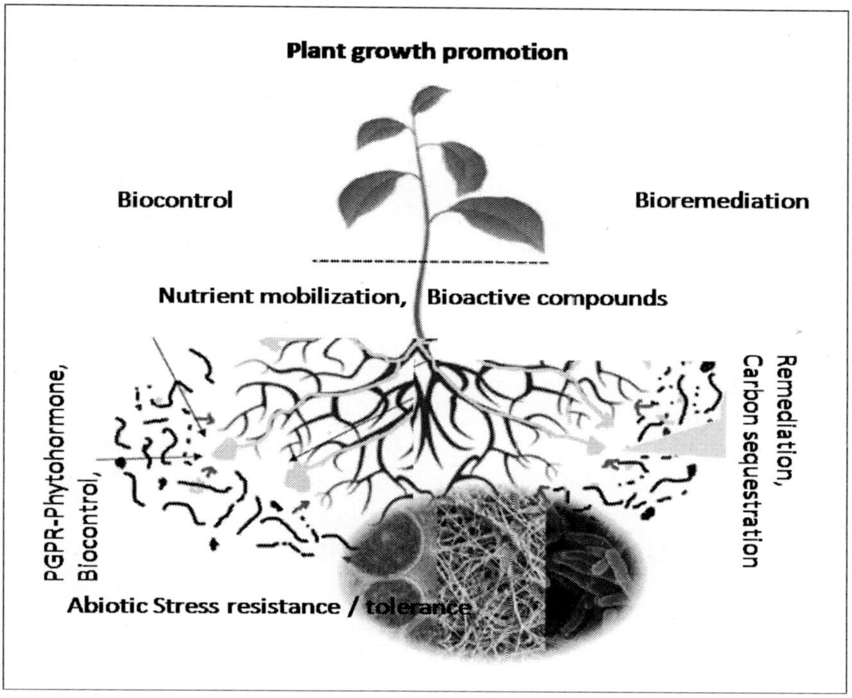

Fig. 2. Beneficial effects of plant-microbial interactions in soil

In this context, investigations of the plant growth promotion mechanisms by plant growth promoting rhizobacteria (PGPR) are increasing at a rapid rate in order to use the best PGPR strain as commercial biofertilizer. The use of PGPR to enhance successful adaptation and survival of pulses (plants) will gear toward ensuring a sustainable crop yield and improvement of soil fertility and structure. This is a good approach to stress management (Alori *et al*., 2017) because these microbes through deaminase enzyme production (Saleem *et al*., 2015), nodulation, and other physiological activities at the rhizosphere help the plants tolerate stress.

5. POTENTIAL OF PGPR

Free-living beneficial bacteria dwelling in the rhizosphere that exert beneficial activities are known as plant growth promoting rhizobacteria (PGPR). Some of the mare facultative endophytes that further invade intercellular spaces of host tissues and thrive as endophytes to establish a mutually beneficial association. PGPR living outside the plant cell are differently associated with plant roots and directly relate to the underlying mechanisms of plant–microbe interactions. The majority of the PGPR colonizes the root surface and thrives in spaces between root hairs and rhizodermal layers whereas, some are not physically in contact with the roots (Gray and Smith, 2005). Phenols, flavonoids, and organic acids secreted by roots to act as chemical signals for bacterial chemotaxis, secretion of exopolysaccharides, quorum sensing and biofilm formation during rhizosphere colonization (Bauer and Mathesius, 2004). In other words, these rhizosphere dwelling microbes, in order to survive in the rhizosphere, lengthened their biological activities that support the survival and growth of plants (Babalola, 2010).

In this context, PGPR contribute to sustaining the intrinsic resistance of plant to several environmental challenges. A number of rhizobacteria with the enzyme machinery necessary for the breakdown of the exudates of plants are able to protect the plant from stress arising from water scarcity and salt effect (Fig. 3). Under stress conditions, PGPR produce a variety of substances such as deaminase enzyme (Saleem *et al*., 2015), plant hormone–indole acetic acid (Gujral *et al*., 2013), siderophore, PO_4^{2-} solubilizing enzyme (Kumariand Khanna, 2016), salicylic acid, and microbiocidal/biostatic enzyme (Moustaine *et al*., 2017). These substances directly protect the plant against abiotic stress or trigger sequence of physiological changes inside the plant that would ultimately restore the vital plant functions and lead to increase in plant growth and yield (Fig. 3).

Antioxidant enzymes, like catalase (CAT) and peroxidase (POD), have the ability to eliminate free radicals and prevent cell membranes and DNA content from further damage (Munns, 2002). Certain PGPR can raise the levels of reactive oxygen species (ROS) scavenging enzymes in plants and in this article we focus on legumes. In addition to the PGPR there are other microbes viz., VAM, fungi and actinomycetes found in the rhizosphere zone also enhances the ability of the plants in mineral nutrition and protection.

Studies have shown that some rhizobacteria contribute to plants nutritionally by trapping and integrating nitrogen into the plant via nitrogen fixation (Attitalla, 2012). Further, some endophytic bacteria contribute not only to the nutrition of the plant but also to enhancing plants' survival rate and adaptation to their environment (Khan *et al*., 2016).

Some of these organisms are known for biofilm formation and secretion of polysaccharide substances which confer stability to plants during stress (Kasim *et al*., 2016). As microbes contribute significantly in bio-hydrometallurgy by the generation of metal-rich solution through the biological oxidation of sulfur containing ore, they also play a role in the immobilization of metals or cations and render them non-bioavailable using polymeric substances and other chemicals they produced (siderophores). Thus, the growth and survival of plants under stress conditions is greatly aided by PGPR living both within and around the plant surfaces. This can be harnessed to enhance pulse production under abiotic stress conditions prevailing in majority of the pulse growing areas.

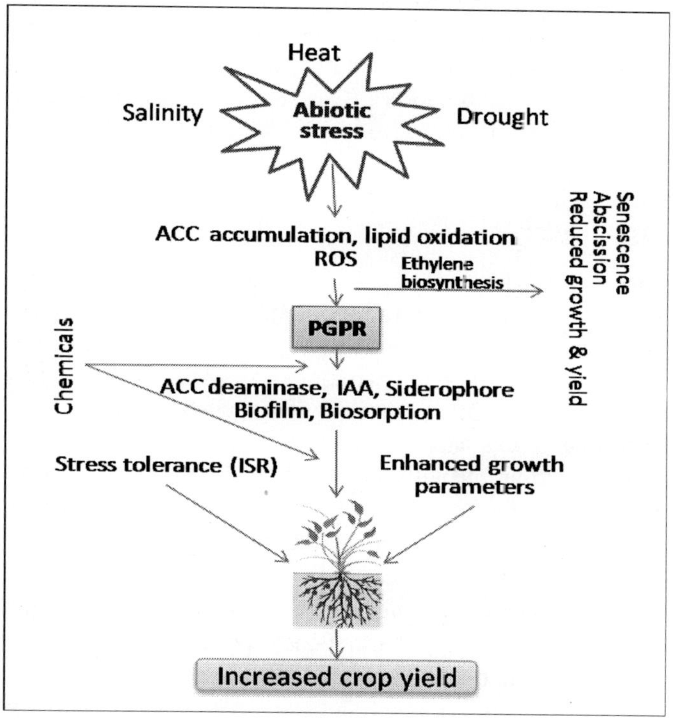

Fig. 3. An overview of plant-PGPR interactions and its role in alleviation of abiotic stresses in legumes (plants)

6. PGPR IN ABIOTIC STRESS MANAGEMENT

A healthy plant is a product of a healthy relationship between the plants and the growth-promoting microbes in a given environment. As discussed earlier, PGPR are the key players in the fight for sustainable plant development amidst stress conditions arising from climatic as well as man made activities. Review of literatures indicated that many PGPR have been investigated for their role in improving plant-water relations, ion homeostasis and photosynthetic efficiency in plants under salt stress. However, their amelioration mechanisms are intricate and often not completely understood in different crops under varying conditions. These mechanisms are regulated by a complex network of signaling events occurring during the plant–microbe interaction and consequently ensuing stress alleviation (Smith *et al.*, 2017). The classes of microbes belonging to these group/genera are *Micrococcaceae* HW-2 (Hong *et al.*, 2016), *Bradyrhizobium*, *Bacillus* (Kasim *et al.*, 2016), *Microbacterium*, *Pseudomonas*, *Curtobacterium*, *Variovorax*, *Paenibacillus* (Yolcu *et al.*, 2011), *Pantoea* (Damam *et al.*, 2014), and many others. Here we discuss the beneficial effect of some of the PGPR on major pulses against salt and drought induced stresses.

6.1 Salt-induced challenges in legumes

Salinity impairs plant growth by causing osmotic imbalance and ion toxicity. Several studies have revealed that PGPRs are capable of maintaining optimum biological function of the plant in the midst of stress inducers that are ascribed to its ability to produce indoleacetic acid, solubilize phosphorus, colonize the plant root, and produce enzyme deaminase (ACC-deaminase). Moreover, improved leaf water status, particularly under salinity has also been reported as an effect of PGPR (Ahmad *et al.*, 2013; Naveed *et al.*, 2014). A summary of beneficial PGPR and the mechanisms responsible for alleviating salinity stress in major pulse crops is given in Table 2.

6.1.1 Osmotic balance

Plant growth promoting rhizobacteria regulate water potential and stomatal opening by affecting hydraulic conductivity and transpiration rate. The co-inoculation of *Pseudomonas* and endomycorrhizae on cowpea plant undergoing salt-induced stress showed a decrease in the mycorrhizal infection of the plant. Treatment of 6000 ppm sodium chloride in the presence of the fungi and bacteria increased the carotenoid concentration (0.449 mg/g). Also, osmolytes (proline and sugars) increased in the presence of endomycorrhizae inoculation in the plant. Irrigation with tap water in the midst of endomycorrhizae and *Pseudomonas fluorescens* gave higher fresh and dry weight, pod length, seed number, and protein content of the cowpea plant (Manaf and Zayed, 2015) compared to the one irrigated with salt water.

Co-inoculation of bean (*Phaseolus vulgaris*) with *Rhizobium tropici* and *Paenibacillus polymyxa* strain modified to over express trehalose 6-phosphate gene resulted in increased nodulation, N content and plant growth. A microarray analysis of nodules revealed up regulation of stress tolerance genes suggesting that extracellular trehalose, which functions as an osmoprotectant can induce salinity tolerance (Figueiredo *et al.*, 2008).

6.1.2 Biofilm formation

In some of the plants microbial exo-polysaccharide formation or the formation of biofilm during environmental stress at the plant rhizosphere is a contributory factor toward plant adjustment to salinity or salt-induced osmotic shock. The polysaccharide molecules produced by microbes in a bid to protect themselves from desiccation and ionic toxicity have a charged part which reacts with dissociated sodium ions and chelates them. This reduces the toxicity and abundance of these ions in the rhizosphere, making the soil suitable for root proliferation (Arora *et al.*, 2010). Rhizobium naturally adapted to desert environments has been shown to perform well in establishing nodules on the root at a high level of salt–soil content and to contribute to growth and better performance of the legume (Sobti *et al.*, 2015).

6.1.3 Production of ACC deaminase

Ethylene production is a plant response mechanism to salinity stress but at the expense of growth and development. Many studies have shown enhanced stress tolerance and growth promotion in plants conferred by soil bacteria producing ACC deaminase (Ahmad *et al.*,

2011; Saikia *et al.*, 2018). PGPR that secrete 1-aminocyclopropane-1-carboxylase (ACC) deaminase restrict ethylene biosynthesis by converting ACC, the precursor of ethylene to ammonia and a-ketobutyrate (Glick *et al.*, 2007). Some of these studies involving pulses are summarized below however, most of these studies were carried out using suitable or sensitive crops to understand the effect of PGPR-plant interactions under stress.

The production of deaminase by rhizobacteria encouraged the root growth of velvet beans undergoing drought stress by catalyzing the disintegration of 1-aminocyclopropane-1-carboxylate to generate nutrients for the microbes and in turn create a shortage of the ingredient (ACC) necessary for the production of ethylene. This helped to limit the action of this hormone (Saleem *et al.*, 2015). The inoculation of *Bradyrhizobium japonicum* together with *Bacillus amyloliquefaciens* enabled better nodule formation, N-fixation and plant viability (Masciarelli *et al.*, 2014). Chickpea plant affected by an increase in salinity was rescued by co-inoculation with two deaminase enzyme-producing microbes (Mesorhizobium MBD26 and Rhizobacteria RHD18) with an observed nodulation capacity of 49 nodules perplant, 201 mg weight of nodules, 12.28 mg per plant nitrogen, and a rise in 31.2% of the above the soil plant part dry weight (shoot dry weight). The result was further increased to 53 nodules and 116.9% grain produced, with N2 level spanning between 9.59 and 27.36 mg per plant investigated (Chaudhary and Sindhu, 2017).

Sometimes the beneficial effect of PGPR increases in a microbial consortia. For example, root inoculation of faba bean with *Rhizobium leguminosarum* and mycorrhizal fungi ameliorate the alkalinity effect on plant by supporting increase in the number of nodules formed (Cordovilla *et al.*, 1999). They boost nitrogenase enzyme activities for efficient nitrogen fixation and improve mycorrhizal dominance in the root to improve productivity and resistance to alkalinity stress (Abd-Alla *et al.*, 2014).

Table 2. A summary of beneficial PGPR and the mechanisms responsible for alleviating salinity stress in major pulse crops

Type of stress	Crop	Microbes involved	Mechanism of action	Significance	References
Salt tolerance	Blackgram	*Ochrobactrum pseudogrignonense*	Osmoregulation, ACC deaminase, ROS	Improves germination, shoot and root length	Saikia *et al.* (2018)
Salinity & water stress	Common bean (*Phaseolus vulgaris*)	*Rhizobium radiobacter, R. etli*	Osmoregulation and ACC deaminase	Improves nodulation, biomass production and seed yield	Yannia *et al.* (2016)
Salt tolerance	Faba bean (*Vicia faba*)	*R. leguminosarum*	Osmoregulation and ACC deaminase	Improves nodulation, biomass production	Cordovilla *et al.* (1999)
Salinity	Greengram (*Vigna radiata)*	*Pseudomonas syringae, Pseudomonas fluorescens, Rhizobium phaseoli*	ACC deaminase	IAA production, siderophore activity nitrogenase activity, P- solubilization	Ahmad *et al.* (2011)

Saline irrigation conditions,	Greengram (*Vigna radiata)*	*E. cloacae, B acillusdrentensis, Rhizobium, B. pumilus*Sol-1, *Alcaligenes sp.* Mal-4, *Providenci-avermicola* Ama-2, *Brevundimonas* Kro13, *Kluyveraascor bata* UD165, *Pseudomonas putida, Ochrobactrum, B. cereus*	Acetylene reducing activity, ACC deaminase activity, nitrogenase activity, IAA, siderophore production	N-fixing, P -solubilizing, improve gaseous exchange, water relations,	Akhtar and Ali (2011) and Mahmood *et al.* (2016)

These studies highlighted that PGPR has the ability to promote plant growth and enhance their tolerance level towards salt by improving plant physiological response and antioxidant potential. These PGPR can be included in pulse production strategy to alleviate salinity stress.

6.2 Moisture stress

Under rainfed conditions, drought has been identified as the main constraint leading to the reductions in crop yields. During moisture stress conditions, the plant-water relation at the cellular level gets destabilized, thus affecting plant metabolic and physiological activities (Nagarajan and Nagarajan, 2009). In broader sense, drought interfere with plant growth, nutrient and water relations, photosynthesis, nodulation, assimilate portioning and reduction in crop yield (Farroq *et al.*, 2009). Reduced absorption of photosynthetically active radiation, radiation use efficiency and decreased harvest index are the major yield-reducing factors under low soil moisture (Earl and Davis, 2003). At the same time plants respond to water shortages with several morphological and physiological modifications. Some of them are induced by PGPR. The establishment of a correlation between PGPR and drought resistance has been reported in several crops, including chickpea, pigeonpea, greengram and wheat (Ngumbi and Kloepper, 2016). In general, PGPR induce tolerance to drought stress in crops by production of phytohormones, producing volatile compounds, ACC deaminase, osmolyte and exopolysaccharides, and triggering antioxidant activities (Fig. 4).

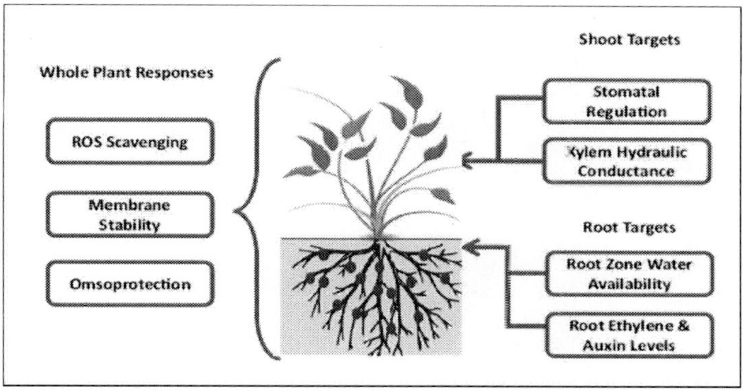

Fig. 4. Summary of key mechanism leading to plant tolerance to moisture stress

A summary of plant growth promoting effects of PGPR on different pulses and the mechanism is summarized in Table 3. It indicates that roots play an important role in water stress tolerance by reduction in leaf expansion and promotion of root growth. In general, deeper and more profuse root systems are able to tap extra water from the soil profile and alleviate drought effects. Root traits like root length, number and root depth, have long been seen as important traits to improve crop adaptation to water stress (Rajendran *et al.*, 2011). In this context, stress-tolerant PGPR helps to change the plant root architecture leading to an increase in root surface area and improved water and nutrient uptake which is an important adaptive measure in plants during a drought (Vacheron *et al.*, 2013). Production of plant hormones (indole acetic acid, gibberellin) stimulates growth in lentil, and chickpea leading to alleviation of moisture stress. The increase in root number and tissue mass promoted the wider reach of plant roots for water and nutrient absorption, which enabled them to thrive in water scarce environments.

Table 3. A summary of beneficial PGPR and the mechanisms responsible for alleviating moisture stress in major pulse crops

Type of stress	Crop	Microbes involved	Mechanism of action	Significance	References
Drought	Pea	*Variovorax paradoxus*	Decreased ethylene due to ACC deaminase (ACCd)	Sustain plant growth	Belimov *et al.* (2009)
Drought	Greengram (*Vigna radiata*)	*Pseudomonas aeruginosa*	IAA, Up regulation of DREB2A, CAT1, DHN, Increased activity of SOD, peroxidase (POX), CAT	Reduce the crop loss, maintain growth	Sarma and Saikia (2014)
		Achromobacter xylosoxidans	ACCd	Influence plant homeostasis	Ma *et al.* (2009)
		Bacillus circulans, Cladosporium herbarum	ACCd	Phosphate solubilization	Oteino *et al.* (2015)
		Pseudomonas aeruginosa, Bacillus subtilis	ROX and ACCd	Prevent root knot formation	Ngumbi and Kloepper (2016)

Drought tolerance	Greengram (*Vigna radiata*)	*Pseudomonas fluorescens* strain Pf1 *Bacillus subtilis*EPB5, EPB22, and EPB 31	Stress-related enzymes, Increased proline content	Improves growth under stress	Saravanakumar *et al.*(2011)
Mitigation of drought	Blackgram (*Vigna mungo*)	*Consortium of Ochrobactrum. pseudogrignonense RJ12, Pseudomonas sp. RJ15, and B. subtilis RJ46*	ACCd	Improves germination, shoot and root length	Saikia *et al.*, 2018
Water deficit	Chickpea (*Cicer arietinum*) and pigeonpea (*Cajanus cajan*)	*E. ludwigii* SRI-229	IAA, ROX	Improved nodulation	Gopalakrishnan *et al.* (2016)
Moisture stress	Chickpea (*Cicer arietinum*)	*Pseudomonas aeruginosa*	ACCd	Positively stimulate potassium and phosphorus uptake	Ahemad and Kibret (2014)
		Pseudomonas putida MTCC5279 (RA)	Reduced/ controlled the expression of stress response gene	maintained water content, osmolyte, membrane structure, and germination rate of the plant	Tiwari *et al.* (2016)
		Pseudomonas with *Mesorhizobium ciceris*	ACCd activity	Increased growth	Sharma *et al.*(2013)
		Pseudomonas jessenii PS06, *Mesorhizobium ciceri* C-2/2, *Pseudomonas putida*, *Burkholderia*, *Bacillus Amyloliquefacien*	IAA production, siderophore production, phosphate solubilization,	seed yield and nodule fresh weight,	Paul and Sarma (2006); Joseph *et al.*(2012)
Moisture stress	Common bean (*Phaseolus vulgaris*)	*Paenibacillus polymyxa*	ACCd, IAA, ROX	Alleviate adverse effects of drought stress and maintain plant growth	Ngumbi and Kloepper (2016)
		Rhizobium leguminosarum	Ion homeostasis and increased nutrition	Phosphate solubilization	Ahemad and Kibret (2014)

Several rhizobial species were isolated from *V. radiate* grown in different agro-climatic zones and studied for its tolerance to drought. The results showed that some isolates (PM-2, PDM-54, K-851 and PM-5) promoted plant biomass, maximum nodule numbers and higher nitrogenase activity under moisture stress (Neeraj *et al.*, 2009). Apart from the ectophytic PGPR, endophytic microbes helps the plants to adapt to water stress, for example, the presence of these organisms aided the physiological functions of the plant and enabled it to tolerate water scarcity in the soil. The studies showed 28% increase in biomass (shoot and roots) as against the control. Further, studies revealed production of poly-sugar substance known as trehalose by the endophytes which is capable of protecting biologically produced compounds and molecules from breakdown during water induced osmotic tension (Khan *et al.*, 2016).

The symbiotic N_2 fixers (*Sinorhizobium meliloti* and *Sinorhizobium medicae*) possess the ability of protecting plants from senescence of the leaves under drought stress (Staudinger *et al.*, 2016). PGPR promoted nodulation in pulses helps to recover from drought stress faster than the non-nodulated plants as a result of the shifts in the carbon partitioning from starch to sugars and thus the enhanced allocation of reserves to osmolytes during drought, enabling to stay green with the ability of fast recovery after re-watering (Staudinger *et al.*, 2016). The symbiont (*Bradyrhizobium sp.*) aided the tolerance of cowpea plant to water scarcity stress and boosted the quality of NO^{3-} (nitrate) and amino acid (proline) in the inoculated plant (Barbosa *et al.*, 2013). Similarly, the rhizobial inoculants aid in the delay of the process of senescence by promoting the accumulation of potassium ions, reduction in the protein mediating ethylene production, and induction of cytokinin production which inhibit senescence and accumulation of sugars and amino acids that added in the plant (*Medicago truncatula*) survival during stress.

Genistein treatment of soybean seed together with a bioinoculant (*B. japonicum*) greatly encouraged the resistance of soybean to drought stress and decreased its detrimental effect on the formation of nodules by the microbe (Hasanah and Rahmawati, 2012). Similar to this, bioinoculation of legume plant with *Rhizobium* helps to encourage the development of many leaves in the plant due to the greater numbers of root nodules in the plant roots that aid in nitrogen fixation (Ferreira *et al.*, 2011). Groundnut inoculated by *Bradyrhizobium* under drought conditions had a large quantity of amino acids which was derived from the nitrogenase-catalyzed conversion of atmospheric N to NH_4^+ ions necessary for amino acid and protein formation in the plant (Delfini *et al.*, 2010). The observed increase in protein level of legume-inoculated plants is also implied for pulse crops. Thus, it is evidenced from the review that PGPR helps to maintain leaf turgor pressure, improves root architecture, N-fixation, enhances anti-oxidant activity, and restores photosynthetic activity even under moisture stress.

7. CONCLUSION

With the advent of global climate change and increased pressure to achieve food security, strenuous efforts are needed to increase global agricultural production. Pulses are important food crops required to meet the food requirement including nutritional demand. But the production is stagnating due to abiotic stress among other factors. Significant improvement in plant tolerance to abiotic stress will improve yield and production of crops to feed humans and livestock. This can be achieved by two approaches. One approach is via search, selection, and engineering of plant species capable of resisting salinity and drought stress. The other method is the use of plant growth-promoting rhizobacteria that will go a long way in supporting the plant to develop both intrinsic and extrinsic ability to tolerate stressful conditions and sustain yield.

In traditional approach, abiotic stress management is sought at genetic level by use of biotechnological tools and identification of tolerant genotypes. But use of PGPR and agronomical tools and techniques can offer an adequate and rapid solution for restoring and improving crop production under stress conditions in a sustainable manner. An integrated abiotic stress management strategy of co-integration of external application of proline, caffeic acid, nutrients, synthetic plant hormones, and microbes (PGPR) could aid greatly in ensuring continuous and efficient agricultural practices that will manage the stresses and boost the yield of pulse crops.

8. FUTURE THRUST

Researchers from across the globe have persistently engaged in understanding the crop response to abiotic stress and the mechanisms behind the alleviation of such stress due to plant-PGPR interaction. These studies proved that use of PGPR is a viable option for crop cultivation under abiotic stress conditions but the stress conditions are being modified by the emerging climate change. Further, the increased pulse production essentially should come by improved productivity rather than area expansion. Therefore, future research should focus on managing plant-microbe interactions at root zone environments with respect to improving the ability of plants adapt to extreme environments. Rhizo-engineering and screening of bacterial strains through molecular techniques like proteomics and docking methods will be the focused area of researchers to identify problem specific PGPR in the coming years.

The role of PGPR-based bio-formulations has shown great potential toward sustainable agriculture and seen as an alternative to chemicals. Majority of researches are confined either to laboratory or greenhouse level, hence these should be scaled up to the field level (Wallenstein, 2017). Moreover, bulk of the studies focused on isolation and inoculation of PGPR into rhizosphere to get better yield in selected crops viz., wheat, rice, maize some vegetables, fruits, and herbs. More studies are needed to uncover the beneficial interactions of pulse-PGPR for progressing it to field level use.

At the same time more efforts are required to explore ways and means for better utilization of PGPR at field scale and its commercialization to improve pulse production. However, the approaches focused for a long time on each organism individually rather than an integrated metaorganism approach in an ecological perspective. So, microbial formulation is also an important parameter to be focused in the coming years, like the type of formulation and their acceptance at physiological and ecological level.

REFERENCES

Abd-Alla, M.H., El-Enany, A.E., Nafady, N.A., Khalaf, D.M. and Morsy, F.M. (2014). Synergistic interaction of Rhizobium leguminosarum bv. Viciae and arbuscular mycorrhizal fungi as a plant growth promoting biofertilizers for faba bean (*Vicia faba* L.) in alkaline soil. *Microbiological Research* 169, 49–58.

Ahemad, M. and Kibret, M. (2014). Mechanisms and applications of plant growth promoting rhizobacteria: current perspective. *Journal of King Saud University Science* 26, 1–20.

Ahmad, M., Zahir, Z.A., Asghar, H.N. and Asghar, M. (2011). Inducing salt tolerance in mung bean through coinoculation with rhizobia and plant-growth-promoting rhizobacteria containing 1-aminocyclopropane-1-carboxylate deaminase. *Canadian Journal of Microbiology* 57, 578–589.

Ahmad, M., Zahir, Z.A. and Khalid, M. (2013). Efficacy of Rhizobium an Pseudomonas strains to improve physiology, ionic balance and quality of mung bean under salt-affected conditions on farmer's fields. *Plant Physiology and Biochemistry* 63, 170–176.

Akhtar, S., and Ali, B. (2011) Evaluation of rhizobacteria as non-rhizobial inoculants for mung beans. *Australian Journal of Crop Science 5*, 1723–1729.

Alori, E.T., Dare, M.O. and Babalola, O.O. (2017). Microbial inoculants for soil quality and plant health. *In*: Sustainable agriculture reviews. Springer, Switzerland. pp. 281–307

Arora, M., Kaushik, A., Rani, N. and Kaushik, C.P. (2010). Effect of cyanobacterial exopolysaccharides on salt stress alleviation and seed germination. *Journal of Environmental Biology* 31(5), 701–704

Attitalla, I.H. (2012). Traits biologically interacting Azospirillum lipoferum strain R23. *Adv Biol Res* 6(2), 87–94

Babalola, O.O. (2010). Beneficial bacteria of agricultural importance. *Biotechnology Letters* 32(11), 1559–1570.

Barbosa, M.A.M., da Silva Lobato, A.K., Tan, D.K.Y., Viana, G.D.M., Coelho, K.N.N., Barbosa, J.R.S., dos Santos Moraes, M.H., Lobo da Costa, R.C., dos Santos Filho, B.G. and de Oliveira Neto, C.F. (2013). Bradyrhizobium improves nitrogen assimilation, osmotic adjustment and growth in contrasting cowpea cultivars under drought. *Australian Journal of Crop Science* 7(13), 1983–1989

Bauer, W.D. and Mathesius, U. (2004). Plant responses to bacterial quorum sensing signals. *Current Opinion in Plant Biology* 7, 429–433.

Belimov, A.A., Dodd, I.C., Hontzeas, N., Theobald, J.C., Safronova, V.I. and Davies, W.J. (2009). Rhizosphere bacteria containing 1-aminocyclopropane-1-carboxylate deaminase increase yield of plants grown in drying soil via both local and systemic hormone signalling. *New Phytologist* 181, 413–423.

Blum, A. (2017). Osmotic adjustment is a prime drought stress adaptive engine in support of plant production: osmotic adjustment and plant production. *Plant Cell and Environment* 40, 4–10.

Borlaug Institute for South Asia. (2015). Major Accomplishments 2012–2014. BISA Report Series 1. New Delhi, India. 38pp.

Chaudhary, D. and Sindhu, S.S. (2017). Amelioration of salt stress in chickpea (*Cicer arietinum* L.) by coinculation of ACC deaminase containing rhizospheric bacteria with Mesorhizobium strains. *Legume Research* 40(1), 80–86

Cordovilla, D.M., Berrido, S.I., Ligero, F. and Lluch, C. (1999). Rhizobium strain effects on the growth and nitrogen assimilation in *Pisum sativum* and *Vicia faba* plant growth under salt stress. *Journal of Plant Physiology* 154, 127–31.

Cramer, G.R., Urano, K., Delrot, S., Pezzotti, M. and Shinozaki, K. (2011). Effects of abiotic stress on plants: A systems biology perspective. *BMC Plant Biology* 11, 163.

Damam, M., Gaddam, B. and Kausar, R. (2014). Effect of plant growth promoting rhizobacteria (PGPR) on Coleus forskohlii. *International Journal of Current Microbiology and Applied Sciences* 3(9), 266–274.

DES. (2019). Crop production statistics (2012-19). Directorate of Economics and Statistics, GOI, New Delhi.

Delfini, R., Belgoff, C., Fernández, E., Fabra, A. and Castro, S. (2010). Symbiotic nitrogen fixation and nitrate reduction in two peanut cultivars with different growth habit and branching pattern structures. *Plant Growth Regulator* 61, 153–159

Earl, H. and Davis, R.F. (2003). Effect of drought stress on leaf and whole canopy radiation use efficiency and yield of maize. *Agronomy Journal* 95, 688–696.

Enebe, M.C. and Babalola, O.O. (2018). The influence of plant growth-promoting rhizobacteria in plant tolerance to abiotic stress: a survival strategy. *Applied Microbiology and Biotechnology* 102, 7821–7835.

FAOSTAT. (2013). Statistical database. http://faostat.fao.org/site/339/ default.aspx

Farooq, M., Wahid, A., Kobayashi, N., Fujita, D. and Basra, S.M.A. (2009). Plant drought stress: effects, mechanisms and management. *Agronomy for Sustainable Development* 29, 185–212.

Ferreira, E.P.B., Martins, L.M.V., Xavier, G.R. and Rumjanek, N.G. (2011). Nodulation and grain yield by cowpea (*Vigna unguiculata* L.Walp.) inoculated with rhizobia, isolates. *Revista Caatinga* 24, 27–35.

Figueiredo, M.V.B., Burity, H.A., Martinez, C.R. and Chanway, C.P. (2008). Alleviation of drought stress in the common bean (Phaseolus vulgaris L.) by coinoculation with Paenibacillus polymyxa and Rhizobium tropici. *Applied Soil Ecology* 40, 182–188.

Gayathri, I. and Smith, D.L. (2017). Plant Growth Promoting Rhizobacteria in Amelioration of Salinity Stress: A Systems Biology Perspective. *Frontiers in Plant Science* 8, 1768. doi: 10.3389/fpls.2017.01768

Glick, B.R. (2012). Plant growth-promoting bacteria: Mechanisms and applications. *Scientifica* 15. doi.org/10.6064/2012/963401

Glick, B.R., Cheng, Z., Czarny, J. and Duan, J. (2007). Promotion of plant growth by ACC deaminase-producing soil bacteria. *European Journal* of *Plant Pathology* 119, 329–339.

Gopalakrishnan, S., Vadlamudi, S., Samineni, S. and Sameer Kumar, C.V. (2016). Plant growth-promotion and biofortification of chickpea and pigeonpea through inoculation of biocontrol potential bacteria, isolated from organic soils. *Springer Plus* 5, 1882

Goufo, P., Moutinho-Pereira, J.M., Jorge, T.F., Correia, C.M., Oliveira, M.R. and Rosa, E.A.S. (2017). Cowpea (*Vigna unguiculata* L. Walp.) metabolomics: osmoprotection as a physiological strategy for drought stress resistance and improved yield. *Frontier in Plant Science* 8, 586.

Gowda, C.L.L., Rao, P.P., Tripathy, S., Gaur, P.M. and Deshmukh, R.B. (2009). Regional shift in chickpea production in India. In: Masood A, Kumar S (eds.), Milestone in

Food Legumes Research. Indian Institute of Pulses Research, Kanpur, 208.024, India. pp.21-35.

Gray, E.J. and Smith, D.L. (2005) Intracellular and extracellular PGPR: commonalities and distinctions in the plant-bacterium signaling processes. *Soil Biology and Biochemistry* 37, 395–412.

Greenway, H. and Munns, R. (1980). Mechanisms of salt tolerance in non halophytes. *Annual Review of Plant Physiology* 31, 149–190.

Gujral, M.S., Agrawal, P., Khetmalas, M.B. and Pandey, R. (2013). Colonization and plant growth promot ion of sorghum seedl ing by endorhizospheric *Serratia* sp. *Acta Biologica Indica* 2(1), 343–352.

Hasanuzzaman, M., Nathan, K. and Fujita, M. (2013). Plant response to salt stress and role of exogenous protectants to mitigate salt induced damages. In: Ahmad P, Azooz MM, Prasad MNV (Eds.), Ecophysiology and Responses of Plant under Salt Stress, Springer, New York. pp. 25–87.

Hidangmayum, A., Ankita Singh, Vivek Kumar and Dwivedi, P. (2018). Abiotic stress responses in cereals and pulses crop and their agronomic practices to enhance tolerance. *EurAsian Journal of BioSciences* 12, 487-493.

Hasanah, Y. and Rahmawati, N. (2012). Soybean production under drought stress with application of *Bradyrhizobium japonicum* induced by genistein, The Proceedings of the 2nd Annual International Conference Syiah Kuala University 2012 & The 8th IMT-GT Uninet Biosciences Conference Banda Aceh, 22–24 November 2012. 2(1), 145–149.

Hong, S.H., Ham, S.Y., Kim, J.S., Kim I-S. and Lee, E.Y. (2016). Application of sodium polyacrylate and plant growth-promoting bacterium, Micrococcaceae HW-2 on the growth of plants cultivated in the rooftop. *International Biodeterioration & Biodegradation* 133, 297–303.

Joseph, B., Ranjan Patra, R. and Lawrence, R. (2012). Characterization of plant growth promoting rhizobacteria associated with chickpea (*Cicer arietinum* L.). *International Journal of Plant Production* 1: 141–152.

Kasim, W.A., Gaafar, R.M., Abou-Ali, R.M., Omar, M.N. and Hewait, H.M. (2016). Effect of biofilm forming plant growth promoting rhizobacteria on salinity tolerance in barley. *Annals of Agricultural Sciences* 61(2), 217–227.

Kasmani, M.R.B., Samavat, S., Mostafavi, M. and Khalighi, A. (2013). The effect of application of humic acid foliar on biochemical parameters of pistachio under drought stress. *New York Science Journal* 6(12), 26–31.

Khan, A.L., Halo, B.A., Elyassi, A., Ali, S., Al-Hosni, K., Hussain, J., Al-Harrasi, A. and Lee, I. (2016). Indole acetic acid and ACC deaminase from endophytic bacteria improves the growth of Solanum lycopersicum. *European Journal of Biotechnology* 21, 58–64.

Klein, A., Keyster, M. and Ludidi, N. (2015). Response of soybean nodules to exogenously applied caffeic acid during NaCl-induced salinity. *South African Journal of Botany* 96, 13–18.

Kumar, K., Solanki, S., Singh, S.N. and Khan, M.A. (2016). Abiotic constraints of pulse production in India. In: Biswas SK, Kumar S, Chand G (Eds.), Disease of Pulse Crops and their Sustainable Management pp. 23–39, Biotech Books, New Delhi, India. (Eds.), Kumar, Shiv Indian Institute of Pulses Research, Kanpur, Uttar Pradesh, India, pp. 21–35.

Kumari, P. and Khanna, V. (2016). Biodiversity of Pseudomonas and Bacillus possessing both bioantagonistic and plant growth promoting traits in chickpea rhizosphere. International Journal of Science and Nature 7(1), 153–158.

Lal, R. (2015). The soil–peace nexus: our common future. *Soil Science and Plant Nutrition* 61, 566–578.

Lobell, D.B., Schlenker, W. and Costa-Roberts, J. (2011) Climate trends and global crop production since 1980. *Science* 333, 616–620.

Ma, Y., Rajkumar, M. and Freitas, H. (2009) Inoculation of plant growth promoting bacterium Achromobacter xylosoxidans strain Ax10 for the improvement of copper phytoextraction by Brassica juncea. *Journal of Environmental Management* 90, 831–837.

Manaf, H.H. and Zayed, M.S. (2015). Productivity of cowpea as affected by salt stress in presence of endomycorrhizae and Pseudomonas fluorescens. *Annals of Agricultural Sciences* 60(2), 219–226.

Mahmood, S., Daur, I. and Al-Solaimani, S.G. (2016). Plant growth promoting rhizobacteria and silicon synergistically enhance salinity tolerance of mung bean. *Frontiers in Plant Science* 7, 876.

Manikavelu, A., Nadarajan, N., Ganesh, S.K., Gnanamalar, R.P. and Babu, R.C. (2006). Drought tolerance in rice: morphological and molecular genetic consideration. *Plant Growth and Regulation* 50, 121–138.

Moustaine, M., Elkahkahi, R., Benbouazza, A., Benkirane, R. and Achbani, E.H. (2017). Effect of plant growth promoting rhizobacterial (PGPR) inoculation on growth in tomato (*Solanum lycopersicum* L.) and characterization for direct PGP abilities in Morocco. *International Journal of Environment, Agriculture & Biotechnology* 2(2), 590–596

Munns, R. (2002). Comparative physiology of salt and water stress. *Plant Cell and Environment* 25, 239–250

Nagarajan, S. and Nagarajan, S. (2009). Abiotic stress adaptation in plants (ed. Pareek, A. *et al.*) pp.1–11.

Nam, N.H., Chauhan, Y.S. and Johansen, C. (2001). Effect of timing of drought stress on growth and grain yield of extra-shortduration pigeonpea lines. *Journal of Agricultural Science* 136, 179–189.

Naveed, M., Hussain, M.B., Zahir, Z.A., Mitter, B. and Sessitsch, A. (2014). Drought stress amelioration in wheat through inoculation with Burkholderia phytofirmans strain PsJN. *Plant Growth Regulators* 73, 121–131.

Neeraj, Gaurav, S.S., Chatterjee, S.C. and Sachin, C.M. (2009). Efficient nitrogen fixing rhizobial isolate infecting *Vigna radiata* L. *Asian Journal of Agricultural Sciences* 1(2), 62-65.

Ngumbi, E. and Kloepper, J. (2016). Bacterial-mediated drought tolerance: current and future prospects. *Applied Soil Ecology* 105, 109–125.

Oteino, N., Lally, R.D., Kiwanuka, S., Lloyd, A., Ryan, D., Germaine, K.J. and Dowling, D.N. (2015). Plant growth promotion induced by phosphate solubilizing endophytic Pseudomonas isolates. *Frontiers in Microbiology* 6, 745.

Paul, D. and Sarma, Y.R. (2006). Plant growth promoting rhizhobacteria (PGPR)-mediated root proliferation in black pepper (*Piper nigrum* L.) as evidenced through GS Root software. *Archives of Phytopathology and Plant Protection* 39, 311–314.

Porter, J.R. (2005) Rising temperatures are likely to reduce crop yields. *Nature* 436, 174.

Rahnama, A., Poustini, K., Tavakkol-Afshari, R. and Tavakoli, A. (2010). Growth and stomatal responses of bread wheat genotypes in tolerance to salt stress. *International Journal of Biology and Life Science* 6, 216–221.

Rainey, K. and Griffiths, P. (2005). Evaluation of *Phaseolus acutifolius* A. Gray plant introductions under high temperatures in a controlled environment. *Genetic Resources and Crop Evolution* 52, 117–120.

Rajendran, R.A., Muthiah, A.R., Manickam, A., Shanmugasundaram, P. and Joel, A.J. (2011). Indices of drought tolerance in sorghum (*Sorghum bicolor* L. Moench) genotypes at early stages of plant growth. *Research Journal of Agricultural and Biological Sciences* 7, 42-46.

Rana, D.S., Dass, A., Rajanna, G.A. and Kaur, R. (2016). Biotic and abiotic stress management in pulses. *Indian Journal of Agronomy* 61, S238-S248.

Ruggiero, A., Punzo, P., Landi, S., Costa, A., VanOoosten, M. and Grillo, S. (2017). Improving plant water use efficiency through molecular genetics. *Horticulturae* 3, 31.

Saikia, J., Sarma, R.K., Dhandia, R., Yadav, A., Bharali, R. and Gupta, V.K. (2018). Alleviation of drought stress in pulse crops with ACC deaminase producing rhizobacteria isolated from acidic soil of Northeast India. *Scientific Reports* 8, 3560.

Saleem, A.R., Bangash, N., Mahmood, T., Khalid, A., Centritto, M. and Siddique, M.T. (2015). Rhizobacteria capable of producing ACC deaminase promote growth of velvet bean (*Mucuna pruriens*) under water stress condition. *International Journal of Agriculture and Biology* 17, 663–667.

Saravanakumar, D., Kavino, M., Raguchander, T., Subbian, P. and Samiyappan, R. (2011). Plant growth promoting bacteria enhance water stress resistance in green gram plants. *Acta Physiologiae Plantarum* 33, 203–209.

Sarma, R.K. and Saikia, R. (2014). Alleviation of drought stress in mung bean by strain *Pseudomonas aeruginosa* GGRJ21. *Plant and Soil* 377, 111–126.

Sehrawat, N., Yadav, M., Bhat, K.V., Sairam, R.K. and Jaiwal, P.K. (2015). Effect of salinity stress on mungbean during consecutive summer and spring seasons. *Journal of Agricultural Science* 60(1), 23–32.

Shaddad, M.A.K., Abd El- Samad, H.M. and Mostafa, D. (2013). Role of gibberellic acid (GA3) in improving salt stress tolerance of two wheat cultivars. *International Journal of Plant Physiology and Biochemistry* 5(4), 50–57.

Sharma, P., Khanna, V. and Kumar P.I. (2013). Efficacy of aminocyclopropane-1-carboxylic acid (ACC)- deaminase-producing rhizobacteria in ameliorating water stress in chickpea under axenic conditions. *African Journal of Microbiology Research* 7, 5749-5757.

Simoes-Araujo, J.L., Rumjanek, N.G. and Margis-Pinheiro, M. (2003). Small heat shock proteins genes are differentially expressed in distinct varieties of common bean. *Brazilian Journal of Plant Physiology* 15, 33–41.

Smertenko, A., Draber, P., Viklicky, V. and Opatrny, Z. (1997). Heat stress affects the organization of microtubules and cell division in *Nicotiana tabacum* cells. *Plant Cell and Environment* 20, 1534–1542.

Smith, D.L., Gravel, V. and Yergeau, E. (2017). Editorial: signaling in the phytomicrobiome. *Frontiers in Plant Science* 8, 611.

Sobti, S., Belhadj, H.A. and Djaghoubi, A. (2015). Isolation and characterization of the native rhizobia under hyper-salt edaphic conditions in Ouargla (Southeast Algeria). *Energy Procedia* 74, 1434–1439.

St Clair, S.B. and Lynch, J. (2010). The opening of Pandora's box: climate change impacts on soil fertility and crop nutrition in developing countries. *Plant and Soil* 335, 101–115

Staudinger, C., Mehmeti-Tershani, V., Gil-Quintana, E., Gonzalez, E.M., Hofhans, F., Bachmann, G. and Wienkoop, S. (2016). Evidence for a rhizobia-induced drought stress response strategy in Medicago truncatula. *Journal of Proteomics* 136, 202–213.

Tiwari, S., Lata, C., Chauhan, P.S. and Nautiyal, C.S. (2016). *Pseudomonas putida* attunes morphophysiological, biochemical and molecular responses in Cicer arietinum L. during drought stress and recovery. *Plant Physiology and Biochemistry* 99, 108–117.

Vacheron, J. (2013). Plant growth-promoting rhizobacteria and root system functioning. *Frontiers in Plant Science* 4, 356.

Velmurugan, A., Swarnam, T.P., Ambast, S.K., Ravisankar, N. and Subrmani, T. (2014). Land degradation and its spatio-temporal changes induced by natural events in Andaman islands. *Journal of the Andaman Science Association* 19(1), 65-74.

Wahid, A., Gelani, S., Ashraf, M. and Foolad, M.R. (2007). Heat tolerance in plants: an overview. *Environment and Experimental Botany* 61, 199–223.

Wallenstein, M.D. (2017). Managing and manipulating the rhizosphere microbiome for plant health: A systems approach. *Rhizosphere* 3, 230–232.

Wang, W., Vinocur, B. and Altman, A. (2003). Plant responses to drought,salinity and extreme temperatures: towards genetic engineering for stress tolerance. *Planta* 218, 1-14.

Yannia, Y., Mohamed Zidanb, M., Dazzoc, F., Rizka, R., Mehesena, A., Abdelfattaha, F. and Elsadanya, A. (2016). Enhanced symbiotic performance and productivity of drought stressed common bean after inoculation with tolerant native rhizobia in extensive fields. *Agriculture Ecosystem and Environment* 232, 119–128.

Yolcu, H., Turan, M., Lithourgidis, A., Cakmakçi, R. and Ali, K.O.C. (2011). Effects of plant growth-promoting rhizobacteria and manure on yield and quality characteristics of Italian ryegrass under semi-arid conditions. *Australian Journal of Crop Science* 5(13), 1730–1736.

Zandalinas, S.I., Rivero, R.M., Martínez, V., Gómez-Cadenas, A. and Arbona, V. (2016). Tolerance of citrus plants to the combination of high temperatures and drought is associated to the increase in transpiration modulated by a reduction in abscisic acid levels. *BMC Plant Biology* 16, 105.

CHAPTER 9

Mechanism of plant growth-promoting rhizobacteria in grain legumes for biotic and abiotic stress management

Karivaradharajan Swarnalakshmi*, Vandana Yadav, Deepti Tyagi, Dolly Wattal Dhar and Annapurna Kannepalli

Division of Microbiology, ICAR-Indian Agricultural Research Institute, PUSA, New Delhi 110012, India.

*Corresponding author e-mail: swarnalakshmi.k@icar.gov.in

Abstract

Grain legumes are important source of human nutrition and grown under low-input, diverse farming systems under rainfed conditions. These crops benefit from a root nodulating symbiotic rhizobia that convert atmospheric nitrogen into plant usable form. They are also colonized by the naturally occurring plant-growth promoting rhizobacteria (PGPR) present in the soil. PGPRs help in improving seed germination, root development, plant growth and protect plants against broad spectrum of soil borne pathogens. Among various PGPRs, Bacillus and Pseudomonas are well known biocontrol agents. They can produce various antagonistic substance viz cell wall degrading enzymes, siderophores, HCN, antibiotics against pathogen. They also induce systemic resistance in host plant against pathogens. Similar to pathogen-induced systemic acquired resistance (SAR), PGPRs trigger induced systemic resistance (ISR) in plants via salicylic acid-dependent SAR pathway. PGPRs elicit induced systemic tolerance (IST) that enhances plant tolerance against biotic and abiotic stress like drought, salinity, high temperature. This chapter discusses various mechanisms of PGPRs that enhance plant resistance under biotic and abiotic stress conditions with special reference to grain legumes.

Keywords: Grain legumes, PGPR, Biocontrol, Stress alleviation

1. INTRODUCTION

Grain legumes (Family Leguminosae), also called pulses, are high in protein content (20-25%) and a good source of vitamins, minerals and the important micronutrients. They are mainly cultivated on marginal lands with minimum agro-inputs and a diverse range of soil microorganisms are vital components of low-input farming. Role of PGPRs (Plant growth promoting rhizobacteria) for improving the growth and yield of agronomically important crops widely reported. In India, the exploitation of PGPRs for improving production and productivity of grain legumes are being popularized in the last decade. PGPRs live in the rhizosphere region and exist either as an extracellular PGPR found in rhizosphere or intracellular PGPR, present inside the root cells or nodular structure. The various root exudates components *viz* sugars, organic acids and amino acids released by the plant roots makes the rhizosphere soil rich in nutrients in comparison to bulk soil resulting in the higher active microbial population in and around the root region and this phenomenon is called as rhizosphere effect. PGPRs represent a wide range of microorganisms belong to the phyla bacteroidetes, actinobacteria, firmicutes and proteobacteria which live in association with plants. PGPRs exert beneficial traits such as enhancement of seedling germination and early plant growth through array of mechanisms. Inoculation of plant-growth promoting rhizobacteria (PGPR) that aggressively colonize legume rhizosphere protects the plants from pathogenic infections. Several reviews discussed their role in growth stimulation and hence, the present manuscript is an attempt to review the role of PGPRs and their applications as biocontrol agent in leguminous crops.

Different bacterial species such as *Alcaligenes*, *Azospirillum, Arthrobacter, Acinetobacter, Bacillus, Burkholderia, Enterobacter, Erwinia, Flavobacterium, Methylobacterium, Paenibacillus, Pseudomonas* as well as various rhizobial species are reported to exhibit PGPR effect among them *Bacillus* and *Psudomonas* sp are most versatile and influence plant growth directly or indirectly (Glick *et al.,* 1999; Persello-Cartieaux *et al.,* 2001). Direct growth promotion includes nitrogen fixation (Zahran, 2001), phosphate solubilization (Wakelin *et al.,* 2004), phytohormone production (Lee *et al.,* 2004) or enhancement in the availability of minerals (Sturz *et al.,* 2000). Indirect growth promotion of PGPR is attributed to the prevention of the deleterious effects of phytopathogens (Glick *et al.,* 1999) by producing antagonistic substances such as phenazine, diacetylphloroglucinol (DAPG), HCN, 2-3 butanediol, acetoin (Ryu *et al.,* 2003) and siderophores (Costa and Loper, 1994). The lytic enzymes *viz*. chitinase and glucanase produced by these PGPRs can degrade cell-wall of fungal pathogens. The bacterial metabolites exhibit a direct inhibitory action against different pathogens of various leguminous crops is listed in Table 1. The PGPRs can also induce systematic resistance throughout the entire plant system (Kloepper and Ryu, 2006). Among the various PGPRs, the *Pseudomonas* and *Bacillus* strains were used as the potential biocontrol agents (BCA). Another crucial contribution of PGPRs is to enhance the plant's resistance or tolerance to abiotic stresses like pH, drought, salinity and temperature. However, the ways by which PGPRs influence plant growth directly may differ from species to species or can be strain specific.

Table 1. Plant growth promoting rhizobacteria (PGPR) controlling root diseases of legume crops

Legume and associate disease	PGPR strains used	References
Pigeonpea		
Wilt	*Bacillus brevis*	Bapat and Shah (2000)
	Pseudomonas fluorescens	Siddiqui *et al.* (1998)
	Pseudomonas aeruginosa RRLJ04, *Bacillus cereus* BS03	Dutta *et al.* (2014)
	Pseudomonas aeruginosa PNA1	Anjaiah *et al.* (2003)
	Pantoea dispersa	Maisuria *et al.* (2008)

	Pseudomonas fluorescens, Pseudomonas aeruginosa	Siddiqui and Dhakeel (2009)
	Bacillus subtilis AF-1	Harish *et al.* (1998)
	Pseudomonas aeruginosa RRLJ 04, *Bacillus cereus* BS 03	Dutta *et al.* (2014)
Chickpea		
Wilt	*Pseudomonas aeruginosa* PNA1	Anjaiah *et al.* (2003)
	Pseudomonas fluorescens	Inam-ul-Haq *et al.* (2003)
	Pseudomonas fluorescens Pf4-92	Saikia *et al.* (2009)
	Pseudomonas seruginosa, Bacillus subtilis	Karimi *et al.* (2012)
Charcoal rot	*Pseudomonas fluorescens*	Srivastava *et al.* (2001)
Root Rot	*Bacillus pumilus, Pseudomonas alkaligenes*	Akhtar and Siddiqui (2008)
	Paenibacillus illinoisensis, Bacillus subtilis, Pseudomonas psychrotolerans	Inam-ul-Haq *et al.* (2015)
	Pseudomonas straita	Akhtar and Siddiqui (2008)
	Pseudomonas aeruginosa	Izhar *et al.* (1995)
	Pseudomonas alcaligenes, Pseudomonas putida	Akhtar and Siddiqui (2008)
	Pseudomonas aeruginosa	Siddiqui and Akhtar (2007)
	Pseudomonas fluorescens	Manjunatha *et al.* (2013)
Pea		
Root rot	*Bacillus subtilis*	El-Mohamedy and El- Baky (2008)
	Pseudomonas fluorescens PRA25	Xi *et al.* (1996)
	Bacillus subtilis	Hwang and Chakravarty (1992)
	Pseudomonas fluorescens	Hamid *et al.* (2013)
	Pseudomonas putida, Bacillus pumilus	Akhtar and Azam (2014)
Damping-off and Root rot	*Pseudomonas cepacia* AMMD, *Pseudomonas fluorescens*	Parke *et al.* (1991)
Soybean		
Damping-off	*Bacillus megaterium*	Zheng and Sinclair (2000)
Root rot	*Pseudomonas aeruginosa*	Inam-ul-Haq *et al.* (2012)

	Bacillus megaterium, Azotobacter chroococcum, Azospirillum brasilense	El-Barougy *et al.* (2009)
Charcoal root rot	*Bacillus* sp. BIN	Yalda *et al.* (2013)
Cowpea		
Root rot	*Pseudomonas fluorescens* BICC602	Mukherjee and Babu (2013)
	Pseudomonas fluorescens	Sendhilvel *et al.* (2005)
Damping-off	*Pseudomonas fluorescens* bv. V	Nandi *et al.* (2013)
Bean		
Halo blight	*Pseudomonas fluorescens*	Alstrom (1991)
Damping-off	*Burkholderia cepacia*	Devi *et al.* (2012)
Wilt	*Bacillus subtilis*	Martins *et al.* (2013)
Gray mold	*Pseudomonas aeruginosa* 7NSK2	Geert De Meyer (1997)
	Pseudomonas putida BTP1	Ongena *et al.* (2005)
Lentil		
Wilt	*Pseudomonas alcaligenes, Bacillus pumilus*	Akhtar *et al.* (2010)
Groundnut		
Charcoal rot	Fluorescent Pseudomonads	Bhatia *et al.* (2008)
Greengram		
Root rot	Fluorescent Pseudomonads	Noreen *et al.* (2015)
	Burkholderia sp. TNAU-1	Satya *et al.* (2011)
	Bacillus subtilis	Shahid and Khan (2016)
	Pseudomonas aeruginosa	Khatoon *et al.* (2014)
	Pseudomonas aeruginosa, Bacillus subtilis	Ali Siddiqui *et al.* (2001)
	Pseudomonas fluorescens, Bacillus subtilis	Sarkar and Bhattacharyya (2008)
	Pseudomonas fluorescens	Thilagavathi *et al.* (2007)
	Pseudomonas fluorescens PF1	Jayashree *et al.* (2000)
Faba bean		
Root rot	*Bacillus* sp.	Belete *et al.* (2015)
	Bacillus megaterium	Abdel-Monaim (2013)
	Bacillus megaterium	Farfour and Al-Saman (2014)
Clusterbean		
Root rot	*Pseudomonas, Bacillus sp.*	Choudhary and Sindhu (2015)
Blackgram		
Root rot	*Pseudomonas fluorescens*	Latha *et al.* (2000)

2. MECHANISMS OF PGPRs AS BCA

2.1 Antibiosis

A growing concern of soil health and environmental hazards necessitated the use of biocontrol agents for disease suppression. The occurrence of antagonistic soil bacteria and their suppressive ability to soil-borne pathogens are referred to as disease-suppressive soils. The presence of fluorescent pseudomonads, bacilli and streptomyces play a substantial role in disease suppression and the antagonistic activity may be associated with the production of bacterial metabolites that reduce the population or activities of pathogens or deleterious rhizospheric microflora (Glick, 1995; Kloepper 1996; Stutz *et al.,* 1986). Many bacterial strains produce secondary metabolites and conditions favouring one compound do not support others. Fluorescent pseudomonads have been widely studied amongst rhizospheric microflora with respect to the production of antibiotics and the first antibiotic produced by this bacterium with a biocontrol potential is phenazine (*phzC*) derivative. Phenazines are heterocyclic, nitrogen-containing pigments with broad-spectrum antibiotic activity and are synthesised by *Pseudomonas*, *Streptomyces*, *Nocardia*, *Sorangium*, *Brevibacterium* and *Burkholderia* species (Turner and Messenger, 1986). The phenazine derivatives secreted by *Bacillus cereus*, *Bacillus brevis* and *Pseudomonas aeruginosa* controlled the wilt disease caused by *Fusarium oxysporum f. sp. udum* in pigeonpea (Dutta *et al.*, 2014). Phenazine and Diacetyl Phloro Glucinol (DAPG) producing pseudomonads showed biocontrol activity against *Fusarium* suppressive soils (Chin-A-Woeng *et al.*, 1998; Schouten *et al.*, 2004). Production of phenazine-1-carboxylate defective mutant by transposon insertion reduced disease suppressive activity (Pierson and Pierson, 1996). The phenazine biosynthesis in *Pseudomonas* sp. includes five genes *phzFABCD* (Pierson *et al.,* 1995). The biosynthetic genes for DAPG (*phlD*) in fluorescent *Pseudomonas* sp. include six genes of *phlF* and *phlABCDE* that function in the regulation, synthesis and export (Bangera and Thomashow, 1999). Ten genes (*pltA to pltG, pltL, pltM,* and *pltR)* required for the biosynthesis of pyoluteorin were identified within a 24-kb genomic region of *Pseudomonas fluorescens* Pf-5 (Nowak-Thompson *et al.,* 1999). Biosynthesis of DAPG and suppression of pyoluteorin by *Pseudomonas fluorescens* CHAO is stimulated by glucose; however, depletion of glucose induces synthesis of pyoluteorin. This showed a degree of flexibility of antagonism under different environments (Duffy and Defago, 1999). The volatile components 2,3 butanediol and 3-hydroxy-2-butanone (acetoin) released exclusively from PGPR strains were shown to trigger the highest level of growth promotion and induced diseased resistance (Ryu *et al.,* 2003).

Plant growth-promoting rhizobacteria can suppress the growth of soil pathogens through cyanide production. Hydrogen cyanide (HCN) is a secondary metabolite produced by Gram-negative rhizobacteria and acts as a metabolic inhibitor, which avoids predation or competition. HCN is a co-product of ethylene biosynthesis from ACC, catalysed by ACC oxidase. In *Pseudomonas fluorescens,* HCN is produced from glycine by the enzyme HCN synthase under microaerophilic conditions (Castric, 1977). However, cyanogenesis in other bacteria differs significantly (Owen and Zdor, 2001). HCN inhibits Cytochrome C Oxidase of the respiratory system, suppresses the growth of pathogenic fungi (Martinez-Viveros *et al.,* 2010) and influences fungus-bacterium ecological interactions. The HCN producing *Pseudomonas* is also active against root-knot nematode (Siddiqui *et al.,*

2006). The structural genes of HCN synthase are coded by *hcnABC* and its synthesis is regulated by global activator - GacA (Reimmann *et al.,* 1997) and anaerobic regulator ANR (Zimmermann *et al.,* 1991). Mutation analysis and complementation study in *Pseudomonas fluorescens* CHAO demonstrated that HCN production alone accounts for 60% of the biocontrol activity (Voisard *et al.,* 1989).

2.2 Siderophores

Iron is abundant in Earth's crust, but most of it is in the form of insoluble ferric hydroxide, which is unavailable in neutral to alkaline soils, rendering iron deficiency in plants. Globally, 30% of the cultivated soils are calcareous with low iron availability. Some PGPRs produce siderophores (Neilands and Nakamura, 1991) or ferric iron (Fe^{3+}) specific carriers that sequester iron from the soil solution and organic iron complexes. Thus, they mediate nutritional competition for iron and inhibit the growth of plant pathogens which have low affinity for iron (Kloepper *et al.,* 1980). Siderophore mechanism is relevant under iron-deficient conditions and these become less effective under acidic pH (below 6), in which iron availability increases (Neilands and Nakamura, 1991). Different structural types of siderophores such as catecholates, hydroxamates and citrate based polycarboxylates are produced by PGPRs (Holden and Bachman, 2015). Examples of siderophores produced by various PGPR include enterobactin by *Escherichia coli*, azotobactin (*Azotobacter vinelandii*), pseudobactin (*Pseudomonas B 10*), pyoverdin (fluorescent *Pseudomonad*), bacillibactin (*Bacillus subtilis*), rhizoferrin (*Rhizobium*), ferrioxamine B (*Streptomyces pilosus*), vibriobactin (*Vibrio cholerae*), yersiniabactin (*Yersinia pestis*), erythrobactin (*Saccharopolyspora erythraea*) or ornibactin (*Burkholderia cepacia*) have been reviewed (Ghosh *et al.,* 2020; Khan *et al.,* 2018). In gram-negative bacteria, siderophore biosynthesis is regulated directly by TonB –dependent receptor genes (Cascales *et al.,* 2007) or indirectly by iron-sensitive repressor, Fur (Ferric Uptake Regulator) along with its co-repressor Fe^{2+} (Chen *et al.,* 2007). Siderophores form complexes with iron (1:1) and iron-uptake is mediated by TonB- dependent outer membrane proteins. Insoluble ferric (Fe^{3+}) is reduced to soluble ferrous (Fe^{2+}) by esterase activity which is released into the cell. The energy required for translocation of the ferric - siderophore complex into the periplasmic spaces is driven by the protonmotive force / ATPase activity. In iron-deficient environment, *Pseudomonas putida* strain GRP3A is able to produce siderophores and make iron available to plants. Peroxidase levels are also increased along with siderophore production under stress conditions (Li *et al.,* 2018). Siderophore production under iron stress conditions provides an added advantage to rhizobia, resulting in the exclusion of pathogens due to iron starvation.

2.3 Hydrolytic enzyme production

An additional mechanism of biocontrol agents that reduces plant diseases is hyper parasitism of fungal pathogens by excreting cell wall hydrolases. Digestion of the parasite cell wall is accomplished by several excreted enzymes including proteases, chitinases and β-1, 3-glucanases. Individually, all these enzymes display antifungal activity, but they often act synergistically with antibiotics (Lorito *et al.,* 1994; Jadhav *et al*., 2017). Extracellular chitinase and laminarinase synthesised by *Pseudomonas stutzeri* can digest and lyse mycelia of *Fusarium solani* (Lim *et al.,* 1991). Cell wall lysis of *Sclerotium rolfsii* by extracellular chitinase of *Serratia marcescens* has been reported (Ordentlich *et al.,* 1988). Chitinase synthesis by *Serratia plymuthica* C48 inhibits spore germination and germ-tube elongation in *Botrytis cinerea* (Frankowski *et al.,* 2001). Similarly, *Pseudomonas fluorescens* strain produced lytic enzymes *viz.*, chitinase and glucanase (Nagarajkumar *et al.,* 2004); hence, *Pseudomonas* sp. can be an effective biocontrol agent against phytopathogenic fungus *Rhizoctonia solani* and *Pythium aphanidermatum*. The bacterium also promotes nodulation in legumes in coordination with Rhizobia. Production of β-1, 3-glucanase by *Paenibacillus* sp. and *Streptomyces* sp. showed antagonistic potential against *Fusarium oxysporum f* sp *cucumerinum* (Singh *et al.,* 1998). *Pseudomonas cepacia* synthesises β -1,3-glucanase that destroys the cell walls of *Rhizoctonia solani*, *Scelerotium rolfsii* and *Pythium ultimum* (Fridlender *et al.,* 1993).

2.4 Competition for nutrients and niches

Competition for nutrients supplied by root exudates is probably a significant factor in most interactions between PGPR and pathogens. Populations of bacteria established on a plant roots act as a sink for the nutrients in the rhizosphere. This in turn reduces the availability of nutrients for pathogens and their subsequent colonisation of the roots. This mechanism is often used by fluorescent pseudomonads due to their nutritional versatility and high growth rates in the rhizosphere (Sorensen, 1997); however, this mechanism is often overlooked, because of the difficulty to study them in biological systems.

2.5 Induced systemic resistance

The use of PGPR as inducers of systemic resistance in crop plants against different pathogens has been demonstrated under field conditions (Wei *et al.,* 1991). Two general plant defence mechanisms are Induced Systemic Resistance (ISR) and Systemic Acquired Resistance (SAR) and ISR is rhizobacteria-mediated systemic resistance that does not involve any damage to plants. *Pseudomonas fluorescens* (S97) is able to elicit ISR against halo blight disease-causing strain of *Pseudomonas syringae pv. phaseolicola* in bean (*Phaseolus vulgaris L*) (Alstrom, 1991). On the other hand, SAR is induced by foliar pathogens and activates resistance mechanisms in uninfected parts, hence, in SAR, a first infection predisposes the plant to resist further attacks.

Various strains of *Pseudomonas* sp., induce broad-spectrum disease resistance in plants in a species-specific manner. In ISR, there is an increased sensitivity towards hormones rather than their production. Signalling molecules like jasmonic acid (JA), salicylic acid (SA) and ethylene (ET) accumulation coordinate the defence response to induce resistance. Jasmonic acid is produced through octadecanoid pathway in which linolenic acid (18:3) is converted to oxylipin molecules such as JA (Van Spronsen *et al.,* 2003) in response to pathogen invasion in plants. This signalling molecule activates defence-related genes like defensins, thionins and pectinase inhibitors (Pieterse *et al.*, 1998). In addition, JA also plays an important role in the onset of senescence, root formation and ethylene synthesis.

Salicylic acid induces defence against pathogens feeding on live tissues and SA induced genes encoding pathogenesis-related proteins (PRs) (Van Loon *et al.,* 2006) show anti-microbial activity against pathogens (Chandrasekaran and Chun, 2016). *Pseudomonas fluorescens* treated plants trigger ISR by activating the genes encoding pathogenesis-related proteins such as chitinase, β-1,3 glucanases, thaumatin like proteins (TLP) and peroxidases. Further, phenyl propanoid pathway enzymes such as phenylalanine ammonia lyase (PAL), peroxidase (PO) and polyphenol oxidase (PPO) were induced in ISR and their products like phenolics and lignin exhibit enhanced resistance to pathogens (Kandan *et al.,* 2002). ISR induced by endophytic bacterium *Bacillus pumilus* strain SE34 in pea against *Fusarium oxysporum f. sp. pisi* leads to modifications in host plants like strengthening of epidermal and cortical cell walls and deposition of barriers containing callose and phenolic compounds (Benhamou *et al.,* 1996). De Meyer *et al.* (1999) reported that the minute amount of the salicylic acid produced by the rhizosphere *Pseudomonas aeruginosa* 7NSK 2 activated the SAR pathway and also increased the levels of phenylalanine ammonia lyase (PAL) in bean roots. SA produced by this strain grown on iron deficient medium and N-alkylated benzylamine derivative (NABD) produced by *Pseudomonas putida* (BTP1) are the elicitors of the ISR against the *Botrytus cinerea* infection in beans (De Meyer and Hofte, 1997; Ongena *et al.,* 2008). *Bacillus subtilis* AF1 increases the PAL content and decreases the incidence of fusarium wilt in pigeon pea (Podile and Laxmi, 1998).

3. ABIOTIC STRESS MANAGEMENT

The use of PGPRs not only enhances nutrient uptake by the plants but also helps in the alleviation of abiotic stresses like drought, salinity, high temperature and metal toxicity. Rhizobacteria belonging to genera like *Achromobacter, Azospirillum, Bacillus, Burkholderia, Enterobacter, Microbacterium, Methylobacterium, Rhizobium, Pseudomonas, Pantoea, Paenibacillus, etc.* are known to alleviate abiotic stress in plants. Inoculation of PGPR can modify plants response at gene level under abiotic stress conditions (Sergeeva *et al.,* 2006). A *Pseudomonas sp.* can influence physiological changes in plants that modulate their growth and development. Salinity stress responses in *Azospirillum* sp. has been studied in detail and the bacterium has been shown to accumulate compatible solutes such as glutamate, proline, glycine betaine, trehalose, *etc.* in order to adapt to fluctuations in soil salinity or osmolarity (Tripati *et al.,* 1998) and proline seems to play a major role in osmoadaptation. Kulkarni and Nautiyal (2000) conducted a study to examine the growth response of a *Rhizobium* sp. NBRI 330 isolated from root nodules of *Prosopis juliflora* growing in alkaline soil and the individual stress survival limits of the strain was 32% NaCl for 8 h and at 55°C up to 3 h.

Several soil bacteria produce osmolytes to protect against frequent fluctuations in osmotic conditions as *Paenibacillus polymyxa* and *Bacillus subtilis* produce glycine betaine. This compound lowers the water potential outside the cell wall and inhabits intercellular spaces. The increased concentration of this compound is sensed by the plant as a local dehydration and consequently, dehydration genes such as ERD 15 and RAB 18 are activated in plants in order to tolerate drought (Lucht and Bremer, 1994). In drought conditions, *Pseudomonas aeruginosa* GGRJ21 can alleviate the water stress response in mungbean plants by accelerated production and accumulation of antioxidants, and upregulation of drought stress-responsive genes (Sarma and Saikia, 2014).

Water availability controls the production and consumption of proteins and polysaccharides by the bacteria which can indirectly influence soil structure (Roberson and Firestone, 1992). Survival of *Pseudomonas* under water stress conditions is mediated through exopolysaccharide (EPS) production (Chenu and Roberson, 1996). EPS enhances water retention and regulates the diffusion of organic carbon sources and help in the formation of soil aggregates (Costa *et al.,* 2018). Inoculation of *Pseudomonas putida* GAP-P45 in sorghum showed the highest level of EPS production under water stress conditions (Sandhya *et al.,* 2009). Ethylene levels can enhance under stress conditions which can lead to stunted root and shoot growth. Inoculation with *Pseudomonas* sp. containing ACC deaminase which hydrolyses precursor of ethylene *i.e.* ACC, which in turn, can alleviate the imbalance of salinity stress on growth, yield and ripening of pea plants (Arshad *et al.,* 2008). Such a study can also increase nodulation and plant growth in greengram as well (Ahmad *et al.,* 2011).

PGPRs elicit induced systemic tolerance (IST) that enhances plant tolerance against ROS (Reactive Oxygen Species) which are generated under biotic and abiotic stress conditions. Under steady state conditions, low levels of ROS are produced during cell metabolism which in turn can regulate plant development (Miller *et al.,* 2010), or these are detoxified by antioxidative defence mechanisms (Foyer and Noctor, 2005). However, under stress conditions, enhanced level of ROS such as singlet oxygen ($^1O_2$), hydrogen peroxide (H_2O_2), superoxide radical ($O_2^{\cdot -}$) and hydroxyl radicals ($^{\cdot}OH$) (Azevedo Neto *et al.,* 2005) can affect the equilibrium between ROS generation and scavenging, which, in- turn can damage nucleic acids, proteins, lipids and chlorophyll (Bhattachrjee, 2005). Microbial synthesis of ROS-scavenging enzymes such as superoxide dismutase (SOD), peroxidases (POX) and catalase (CAT) prevent ROS toxicity and enhance the plant tolerance to stress conditions (Mittler, 2017; Azevedo Neto *et al.,* 2005).

4. MOLECULAR TECHNIQUES USED IN PGPR STUDY

For a long time, screening and characterization of PGPRs has been focused on culture based biochemical methods. The 16S-23S intra genomic spacer (IGS) and repetitive sequence based-PCR (rep-PCR) that target highly conserved inverted repeats have been used for species level identification. In rep-PCR method, primers specific for either Repetitive Extragenic Palindromic (REP-PCR) elements or Enterobacterial Repetitive Intergenic Consensus (ERIC-PCR) sequences are used to study the genetic diversity of the PGPRs. To improve bacterial species identification with higher resolution, Multilocus Sequence Analysis (MLSA) technique that target partial sequencing of several proteins (*atp*D, recA, *rpo*A, *rpo*B, *thr*C, *dna*K, *dna*J, *gln*II, *gap*, *gln*A, *glt*A, *gyr*B and *pnp*) with conserved functions is also used (Glaeser *et al.,* 2015). The genetic analysis of 16S ribosomal DNA was widely used method for identification of PGPRs at the genus level. Restriction Fragment Length Polymorphism (RFLP) analysis of 16S rRNA (ARDRA-Amplified Ribosomal DNA Restriction Analysis of 16S rDNA) gene used to PGPR grouping. Random Amplified Polymorphic DNA (RAPD) analysis which uses arbitrary primers for random amplification of genome sequence was employed to study PGPR diversity. Culture independent metagenomic approach including genomic, metagenomic, transcriptomic and proteomic data is available for the genetic analysis of PGPRs. Recently, the whole genome metagenomic and partial genome analysis of PGPR species are widely used for identification and characterization of PGPR traits.

5. OTHER MECHANISMS

Plants exhibit strategies like ISR, SAR and IST to combat stress at their levels and PGPRs can help to enhance these traits by influencing pathways like ethylene production and jasmonic acid signalling in plants. Table 2 provide a detailed account of the genes activated during the direct and indirect influence of PGPRs in legumes. The magnitude and nature of plant growth promotion (PGP) activity vary from one PGPR to another, thus, to study these in detail, specific genes responsible for each PGPR trait need to be targeted and compared with the expression studies across PGPR community.

Table 2. Genes activated during indirect mode of action of PGPR

PGP traits and related genes	PGPR strains	References
Antibiosis		
Hydrogen cyanide biosynthesis *hcn*AB, *hcn*BC	*Pseudomonas* sp.	Ramette *et al.* (2003); Meyer *et al.* (2011)
Antibiotics: *str*A (streptomycin)	*Streptomycetes* sp.	Huddleston *et al.* (1997)
Polyketide synthesis KSα (α-ketoacyl synthase)	*Streptomyces* sp.	Metsa-Ketela *et al.* (1999)
Response to stress		
Ethylene regulation *acd*S (acc deaminase)	*Pseudomonas* sp., *Rhizobium* sp., *Achromobacter* sp.	Hontzeas *et al.* (2005); Nascimento *et al.* (2012)
Induced systemic resistance (ISR) JA pathway genes -coi1(coronatine insensitive 1) and *PDF1.2* PeBA1 (protein elicitor, can induce transcription of defence genes, ROS production, etc.)	*Bacillus* sp. Rhizobacteria *Bacillus amyloliquefaciens* NC6	Wang *et al.* (2009) Ryu *et al.* (2004) Wang *et al.* (2016)
Systemic Acquired Resistance (SAR) NPR1 (encodes for defence genes) Pathogenesis related (PR) genes: PR1a, PR1b, PR2 (encodes for β-1,3-glucanase), PR3 and PR4 (encode for chitinase), PR9 (encodes peroxidase)	*Bacillus* sp. *Bacillus* sp. *Bacillus* sp.JS	Cao *et al.* (1997); Wang *et al.* (2009); Beneduzi *et al.* (2012); Kim *et al.* (2015)
Induced systemic tolerance (IST) SOS1 and SOS4 (SOS pathway genes), APX (ascorbate peroxidase), MnSOD (Manganese-dependent Superoxide Dismutase), CAT (catalase) and POD (peroxidase)	*Dietzia natronolimnaea*	Bharti *et al.* (2016)

6. CONCLUSION AND FUTURE PROSPECTS

Application of agrochemicals also leads to pathogen resistance besides affecting non-target beneficial microorganisms associated with plant. PGPRs inoculation can also assure the continuous supply of PGR than the one-time application of synthetic chemicals. The PGPRs showing growth promotion or disease control under *in-vitro* conditions will often show inconsistent response under *in-situ* conditions. The strain efficacy is often related to their establishment and population density of introduced strain in the rhizosphere. With the advent of molecular techniques, it is possible to predict their establishment and plant response. However, investigation on root exudate mediated rhizobacterial community can provide insight into rhizosphere microbiology. Knowledge of rhizosphere processes mediated by root exudates is scanty due to their complex underground nature. Various environmental factors, plant genotype and soil types can affect their performance under field conditions. An understanding of these aspects can aid in the utilisation of microbial inoculants for the betterment and benefit of sustainable legume cultivation.

Acknowledgement: The funding support provided by the ICAR-AINP, Government of India is duly acknowledged

REFERENCES

Abdel-Monaim, M.F. (2013). Improvement of biocontrol of damping-off and root rot/wilt of faba bean by salicylic acid and hydrogen peroxide. *Mycobiology* 41, 47–55.

Ahmad, M., Zahir, Z.A., Asghar, N. and Asghar, M. (2011). Inducing salt tolerance in mung bean through coinoculation with rhizobia and plant-growth-promoting rhizobacteria containing 1-aminocyclopropane-1-carboxylate deaminase. *Canadian Journal of Microbiology* 57(7), 578-589.

Akhtar, M.S. and Azam, T. (2014). Effects of PGPR and antagonistic fungi on the growth, enzyme activity and *Fusarium* root-rot of pea. *Phytopathology and Plant Protection* 47, 138–148.

Akhtar, M.S. and Siddiqui, Z.A. (2008). *Glomus intraradices*, *Pseudomonas alcaligenes*, and *Bacillus pumilus*: effective agents for the control of root-rot disease complex of chickpea (*Cicer arietinum L.*). *Journal of General Plant Pathology* 74, 53–60.

Akhtar, M.S. and Siddiqui, Z.A. (2009). Use of plant growth-promoting rhizobacteria for the biocontrol of root-rot disease complex of chickpea. *Australasian Plant Pathology* 38, 44–50.

Akhtar, M.S., Shakeel, U. and Siddiqui, Z.A. (2010) Biocontrol of *Fusarium* wilt by *Bacillus pumilus*, *Pseudomonas alcaligenes*, and *Rhizobium sp.* on lentil. *Turkish Journal of Biology* 34, 1–7.

Ali Siddiqui, I., Ehetshamul-Haque, S. and Shahid Shaukat, S. (2001). Use of rhizobacteria in the control of root rot-root knot disease complex of mungbean. *Journal of Phytopathology* 149, 337–346.

Alstom, S. (1991). Induction of disease resistance in common bean susceptible to halo blight bacterial pathogen after seed bacterization with rhizosphere Pseudomonads. *Journal of General Applied Microbiology* 37, 495- 498.

Alstrom, S. (1991). Induction of disease resistance in common bean susceptible to halo blight bacterial pathogen after seed bacterization with rhizosphere pseudomonads. *The Journal of General and Applied Microbiology* 37, 495–501.

Anjaiah, V., Cornelis, P. and Koedam, N. (2003). Effect of genotype and root colonization in biological control of fusarium wilts in pigeonpea and chickpea by *Pseudomonas aeruginosa* PNA1. *Canadian Journal of Microbiology* 49, 85–91.

Arshad, M., Shaharoona, B. and Mahmood, T. (2008). Inoculation with *Pseudomonas* spp. containing ACC deaminase partially eliminates the effects of drought stress on growth, yield, and ripening of pea (*Pisum sativum* L.). *Pedosphere* 18, 611-620.

Azevedo Neto, A.D., Prisco, J.T., Eneas-Filho, J., Medeiros, J. and Gomes-Filho, E. (2005). Hydrogen peroxide pre-treatment induces salt-stress acclimation in maize plants. *Journal of Plant Physiology* 162(10), 1114-1122.

Bangera, M.G. and Thomashow, L.S. (1999). Identification and characterization of a gene cluster for synthesis of the polyketide antibiotic 2, 4-diacetylphloroglucinol from *Pseudomonas fluorescens* Q2–87. *Journal of Bacteriology* 181, 3155–3163.

Bapat, S. and Shah, A.K. (2000). Biological control of fusarial wilt of pigeon pea by *Bacillus brevis*. *Canadian Journal of Microbiology* 46, 125–132.

Belete, E., Ayalew, A. and Ahmed, S. (2015). Antagonistic effect of native *Bacillus* isolates against black root rot of faba bean. *African Crop Science Journal* 23, 249–259.

Beneduzi, A., Ambrosini, A. and Passaglia, L.M.P. (2012). Plant growth-promoting rhizobacteria (PGPR): Their potential as antagonists and biocontrol agents. *Genetics and Molecular Biology* 35, 1044–1051.

Benhamou, N., Kloepper, J.W., Quadt-Hallman, A. and Tuzun, S. (1996). Induction of defense-related ultrastructural modifications in pea root tissues inoculated with endophytic bacteria. *Plant Physiology* 112(3), 919–929.

Bharti, N., Pandey, S.S., Barnawal, D., Patel, V.K. and Kalra, A. (2016). Plant growth promoting rhizobacteria *Dietzia natronolimnaea* modulates the expression of stress responsive genes providing protection of wheat from salinity stress. *Scientific Reports* 6, 34768.

Bhatia, S., Maheshwari, D.K., Dubey, R.C., Arora, D.S., Bajpai, V.K. and Kang, S.C. (2008). Beneficial effects of fluorescent *pseudomonads* on seed germination, growth promotion, and suppression of charcoal rot in groundnut (*Arachis hypogea L.*). *Journal of Microbiology and Biotechnology* 18, 1578–1583.

Bhattacharjee, S. (2005). Reactive oxygen species and oxidative burst: roles in stress, senescence and signal. *Current science* 89, 1113–1121.

Cao, H., Glazebrook, J. and Clarke, J.D. (1997). The *Arabidopsis* NPR1 gene that controls systemic acquired resistance encodes a novel protein containing ankyrin repeats. *CELL* 88, 57–63.

Cascales, E., Buchanan, S.K., Duche, D., Kleanthous, C., Lloubes, R., Postle, K., Riley, M., Slatin, S. and Cavard, D. (2007). Colicin biology. *Microbiology and Molecular Biology Reviews* 71, 158–229

Castric, P.A. (1977). Glycine metabolism by *Pseudomonas aeruginosa*: Hydrogen cyanide biosynthesis. *Journal of Bacteriology* 130(2), 826–831.

Chandrasekaran, M. and Chun, S.C. (2016). Expression of PR-protein genes and induction of defense-related enzymes by *Bacillus subtilis* CBR05 in tomato (*Solanum lycopersicum*) plants challenged with *Erwinia carotovora* subsp. *Carotovora*. *Bioscience, Biotechnology and Biochemistry* 80(11), 2277-2283.

Chen, Z., Lewis, K.A., Shultzaberger, R.K., Lyakhov, I.G., Zheng, M., Doan, B., Storz, G. and Schneider, T. (2007). Discovery of Fur binding site clusters in *Escherichia coli* by information theory models. *Nucleic Acids Research* 35, 6762–6777.

Chenu, C. and Roberson, E.B. (1996). Diffusion of glucose in microbial extracellular polysaccharide as affected by water potential. *Soil Biology and Biochemistry* 28, 877–884.

Chin-A-Woeng, T.F.C., van den Broek, D., de Voer, G., van der Drift, K.M., Tuinman, S., Thomas-Oates, J.E., Lugtenberg, B.J.J. and Bloemberg, G.V. (2001). Phenazine-1-carboxamide production in the biocontrol strain *Pseudomonas chlororaphis* PCL1391 is regulated by multiple factors secreted into the growth medium. *Molecular Plant Microbe Interaction* 14, 969-979.

Choudhary, S.R. and Sindhu, S.S. (2015). Suppression of *Rhizoctonia solani* root rot disease of clusterbean (*Cyamopsis tetragonoloba*) and plant growth promotion by rhizosphere bacteria. *Plant Pathology* 14, 48–57.

Costa, J.M. and Loper, J.E. (1994). Characterization of siderophore production by the biological-control agent *Enterobacter cloacae*. *Molecular Plant-Microbe Interactions* 7, 440–448.

Costa, O., Raaijmakers, J.M. and Kuramae, E.E. (2018). Microbial extracellular polymeric substances: ecological function and impact on soil aggregation. *Frontiers in Microbiology* 9, 1636. https://doi.org/10.3389/fmicb.2018.01636

De Meyer, G. and Hofte, M. (1997). Salicylic acid produced by the rhizobacterium *Pseudomonas aeruginosa* 7NSK2 induces resistance to leaf infection by *Botrytis cinerea* on bean. *Phytopathology* 87, 588–593.

De Meyer, G., Capieau, K., Audenaert, K., Buchala, A., Metraux, J.P. and Hofte, M. (1999). Nanogram amounts of salicylic acid produced by the rhizobacterium *Pseudomonas aeruginosa* 7NSK2 activate the systemic acquired resistance pathway in bean. *Molecular Plant-Microbe Interaction* 12, 450-458.

Devi, S.I., Somkuwar, B., Potshangbam, M. and Talukdar, N.C. (2012). Genetic characterization of *Burkholderia cepacia* strain from Northeast India: a potential bio-control agent. *Advances in Bioscience and Biotechnology* 3, 1179–1188.

Duffy, B.K., and Defago, G. (1999). Environmental factors modulating antibiotic and siderophore biosynthesis by *Pseudomonas fluorescens* biocontrol strains. *Applied Environmental Microbiology* 65(6), 2429-2438.

Dutta, S., Morang, P., Nishanth Kumar, S. and Dileep Kumar, B.S. (2014). Fusarial wilt control and growth promotion of pigeon pea through bioactive metabolites produced by two plant growth promoting rhizobacteria. *World Journal of Microbiology and Biotechnology* 30, 1111–1121.

El-Barougy, E., Awad, N.M., Turky, A.S. and Hamed, H.A. (2009). Antagonistic activity of selected strains of rhizobacteria against *Macrophomina phaseolina* of soybean plants. *American-Eurasian Journal of Agricultural and Environmental Science* 5, 337–347.

El-Mohamedy, R.S.R. and El- Baky, M.M.H. (2008). Evaluation of different types of seed treatment on control of root rot disease, improvement, growth and yield quality of pea plant in Nobaria province. *Research Journal of Agriculture and Biological Sciences* 4, 611–622.

Farfour, S.A. and Al-Saman, M.A. (2014). Root-rot and stem-canker control in faba bean plants by using some biofertilizers agents. *Journal of Plant Pathology & Microbiology* 5, 1–6.

Foyer, C.H. and Noctor, G. (2005). Redox homeostasis and antioxidant signaling: A metabolic interface between stress perception and physiological responses. *The Plant Cell* 17, 1866-1875.

Frankowski, J., Lorito, M., Scala, F., Schmidt, R., Berg, G. and Bahl, H. (2001). Purification and properties of two chitinolytic enzymes of *Serratia plymuthica* HRO-C48. *Archieves of Microbiology* 176**,** 421-426.

Fridlender, M., Inbar, J. and Chet, I. (1993). Biological control of soilborne plant pathogens by a β-1,3 glucanase-producing *Pseudomonas cepacia. Soil biology and Biochemistry 25(9), 1211-1221.*

Geert De Meyer, M.H. (1997). Salicylic acid produced by the rhizobacterium *Pseudomonas aeruginosa* 7NSK2 induces resistance to leaf infection by *Botrytis cinerea* on bean. *Phytopathology* 87, 588–593.

Ghosh, S.K., Bera, T. and Chkrabarty, M. (2020). Microbial Siderophore – A boon to agricultural Sciences. *Biological Control* 144, 104214

Glaeser, S.P. and Kampfer, P. (2015). Multilocus sequence analysis (MLSA) in prokaryotic taxonomy. *Systematic and Applied Microbiology* 38(4), 237-245.

Glick, B.R. (1995). The enhancement of plant growth by free-living bacteria. *Canadian Journal of Microbiology* 41, 109–117.

Glick, B.R., Patten, C.L., Holguin, G. and Penrose, D.M. (1999) Biochemical and genetic mechanisms used by plant growth-promoting bacteria. (Imperial College Press: London)

Hamid A, Bhat NA, Sofi TA, Bhat KA and Asif M (2013). Management of root rot of pea (*Pisum sativum L.)* through bioagents. *African Journal of Microbiology Research* 6, 7156–7161.

Harish, S., Manjula, K. and Podile, A.R. (1998). *Fusarium udum* is resistant to the mycolytic activity of a biocontrol strain of *Bacillus subtilis AF 1. FEMS Microbiology Ecology* 25, 385–390.

Holden, I. and Bachman, M.A. (2015). Diverging roles of bacterial siderophores during infection. *Metallomics* 7, 986-995.

Hontzeas, N., Richardson, A.O., Belimov, A., Safronova, V., Abu-Omar, M.M. and Glick, B.R. (2005). Evidence for horizontal transfer of 1-aminocyclopropane-1-carboxylate deaminase genes. *Applied and Environmental Microbiology* 71, 7556–7558.

Huddleston, A.S., Cresswell, N., Neves, M.C.P., Beringer, J.E., Baumberg, S., Thomas, D.I. and Wellington, E.M.H. (1997). Molecular detection of streptomycin-producing

Streptomycetes in Brazilian soils. *Applied and Environmental Microbiology* 63, 1288–1297.

Hwang, S.F. and Chakravarty, P. (1992). Potential for the integrated control of *Rhizoctonia* root-rot of *Pisum sativum* using *Bacillus subtilis* and a fungicide. *Journal of Plant Diseases and Protection* 99, 626-636.

Inam ul Haq, M., Tahir, M.I, Hayat, R., Khalid, R., Ashfaq, M., Jamil, M. and Ali, Z. (2015). Bioefficacy of rhizobacterial isolates against root infecting fungal pathogens of chickpea (*Cicer arietinum L.*). *Journal of Plant Pathology & Microbiology* 3, 1–8.

Inam-ul-Haq, M., Javed, N., Ahmad, R. and Rehman, A. (2003). Evaluation of different strains of *Pseudomonas fluorescens* for the biocontrol of *Fusarium* wilt of Chickpea. *Pakistan Journal of Plant Pathology* 2, 65–74.

Inam-ul-Haq, M., Mehmood, S., Rehman, H.M., Ali, Z. and Tahir, M.I. (2012) Incidence of root rot diseases of soybean in Multan Pakistan and its management by the use of plant growth promoting rhizobacteria. *Pakistan Journal of Botany* 44, 2077–2080.

Izhar, I., Ehteshamul-Haque, S., Javeed, M. and Ghaffar, A. (1995). Efficacy of *Pseudomonas aeruginosa* and *Bradyrhizobium sp.* in the control of root rot disease in chickpea. *Pakistan Journal of Botany* 27, 451–455.

Jadhav, H.P., Shaikh, S.S. and Sayyed, R.Z. (2017). Role of hydrolytic enzymes of rhizoflora in biocontrol of fungal phytopathogens: an overview In: (Ed) Mehnaz S, Rhizotrophs: Plant growth promotion to bioremediation, Springer. pp.183-203.

Jayashree, K., Shanmugam, V., Raguchander, T., Ramanathan, A. and Samiyappan, R. (2000). Evaluation of *Pseudomonas fluorescens* Pf-1 against blackgram and sesame root-rot disease. *Journal of Biological Control* 14, 55–61.

Kandan, A., Commare, R.R., Nandakumar, R., Ramiah, M., Raguchander, T. and Samiyappan, R. (2002). Induction of phenyl propanoid metabolism by *Pseudomonas fluorescens* against tomato spotted wilt virus in tomato. *Folia Microbiology* 47, 121-129.

Karimi, K., Amini, J., Harighi, B. and Bahramnejad, B. (2012). Evaluation of biocontrol potential of *Pseudomonas* and *Bacillus sp.* against *Fusarium* wilt of chickpea. *Australian Journal of Crop Science* 6, 695–703.

Khan, A., Singh, P. and Srivastava, A. (2018). Synthesis, nature and utility of universal iron chelator–siderophore: a review. *Microbial Research* 212, 103-111

Khatoon, N., Shafique, H.A., Noreen, R., Sultana, V., Badar, R, and Ehteshamul-Haque, S. (2014). Role of endophytic and rhizospheric fluorescent *Pseudomonas* associated with mungbean in suppressing the root rotting fungi of mungbean. *International Journal of Biological Research* 2, 115–123.

Kim, J-S., Lee, J., Lee, C-H., Woo, S.Y., Kang, H., Seo, S-G. and Kim, S-H. (2015). Activation of pathogenesis-related genes by the *Rhizobacterium*, *Bacillus sp.* JS, which induces systemic resistance in Tobacco plants. *The plant pathology journal* 31, 195–201.

Kloepper, J.W. and Ryu, C.M. (2006). Bacterial endophytes as elicitors of induced systemic resistance. In: (Eds) B Schulz, C Boyle, TN Sieber, Microbial root endophytes. Springer: Berlin/Heidelberg, pp 33–52.

Kloepper, J.W., Leong, J., Teintze, M. and Schroth, M.N. (1980). Enhancing plant growth by siderophores produced by plant growth-promoting rhizobacteria. *Nature* 286, 885–886.

Kullarni, S. and Nautiyal, C.S. (2000). Effects of salt and pH stress on temperature -tolerant *Rhizobium* sp NBRI 330 nodulating *Prospis juliflora*. *Current Microbiology* 40, 221-226.

Latha, T.K.S., Rajeswari, E. and Narasimhan, V. (2000). Management of root-rot disease complex through antagonists and chemicals. *Indian Phytopathology* 53, 216–218.

Lee, S., Flores-Encarnacion, M., Contreras-Zentalla, M., Garcia-Flores, L., Escamilla, J.E. and Kennedy, C. (2004). Indole-3-acetic acid biosynthesis is deficient in *Gluconacetobacter diazotrophicus* strains with mutations in cytochrome c biogenesis genes. *Journal of Bacteriology* 186, 5384-5391.

Li, C., Zhu, L., Pan, D., Li, S., Xiao, H., Zhang, Z., Shen, X., Wang, Y. and Long, M. (2018). Siderophore-mediated iron acquisition enhances resistance to oxidative and aromatic compound stress in *Cupriavidus necator* JMP134. *Applied and Environmental Microbiology* 85(1), e01938-18.

Lim, H., Kim, Y. and Kim, S. (1991). *Pseudomonas stutzeri* YPL-1 genetic transformation and antifungal mechanism against *Fusarium solani*, an agent of plant root rot. *Applied and Environmental Microbiology* 57, 510-516.

Lorito, M., Peterbauer, C., Christopher, K.H. and Gary Harman, E. (1994). Synergistic interaction between fungal cell wall degrading enzymes and different antifungal compounds enhances inhibition of spore germination. *Microbiology* 140, 623-629.

Lucht, J.M. and Bremer, E. (1994). Adaptation of *Escherichia coli* to high osmolarity environments. Osmoregulation of the high affinity glycine betaine transport system. *FEMS Microbiology Reviews* 4, 3-20.

Maisuria, V.B., Gohel, V., Mehta, A.N., Patel, R.R. and Chhatpar, H.S. (2008). Biological control of *Fusarium* wilt of pigeonpea by *Pantoea dispersa*, a field assessment. *Annals of Microbiology* 58, 411–419.

Manjunatha, S.V., Naik, M.K., Khan, M.F.R. and Goswami, R.S. (2013). Evaluation of bio-control agents for management of dry root rot of chickpea caused by *Macrophomina phaseolina*. *Crop Protection* 45, 147–150.

Martínez-Viveros, O., Jorquera, M.A., Crowley, D.E., Gajardo, G and Mora, M.L. (2020). Mechanisms and practical considerations involved in plant growth promotion by rhizobacteria. *Journal of Soil Science and Plant Nutrition* 10, 293-319.

Martins, S.J., de Medeiros, F.H.V., de Souza, R.M., de Resende, M.L.V. and Junior, P.M.R. (2013). Biological control of bacterial wilt of common bean by plant growth-promoting rhizobacteria. *Biological Control* 66, 65–71.

Metsa-Ketela, M., Salo, V., Halo, L., Hautala, A., Hakala, J., Mantsala, P. and Ylihonko, K. (1999). An efficient approach for screening minimal PKS genes from *Streptomyces*. *FEMS Microbiology Letters* 180, 1–6.

Meyer, J.B., Frapolli, M., Keel, C. and Maurhofer, M. (2011). Pyrroloquinoline quinone biosynthesis gene pqqC, a novel molecular marker for studying the phylogeny and diversity of phosphate-solubilizing *pseudomonads*. *Applied and Environmental Microbiology* 77, 7345–7354.

Miller, G., Suzuki, N., Ciftci-Yilmaz, S. and Mittler, R. (2010). Reactive oxygen species homeostasis and signalling during drought and salinity stresses. *Plant Cell and Environment* 33, 453–467.

Mittler, R. (2017). ROS are good. *Trends in Plant science* 22(1), 11-19.

Mukherjee, A. and Babu, S.P.S. (2013). *Pseudomonas fluorescens* mediated suppression of *Meloidogyne incognita* infection of cowpea and tomato. *Phytopathology and Plant Protection* 46, 607-616.

Nagarajkumar, M., Bhaskaran, R. and Velazhahan, R. (2004). Involvement of secondary metabolites and extracellular lytic enzymes produced by *Pseudomonas fluorescens* in inhibition of *Rhizoctonia solani*, the rice sheath blight pathogen. *Microbiological Research* 159(1), 73-81.

Nandi, A.S., Adhikari, A., Dutta, S., Chattopadhaya, A. and Nath, R. (2013). Potential effects of plant growth promoting rhizobacteria (*Pseudomonas fluorescens*) on cowpea seedling health and damping off disease control. *African Journal of Biotechnology* 12, 1853–1861.

Nascimento, F.X., Brigido, C., Glick, B.R., Oliveira, S. and Alho, L. (2012). *Mesorhizobium ciceri* LMS-1 expressing an exogenous 1-aminocyclopropane-1-carboxylate (ACC) deaminase increases its nodulation abilities and chickpea plant resistance to soil constraints. *Letters in Applied Microbiology* 55, 15–21.

Neilands, J.B. and Nakamura, K. (1991). Detection, determination, isolation, characterization and regulation of microbial iron chelates. In: Winkelmann G (Ed) Handbook of Microbial Iron Chelates. CRC Press. Boca Raton, Florida. pp, 1–14.

Noreen, R., Ali, S.A., Hasan, K.A., Sultana, V., Ara. J, and Ehteshamul-Haque, S. (2015). Evaluation of biocontrol potential of fluorescent *Pseudomonas* associated with root nodules of mungbean. *Crop Protection* 75, 18–24.

Nowak-Thompson, B., Chaney, N., Wing, J.S., Gould, S.J. and Loper, J.E. (1999). Characterization of the pyoluteorin biosynthetic gene cluster of *Pseudomonas fluorescens* Pf-5. *Journal of Bacteriology* 181(7), 2166-2174.

Ongena, M., Jourdan, E., Adam, A., Schafer, M., Budzikiewicz, H. and Thonart, P. (2008). Amino Acids, iron, and growth rate as key factors influencing production of the *Pseudomonas putida* BTP1 benzylamine derivative involved in systemic resistance induction in different plants. *Microbial Ecology* 55, 280–292.

Ongena, M., Jourdan, E., Schafer, M., Kech, C., Budzikiewicz, H., Luxen, A. and Thonart, P. (2005). Isolation of an n-alkylated benzylamine derivative from *Pseudomonas putida* BTP1 as elicitor of induced systemic resistance in Bean. *Molecular Plant-Microbe Interactions* 18, 562–569.

Ordentlich, A., Elad, Y. and Chet, I. (1988). The role of chitinase of *Serratia marcescens* in biocontrol of *Sclerotium rolfsii*. *Phytopathology* 78, 84-88.

Owen, A. and Zdor, R. (2001). Effect of cyanogenic rhizobacteria on the growth of velvetleaf (*Abutilon theophrasti*) and corn (*Zea mays*) in autoclaved soil and the influence of supplemental glycine. *Soil Biology and Biochemistry* 33, 801– 809.

Parke, J.L., Rand, R.E., Joy, A.E. and King, E.B. (1991). Biological control of *Pythium* damping-off and *Aphanomyces* root rot of peas by application of *Pseudomonas cepacia* or *P. fluorescens* to seed. *Plant Disease* 75, 987–992.

Persello-Cartieaux, F., David, P., Sarrobert, C., Thibaud, M.C., Achouak, W., Robaglia, C. and Nussaume, L. (2001). Utilization of mutants to analyze the interaction between *Arabidopsis thaliana* and its naturally root-associated *Pseudomonas. Planta* 212, 190–198.

Pierson, L.S. and Pierson, E.A. (1996). Phenazine antibiotic production in *Pseudomonas tumefaciens*: role in rhizosphere ecology and pathogen suppression. *FEMS Microbiology Letters* 136(2), 101-108.

Pierson, L.S., Gaffney, T., Lam, S. and Gong, F. (1995). Molecular analysis of genes encoding phenazine biosynthesis in the biological control bacterium *Pseudomonas aureofaciens FEMS Microbiology Letters* 134(2-3), 299–307

Pieterse, C.M.J., van Wees, S.C.M., van Pelt, J.A., Knoester, M., Laan, R., Gerrits, H., Weisbeek, P.J. and van Loon, L.C. (1998). A novel signalling pathway controlling induced systemic resistance in *Arabidopsis*. *The Plant Cell* 10, 1571–1580.

Podile, A.R. and Laxmi, V.D.V. (1998). Seed bacterization with *Bacillus subtilis* AF1 increases phenylalanine ammonia-lyase and reduces the incidence of *Fusarium* wilt in pigeonpea. *Journal of Phytopathology* 146, 255- 259.

Ramette, A., Frapolli, M., Defago, G. and Moenne-Loccoz, Y. (2003). Phylogeny of HCN synthase-encoding hcnBC genes in biocontrol fluorescent *Pseudomonads* and its relationship with host plant species and HCN synthesis ability. *Molecular Plant-Microbe Interactions* 16, 525-535.

Reimmann, C., Beyeler, M., Latifi, A., Winteler, H., Foglino, M., Lazdunski, A. and Haas, D. (1997). The global activator GacA of *Pseudomonas aeruginosa* PAO positively controls the production of the autoinducer *N*-butyryl-homoserine lactone and the formation of the virulence factors pyocyanin, cyanide, and lipase. *Molecular Microbiology* 24, 309–319.

Roberson, B. and Firestone, M.K. (1992). Relationship between desiccation and exopolysaccharide production in a soil *Pseudomonas* sp. *Applied and Environmental Microbiology* 58(4), 1284-1291.

Ryu, C-M., Farag, M.A., Hu, C-H., Reddy, M.S., Kloepper, J.W. and Pare, P.W. (2004). Bacterial volatiles induce systemic resistance in Arabidopsis. *Plant physiology* 134, 1017–1026.

Ryu, C-M., Farag, M.A., Hu, C-H., Reddy, M.S., Wie, H-X., Pare, P.W. and Kloepper, J.W. (2003). Bacterial volatiles promote growth of *Arabidopsis*. *Proceedings of the National Academy of Sciences* 100, 4927–4932.

Saikia, R., Varghese, S., Singh, B.P. and Arora, D.K. (2009). Influence of mineral amendment on disease suppressive activity of *Pseudomonas fluorescens* to Fusarium wilt of chickpea. *Microbiological Research* 164, 365–373.

Sandhya, V., Ali, S.K.Z., Grover, M., Reddy, G. and Venkateswarlu, B. (2009). Alleviation of drought stress effects in sunflower seedlings by the exopolysaccharides producing *Pseudomonas putida* strain GAP-P4. *Biology and Fertility of Soil* 46, 17-26.

Sarkar, M. and Bhattacharyya, P.K. (2008). Biological control of root rot of green gram caused by *Macrophomina phaseolina* by antagonistic microorganisms. *Journal of Mycopathological Research* 46, 233–237.

Sarma, R.K. and Saikia, R. (2014). Alleviation of drought stress in mung bean by strain *Pseudomonas aeruginosa* GGRJ21. *Plant and Soil* 377, 111–126.

Satya, V.K., Vijayasamundeeswari, A., Paranidharan, V. and Velazhahan, R. (2011). *Burkholderia sp.* strain TNAU-1 for biological control of root rot in mung bean (*Vigna radiata L.*) caused by *Macrophomina phaseolina*. *Journal of Plant Protection Research* 51, 273-278.

Schouten, A., van den Berg, G., Edel-Hermann, V., Steinberg, C., Gautheron, N., Alabouvette, C., de Vos, C.H., Lemanceau, P. and Raaijmakers, J.M. (2004). Defence responses of *Fusarium oxysporum* to 2,4-diacetylphloroglucinol, a broad-spectrum antibiotic produced by *Pseudomonas fluorescens*. *Molecular Plant-Microbe Interaction* 17, 1201–1211.

Sendhilvel, V., Buvaneswari, D., Kanimozhi, S., Mathiyazhagan, S., Kavitha, K. and Raguchander, T. (2005). Management of cowpea root-rot caused by *Macrophomina phaseolina* (Tassi) Goid, using plant growth promoting rhizobacteria. *Journal of Biological Control* 19, 41–46.

Sergeeva, E., Shah, S. and Glick, B.R. (2006). Growth of transgenic canola (*Brassica napus* cv. Westar) expressing a bacterial 1-aminocyclopropane- 1-carboxylate (ACC) deaminase gene on high concentrations of salt. *World Journal of Microbiology and Biotechnology* 22, 277–282.

Shahid, S. and Khan, M.R. (2016). Biological control of root-rot on mungbean plants incited by *Macrophomina phaseolina* through microbial antagonists. *Plant Pathology Journal* 15, 27–39.

Siddiqui, I.A., Shaukat, S.S., Sheikh, I.H. and Khan, A. (2006). Role of cyanide production *by Pseudomonas fluorescens* CHA0 in the suppression of root-knot nematode, *Meloidogyne javanica* in tomato. *Word Journal of Microbiology and Biotechnology* 22, 641–650.

Siddiqui, Z.A. and Akhtar, M.S. (2007). Biocontrol of a chickpea root-rot disease complex with phosphate-solubilizing microorganisms. *Journal of Plant Pathology* 89, 67–77.

Siddiqui, Z.A. and Shakeel, U. (2009). Biocontrol of wilt disease complex of pigeon pea (*Cajanus cajan* (L.) Millsp.) by isolates of *Pseudomonas sp. African Journal of Plant Science* 3, 1–12.

Siddiqui, Z.A., Mahmood, I. and Hayat, S. (1998). Biocontrol of *Heterodera cajani* and *Fusarium udum* on pigeonpea using *Glomus mosseae*, *Paecilomyces lilacinus* and *Pseudomonas fluorescens*. *Thai Journal of Agricultural Science* 31, 310–321.

Singh, P.P., Shin, Y.C., Park, C.S. and Chung, Y.R. (1998). Biological control of *Fusarium* wilt of cucumber by chitinolytic bacteria. *Phytopathology* 89(1), 92 -99.

Sorensen, J. (1997). The rhizosphere as a habitat for soil microorganisms. In: (EdS) van Elsas JD, Trevors JT and Wellington EMH. Modern Soil Microbiology, Wellington, New York. pp. 21-45.

Srivastava, A.K., Singh, T., Jana, T.K. and Arora, D.K. (2001). Induced resistance and control of charcoal rot in *Cicer arietinum* (chickpea) by *Pseudomonas fluorescens*. *Canadian Journal of Botany* 79, 787–795.

Sturz, A.V. and Nowak, J. (2000). Endophytic communities of rhizobacteria and the strategies required to create yield enhancing associations with crops. *Applied Soil Ecology* 15, 183–190.

Stutz, E.W., Defago, G. and Kern, H. (1986). Naturally occurring fluorescent pseudomonads involved in suppression of black root rot of tobacco. *Phytopathology* 76(2), 181–185.

Thilagavathi, R., Saravanakumar, D., Ragupathi, N. and Samiyappan, R. (2007). A combination of biocontrol agents improves the management of dry root rot (*Macrophomina phaseolina*) in greengram. *Phytopathologia Mediterranea* 46, 157–167.

Tripathi, A.K., Mishra, B.M. and Tripathi, P. (1998). Salinity stress responses in the plant growth promoting rhizobacteria *Azospirillum* spp. *Journal of Bioscience* 23, 463–471.

Turner, J.M. and Messenger, A.J. (1986). Occurrence, biochemistry and physiology of phenazine pigment production. *Advances in Microbial Physiology* 27, 211-275.

Van Loon, L.C., Rep, M. and Pieterse, C.M.J. (2006). Significance of inducible defense related proteins in infected plants. *Annual Review of Phytopathology* 44, 135–162.

Van Spronsen, P.C., Tak, T., Rood, A.M.M., van Brussel, A.A.A., Kijne, J.W. and Boot, K.J.M. (2003). Salicylic acid inhibits indeterminate-type nodulation but not determinate-type nodulation, *Molecular Plant-microbe Interaction* 16(1), 83-91.

Voisard, C., Keel, C., Haas, D. and Defago, G. (1989). Cyanide production by *Pseudomonas fluorescens* helps suppress black root rot of tobacco under gnotobiotic conditions. *EMBO* 8, 351-358

Wakelin, S.A., Warren, R.A., Harvey, P.R. and Ryder, M.H. (2004). Phosphate solubilization by *Penicillium sp.* closely associated with wheat roots. *Biology and Fertility of Soils* 40, 36–43.

Wang, N., Liu, M., Guo, L., Yang, X. and Qiu, D. (2016). A novel protein elicitor (PeBA1) from *Bacillus amyloliquefaciens* NC6 induces systemic resistance in tobacco. *International journal of biological sciences* 12, 757–67.

Wang, S., Wu, H., Qiao, J., Ma, L., Liu, J., Xia, Y. and Gao, X. (2009). Molecular mechanism of plant growth promotion and induced systemic resistance to tobacco mosaic virus by *Bacillus sp. Journal of Microbiology and Biotechnology* 19, 1250–8.

Wei, G., Kloepper, J.W. and Tuzun, S. (1991). Induction of systemic resistance of cucumber *Colletotrichum orbiculare* by select strains of plant growth-promoting rhizobateria. *Phytopathology* 81, 1508-1512.

Xi, K., Stephens, J.H.G. and Verma, P.R. (1996). Application of formulated rhizobacteria against root rot of field pea. *Plant Pathology* 45, 1150–1158.

Yalda, V., Naser, S. and Azizollah, A. (2013). Biological control of soybean charcoal root rot disease using bacterial and fungal antagonists *in vitro* and greenhouse condition. *Journal of Crop Protection* 2, 139–150.

Zahran, H.H. (2001). Rhizobia from wild legumes: diversity, taxonomy, ecology, nitrogen fixation and biotechnology. *Journal of Biotechnology* 91, 143-153.

Zheng, X.Y. and Sinclair, J.B. (2000). The effects of traits of *Bacillus megaterium* on seed and root colonization and their correlation with the suppression of *Rhizoctonia* root rot of soybean. *BioControl* 45, 223–243.

Zimmermann, A., Reimmann. C, Galimand, M. and Haas, D. (1991). Anaerobic growth and cyanide synthesis of *Pseudomonas aeruginosa* depend on *anr*, a regulatory gene homologous with *fnr* of *Escherichia coli*. *Molecular Microbiology* 5, 1483–1490.

CHAPTER 10

Role of hydrogel in pulses production under rainfed rice fallow

Md. Hedayetullah

Directorate of Research, Bidhan Chandra Krishi Viswavidyalaya, Kalyani, Nadia, West Bengal, Pin-741235, India

*Author email: hedaye.bckv@gmail.com

Abstract

Total 11.65 M ha area remained fallow after harvest of rainy season rice in the country. Soil moisture deficit stress and lack of irrigation facilities are the major reason to remain land fallow. These large areas can be utilized for cultivation of pulse crops with appropriate soil moisture conservation technology. Under such condition, hydrogel can act as a soil moisture enhancer under aberrant weather conditions. It helps soil to keep continuous moisture for survival and growth of the crop. Under rice fallow situation, hydrogel meet up the need of life saving irrigation during reproductive phase of crop. Relay cropping of pulses is generally practiced in rainfed rice fallow areas. Hydrogel gives better results with organic manures and facilitates the long time moisture supply to the crops. In general, 5.0 kg/ha of basal application of hydrogel recommended with farm yard manure or organic manure.

Keywords: Hydrogel, Pulses, Rainfed, Rice fallow

1. INTRODUCTION

Super absorbents materials came to exist in the markets in early 1960's by the American Company, Union Carbide. Super absorbents having the capacity to absorb water 20 times more than their weight is considered as superabsorbent (Abedi-koupai and Sohrab, 2004). It is a highly cross-linked polymer with high water holding capacity. It holds water up to 400 to 2000 times of their weight. In respect of solubility, water soluble and insoluble polymers have been marketed for agricultural use. A water soluble polymer does not form gels and are used as soil conditioners. They are polyethylene glycol, polyvinyl alcohol, polyacrylates and polyacrylamides. Water soluble polymers developed primarily to aggregate and stabilize soils, combat erosion and improve percolation. It has ability to absorb water up to 1000 times or more of their weight and form gel. Due to their tremendous water absorbing ability, they are referred as super absorbents (SA).

Rice fallows are those land which remain fallow during winter season after harvesting of *kharif* rice. Relay cropping are generally practiced in rice fallow areas where no option of life saving irrigation (Hedayetullah *et al.*, 2018). It is also called *utera* / *paira* crop or zero tilled crops. The total area under rice fallow in India is about 11.65 M ha. Eastern part of India covered about 4.22 M ha which comprises eastern Uttar Pradesh, Bihar, Jharkhand and West Bengal. North-east hill region namely Assam area covered 0.54 M ha. Central region of India has 5.01 M ha in the states of Madhya Pradesh, Chhattisgarh and Maharashtra, whereas in coastal region is about 1.52 M ha in Andhra Pradesh, Odisha and Tamil Nadu. These rainfed rice fallow lands (RRFL) offers considerable scope to address the twin problems of food and nutritional insecurities. If these existing rice fallow lands brought under cultivation, it may usher another green revolution in the predominantly rice-fallow states, benefiting millions of small landholders (Joshi *et al.,* 2002). Of the total rice fallow area close to 40% lies in the states of Madhya Pradesh and Chhattisgarh, with Chhattisgarh having a larger share. In absolute terms, these 2 states have 38% of RRFL, which is about half of the country's total chickpea area.

Intensification of existing agricultural systems is need of the hour to take care of the rising demand of foodgrain production in the country (Kumar *et al.* 2016). In this perspective, there is an enormous opportunity to increase the total cropping area through strategic research in rice–fallows (Kar and Kumar, 2009). However, inclusion of the second crop in rice–fallows is a great challenge as post-rainy season that often confronts a series of abiotic and biotic stresses (Kumar *et al.,* 2018). Fast depletion of the soil-moisture after rice harvest, lack of irrigation facilities, poor access to extra early-duration varieties of pulses/oilseeds, late harvesting of rice, uncertainty in rainfall event, poor soil structure and problems of stray cattle, are some of the major constraints in cultivation of winter crops in rice–fallows (Kumar *et al.,* 2016).

Hydrogel reduces water runoff and increases infiltration rates, in addition to increasing water holding capacity of the soil. Because of these properties of hydrogel, it is useful in increasing crop yield under RRFL. At the time of water availability either due to rain or irrigation, hydrogel absorbs water and it releases water back to the soil to meet up the water requirement of crop plants during soil moisture deficit stress. Therefore, application of hydrogel reduces frequency of irrigation in almost all the crops including cereals, pulses, vegetables and flowers, thus reducing time and money spend on irrigation, labour and water costs (Dar *et al*., 2017). The efficiency of the technology is also highly suited for farmers growing pulses under RRFL in different parts of the country.

2. WHAT ARE HYDROGEL AND ITS TYPES?

Hydrogel is a network of polymer chains which are hydrophilic in nature that can swell in water and hold a large amount of water. Hydrogel maintains the cross linking structure of chemical and physical polymer chains. Hydrogel was reported first time by Wichterle and Lim (1960). Hydrogel possesses a degree of flexibility due to presence of hydrophilic groups such as $–NH_2$, -COOH, -OH, $-CONH_2$, -CONH-, and $–SO_3H$ (Bahram *et al.,* 2016). Generally poly-acrylamide hydrogel is divided into two i.e. soluble and insoluble. Hydrogel possesses positive (cationic), negative (anionic) and neutral charge. The charge of hydrogel depends on soil and solutes. Generally, cationic hydrogel bind with clay soil and anionic hydrogel cannot bind with clay and act as dispersants (Narjary *et al.,* 2013). These charges are found in both linear and cross-linked poly-acrylamide hydrogel which formed from acrylamide subunits (Fig.1). Cross linked polymers are synthesized with N, N'-methylene bisacrylamide hence they are more stable for many years and it is non-toxic (Kalhapure *et al.*, 2016) (Fig. 2). The charge determines how they will react with soils and solutes. Briefly, clay components of soil have a negative charge; heavy metals have a positive charge, and other commonly found minerals in soil and water possess either a positive or a negative charge. Therefore, cationic hydrogels generally bind to clay components and act as flocculants, anionic hydrogel cannot directly bind to clay and may act as dispersants. However, anionic hydrogel can bind to clay and other negatively charged particles in the presence of ionic bond, such as calcium (Ca^{+2}) and magnesium (Mg^{+2}).

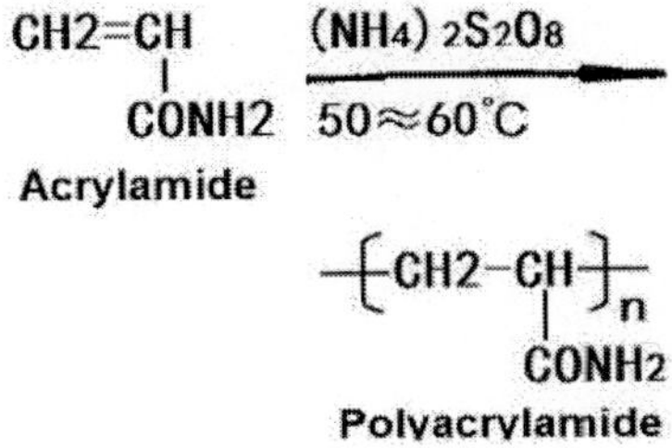

Fig. 1. Formula diagram of acrylamide and poly-acrylamide hydrogel

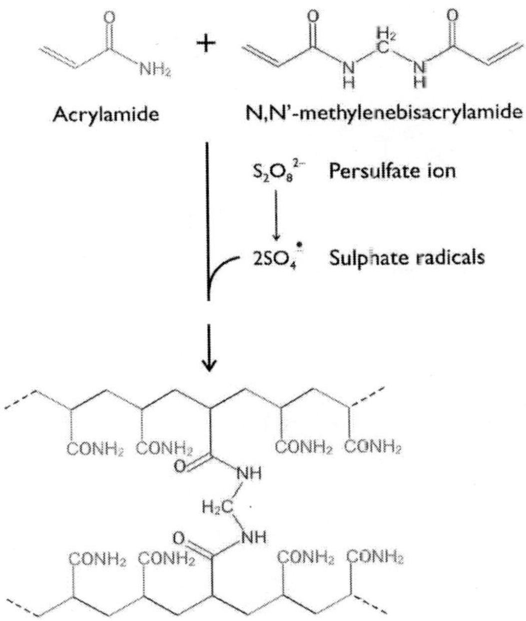

Fig. 2. Formula diagram of acrylamide, N, N'-methylene-bisacrylamide and cross-linked poly-acrylamide hydrogel

3. CHARACTERISTICS OF HYDROGEL

Super absorbent polymers (SAP) potentially influence soil permeability, density, structure, texture, evaporation, and infiltration rates of water. Particularly, hydrogel reduce irrigation frequency and compaction tendency, stop erosion and water runoff, and increase the soil aeration and microbial activity. In arid areas, the use of SAP in the sandy soil, to increase its water-holding capacity seems to be one of the most significant means to improve the quality of plants. The hydrogel helps to uptake of some nutrient elements. Hydrogel enhance the uptake of fertilizers and improves growth rates. SAP can also be used as retaining materials in the form of seed additives, seed coatings, root dips, and for immobilizing plant growth regulator (Neetu *et al.,* 2018). Hydrogel improves porosity, bulk density, and water holding capacity, soil permeability and infiltration rate. It improves the growth and development of microbial activities in the soil (Table 1). Agricultural hydrogel can be used for all crops and all soil types. It is beneficial in nurseries and seedling beds, crops sensitive to moisture stress, crops requiring large quantities of water, and container gardens in pot cultures. Hydrogel swells and retains water up to approximately 350 times its dry weight within its networked structure and releases water slowly to the plant. It requires only 1.5 kg per acre of land and absorbs water 350 times of its dry weight and stable up to 50° C in the soil. Hydrogel products are available in India for agricultural purposes. Many companies are being manufacturing this product in India with different price (Table 2).

Table 1. Effect of hydrogel on soil properties.

Treatments	BD (g/cm^3)	Total porosity (%)	WHC (%)	pH	Dehydrogenase activity (ml H_2/g dry soil/24 hrs)	Total bacteria (10^6 CFU)
Control	1.61	39.13	22.96	7.75	5.1	150
Compost @ 12 t/ha	1.59	39.92	24.18	7.40	29.3	480
Compost @ 24 t/ha	1.58	40.42	25.09	7.36	39.6	510
Hydrogel @ 24 kg/ha	1.56	41.28	26.36	7.30	18.5	320
Hydrogel @ 48 kg/ha	1.54	41.77	27.00	7.27	19.9	360

Source: El-Hady and Abo-Sedera (2006); BD= Bulk density, WHC= water holding capacity, CFU = Colony Forming Unit, t= tone, ha= hectare

Table 2. Agricultural hydrogel products available and price in India

S. No	Trade name	Manufacturing company	Price (Rs/kg)
1	Pusa Hydrogel	IARI, New Delhi	990
2	Water lock 93 N	Acuro Organics Ltd, New Delhi	1000
3	Agro-forestry water absorbent polymer	Technocare Products, Ahmedabad	-
4	Super absorbent polymer	Gel Frost Packs Kalyani, enterprises, Chennai	-
5	Hydrogel	Chemtex Speciality Ltd, Mumbai	650
6	Rain drops	M5 Exotic Lifestyle Concepts, Chennai	-
7	Pusa hydrogel	Kch private ltd., Chennai	990
8	Hydrogel	Chemzest Technoproducts Private Limited, Chennai, TN.	600
9	ALSTA Hydrogel	Chemtex Speciality Ltd., Kolkata	500

Source: Kalhapure *et al.* (2016)

4. METHOD OF APPLICATION OF HYDROGEL

Available in dry powder, the gel should be mixed with approximately 10 times the quantity of the farm soil and basal dose of fertilizers. Seeds to be sown are also mixed with it. The mixture is then applied uniformly in the rows with the help of plough or seed drill. Care must be taken to ensure precision application of the product around or below the seed. Method of application of hydrogel is elaborating below-

- **Field crops:** Prepare an admixture of hydrogel and fine dry soil in 1: 10 ratio and apply along with the seeds/fertilizers or in the opened furrows before sowing. For best results, hydrogel should be close to seeds for better moisture and nutrients uptake.
- **Nursery bed:** Apply 2-3 g/m^2 of nursery bed mix of hydrogel uniformly in the top 2 inches of the nursery bed. In pot culture, mix 3–5 g/kg of soil before planting.
- **Transplanting seedlings:** Thoroughly mix 2-3 g of hydrogel per litre of water to prepare a free-flowing solution; allow it to settle for half an hour. Dip the roots of the seedlings in the solution and then transplant in the field.
- **Seed coating:** Coating of pearl millet seed with 10 and 20 g of hydrogel/kg of seed resulted in the production of significantly higher effective tillers, ear length, test weight, grain and stover yield compared to control and water soaking treatment (Singh, 2012).

5. HYDROGEL APPLICATION RATE

Hydrogel can be mix with fine dry soil in 1:10 ratio and apply along with the seeds/ fertilizers or in the furrows before sowing of seed. Hydrogel use efficiency performs well within rhizosphere. For nursery bed apply hydrogel at the rate of 2 g/m^2 and mix the hydrogel uniformly in the top 5 cm of soil surface. In pot culture mixes the hydrogel at the rate of 5 g/kg of soil before sowing/planting. Rate of application of agricultural hydrogel varied upon the soil texture. The rate of hydrogel in clay soil is 2.5 kg/ha and depth of the soil would be 15-20 cm. The rate of hydrogel in sandy soil is 5 kg/ha and depth of the soil would be 12 cm. Low rate of application (2.5-4.0 kg/ha) also recommended for field crop production.

6. EFFECT OF HYDROGEL ON PULSES PRODUCTION IN RICE FALLOW

6.1 Growth and yield

Crop production under rainfed rice fallow ecosystem is challenging. Irrigation facilities i.e., one life saving irrigation is also rare in winter season for pulses grown under rainfed rice fallow land (RRFL) in India. Due to soil moisture deficit stress crops failure is a common phenomena in RRFL. To alleviate this climatic aberrations hydrogel is the helpful in soil moisture conservation and to eliminate the need of life saving irrigation in pulses under RRFL. Application of SAP could conserve water thereby increasing the soil capacity for water storage, ensuring more available water, relative water content in leaves and increased plant growth under water stress. SAP improved seedling growth and increased absorption of nitrogen and nitrogen wasted through leaching of saline soils was reduced (Martinez *et al.,* 2004). Review of literatures suggest that the amended soil with natural and synthetic SAP improved maize yields by 36%, and 31% and improved dry matter yields by 92% and 81% respectively, than those of the control (Yangyuoru *et al.,* 2012). Use of hydrogel into the pot medium helps the seedling germination (20.4 seedling/pot) compared to control (13.0 seedlings/pot). The beneficial effects were greater when watering frequency was lowest and soil texture was sandier (Mangold and Sheley, 2007). Yield of wheat was found to increase by 8.48% over control with the application of 5 kg/ha of hydrogel in clay loam soil with 100% recommended dose of chemical fertilizers without application of organic manures (Boriboj *et al.*, 2006). Maize yield increased following SAP application by 11.2% under low and 18.8% under medium dose, but significantly at high and very high doses by 29.2 and 27.8% with only half amount (150 kg/ha) of fertilizer as compared to control, which received conventional standard fertilizer dose (Islam *et al.,* 2011). Similarly in pulse crops like chickpea, the hydrogel amendment significantly improves seedling growth (Table 3). The increased water availability with hydrogel amendments is known to improve seed germination and seedling growth, however there are variations in the response of different species to hydrogel products (Akhter *et al*., 2004). Hydrogel application influenced seed yield of chickpea but biological yield and protein content of seeds were not affected by this polymer (Table 4). Application of 18 kg/ha hydrogel observed in a significant increase in seed yield of chickpea compared with control (Farjam *et al.,* 2014).

Table 3. Effect of different levels of hydrogel on seed germination and seedling growth (15 days) of chickpea (*Cicer arietinum*) in sandy loam soil.

Hydrogel level (%)	Seed germination (%)	Shoot length (cm)	Shoot fresh weight (mg)	Shoot dry weight (mg)
0	58.3b*	17.0a	394a	46a
0.1	50.0b	13.7a	321a	57a
0.2	91.6a	15.8a	423a	62a
0.3	75.0ab	21.4a	634a	77a

*Means followed by different letters in a column differ significantly at $P \leq 5\%$ level

Table 4. Effects of hydrogel application on seed yield, biological yield and seed protein content of chickpea

Hydrogel (kg/ha)	Seed yield (kg/ha)	Biological yield (kg/ha)	Protein (%)
0	631.0b*	4899.1a	19.8a
9	618.4b	4613.5a	19.9a
18	715.4a	4526.2a	18.8a

*Different letters in a group of each column indicate significant differences at $P \leq 0.05$ according to Duncan's test.

In a study under rice fallows in Odisha state revealed that highest grain yield of 1029 kg/ha was achieved in greengram variety 'IPM 02-03' with improved management practices (involving line sowing + seed treatment + nano solution + RDF + Hydrogel @ 2.5 kg/ha) which is significantly higher than rest of the practices, and 29.4% higher in comparison to Line sowing + seed treatment + RDF treatment (Chandna *et al.,* 2018). In another study in rice fallow of Odisha where direct seeding of rice followed by greengram under conventional method of establishment with hydrogel application was found superior in terms of yield, economics and energetic (Rath, 2018). An increase in yield and yield related attributes could be because of sufficient availability of water and indirectly nutrients supplied by the SAP to the plants under water stress condition, which in turn lead to better translocation of water, nutrients and photosynthates and finally better plant stand yield (El-Hardy *et al*., 2009). Hydrogel increase water holding capacity ultimately improves soil water content for better plant growth and development (Kalhapure *et al.*, 2016). In an experiment on chickpea under rainfed winter season grown after rice at BCKV, Kalyani, West Bengal showed that application of hydrogel @ 5 kg/ha along with thiourea 500 ppm (T2F5) enhanced acquisition and use efficiency of all the three major nutrients in chickpea. Further, same treatment (T2F5) gave maximum seed yield of chickpea under RRFL and the treatment T_2F_3 observed maximum seed yield of chickpea (Fig 3) and water use efficiency (Fig 4) (Hedayetullah and Sadhukhan, 2019).

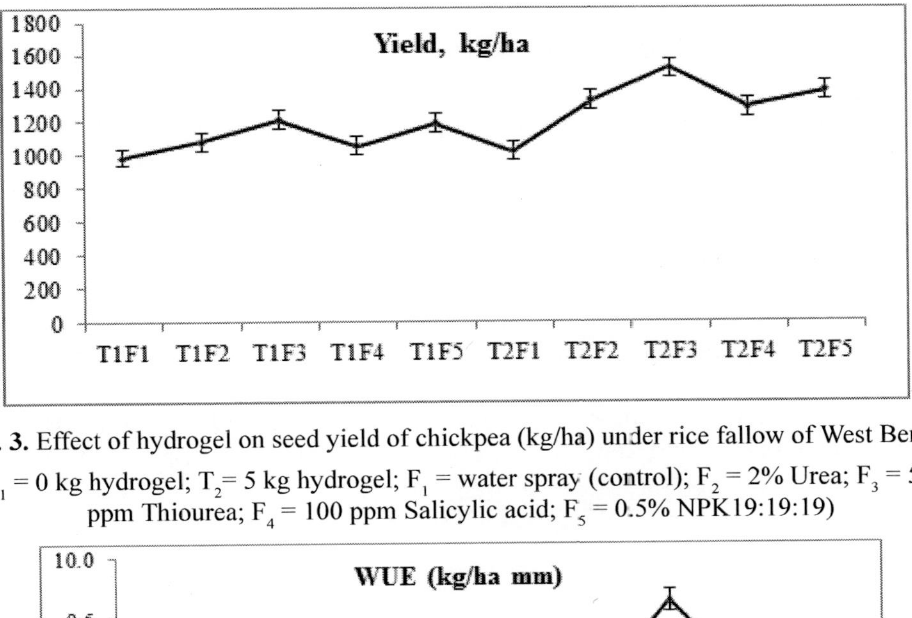

Fig. 3. Effect of hydrogel on seed yield of chickpea (kg/ha) under rice fallow of West Bengal.
(T_1 = 0 kg hydrogel; T_2= 5 kg hydrogel; F_1 = water spray (control); F_2 = 2% Urea; F_3 = 500 ppm Thiourea; F_4 = 100 ppm Salicylic acid; F_5 = 0.5% NPK19:19:19)

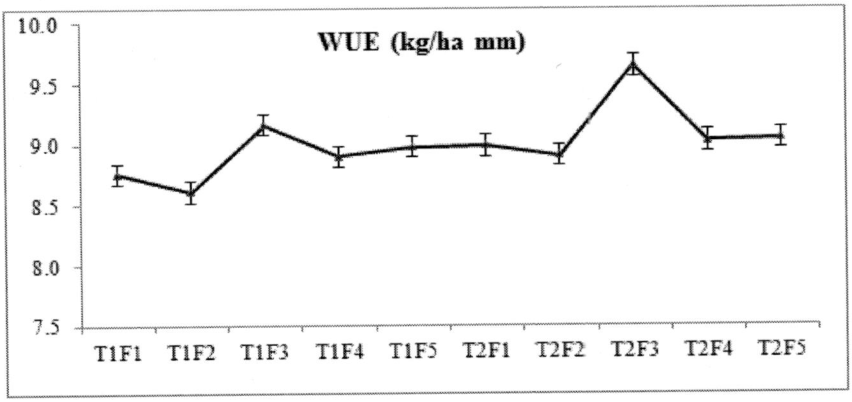

Fig. 4. Relationship of hydrogel in chickpea and WUE (kg/ha mm) under rice fallow of West Bengal.

(T_1 = 0 kg hydrogel; T_2= 5 kg hydrogel; F_1 = water spray (control); F_2 = 2% Urea; F_3 = 500 ppm Thiourea; F_4 = 100 ppm Salicylic acid; F_5 = 0.5% NPK19:19:19)

6.2 Economics

Pulse crop cultivation in winter season with hydrogel 5 kg/ha in rainfed rice fallow situation proven effective in regards to yield. Cultivation of pulse crop by farmers practice yield can be reduced due to non availability of life saving irrigation. In case of conventional practice of taking pulses under RRFL, the grain yield is hardly meeting the cost of cultivation. But, if hydrogel 5 kg/ha is used, the cost of cultivation per hectare will increase by Rs. 2 500/ha. In addition, the efficiency of hydrogel increased many folds if applied with FYM 10 tonnes/ha which cost about Rs. 5 000/. Therefore, the total cost of cultivation increased by Rs. 7 500/ha. The increased yield under these practices may results in significant increase in grain yield of pulses under RRFL. The results of study at BCKV, Kalyani, West Bengal suggest that the net return obtained under following of this practice is 2.2 times of the cost involved (Hedayetullah and Sadhukhan, 2019).

7. FATE OF HYDROGEL IN SOIL

Hydrogel are basically organically produced synthetic polymers. The hydrogel molecules are too voluminous to be absorbed into plant tissue and have zero bio-accumulation potential. Hydrogel can be degraded by both, biotic and abiotic environmental factors. Exposure to ultraviolet radiation, chemical oxidizers, fertilizer salts, mechanical abrasion, and freeze-thaw events can degrade the polymers into smaller fragments and thus losses its original properties. The surface applied hydrogels are more prone to these types of degradation. Studies have confirmed that hydrogel is sensitive to the action of UV rays, and degrades into oligomers. The polyacrylate becomes much more sensitive to aerobic and anaerobic microbiological degradation. Some of the naturally occurring microbes like bacteria (*Bacillus sphaericus* and *Acinetobacter* spp.) and white rot fungi (*Dichomitus squalens*, *Phanerochaete chrysosporium* and *Pleurotus ostreatus*) are well known decomposers of hydrogel in the soil (Narjary *et al*., 2013).

8. LIMITATION TO USE OF HYDROGEL IN AGRICULTURE

(a) Intensive use of hydrogel may affect the soil health and imbalance the environmental ecosystem. Inherent toxicity of the un-reacted monomer acryl amide, acrylic acid, acrylate etc. present in the finished products which is ultimately remains in soil and water system.

(b) Super absorbent polymers materials cannot return to their original monomers, *i.e.* they are scientifically irreversible to toxic initiating materials.

(c) The degradation of hydrogel in soil is very slow by the ionic and micro-organisms.

(d) Long-term use of hydrogel may affect the soil microbiology and plant root development.

(e) Long term research on hydrogel not yet initiated for each and every soil type and crops.

9. ENVIRONMENTAL SAFEGUARD OF HYDROGEL IN AGRICULTURE

(a) Pusa hydrogel does not contain any traceable residual un-reacted monomer and is thus environmentally safe.

(b) Hydrogels of this type degrade completely into carbon dioxide, water and ammonia within one year.

(c) Recommended application rate of Pusa hydrogel is very low i.e. 2.5 to 5 kg/ha for most of the crops which is equivalent to 2.5 to 5 ppm per unit weight of the soil.

(d) Pusa hydrogel is semi-synthetic in nature where in reduced load of monomer component in the end product.

(e) SAP which is purely synthetic in nature is not environment friendly.

(f) Pulse crops require less nitrogen. However, the end product of hydrogel is ammonia which may be detrimental to the crops.

10. CONCLUSION

Total 11.65 m ha area remained fallow after harvest of rainy season rice in the country. Soil moisture deficit stress and lack of irrigation facilities are the major reason to remain land fallow. These large areas can be utilized for cultivation of pulse crops with appropriate soil moisture conservation technology. Under such condition hydrogel can act as a soil moisture balancer under aberrant weather conditions. It helps soil to keep continuous moisture for survival and growth of the crop. Under rice fallow situation, hydrogel meet up the need of life saving irrigation during reproductive phase of crop. Relay cropping of pulses is generally practiced in rice fallow rainfed areas. Hydrogel gives better results with organic manures and facilitates the long time moisture supply to the crops. In general, 5.0 kg/ha of basal application of hydrogel recommended with farm yard manure or organic manure.

REFERENCES

Abedi-koupai, J. and Sohrab, F. (2004). Effect of super absorbent application on water retention capacity and water potential in three soil textures. *Journal of Science and Technology of Polymers* 17, 163-173.

Akhter, J.K., Mahmood, K.A., Malik, A., Mardan, M. and Ahmad, M.M. (2004). Iqbal Effects of hydrogel amendment on water storage of sandy loam and loam soils and seedling growth of barley, wheat and chickpea. *Plant Soil and Environment* 50, 463–469.

Al-Rahim, H.A.A, Hegazy, E.A and El-Mohdy, H.L.A (2007). Radiation synthesis of hyhrogels to enhance sandy soils water retention and increase plant performance. *Journal of Applied Polymer Science* 93, 1360-71.

Bahram, M., Mohseni, N. and Mehdi, M. (2016). An Introduction to Hydrogels and Some Recent Applications. In: Sutapa Biswas Majee (eds) Emerging Concepts in Analysis and Applications of Hydrogels. *Intech Open.* DOI: 10.5772/64301.

Borivoj, S., Rak, L. and Bubenikova, I. (2006). The effect of hydroabsorbent on selected soil biological and biochemical characteristics and its possible use in revitalization. *Ecologia* 25(4), 422–429.

Chandna, P., Mondal, S., Dwivedi, M., Panda, B., Singh, A., Variar, M., Singh, U. and Hellin, J. (2018). Geospatial modelling for targeting improved technologies in rice-fallows. Paper presented in 5 th International Rice Congress (IRC2018), 15 th -17 th October 2018, Marina Bay sands, Singapore.

Dar, S.B., Mishra. D., Zahida, R. and Afshana, B.B. (2017). Hydrogel: To Enhance Crop Productivity Per Unit Available Water Under Moisture Stress Agriculture. *Bull. Env. Pharmacol. Life Science* 6(10), 129-135.

El-Hady, O.A. and Abo-Sedera, S.A. (2006). Conditioning effect of composts and acrylamide hydrogels on a sandy calcareous soil. II physico– biochemical properties of the soil. *International Journal of Agricultural Biology* 8(6), 876–884.

El-Hardy, O.A, El-Kader, A.A. and Shafi, A.M. (2009). Physio-bio-chemical properties of the sandy soil conditioned with acrylamide hydrogels after cucumber plantation. *Australian Journal of Basic* and *Applied* Sciences 3, 3145-51.

Farjam, S., Kenarsari, M.J., Rokhzadi, A. and Yousefi, B. (2014). Effects of inter-row spacing and superabsorbent polymer application on yield and productivity of rainfed chickpea. *Journal of Biodiversity and Environmental Sciences* 5(3), 316-320.

Rath, H. (2018). Effect of stand establishment methods and hydrogel on productivity of rainfed rice (*Oryza sativa* L.) - green gram (*Vigna radiata* L.) cropping system. M.Sc. thesis submitted to Indira Gandhi Krishi Vishwavidyalaya, Raipur, Chhattishgarsh.

Hedayetullah, M., Arnob, R.C., Mainak, G., Hazra, K.K., Nath, C.P., Sadhukhan, R. and Zaman, P. (2018). Paira Chickpea under Rice Fallow in Lowland Ecosystem of West Bengal, India. *Agricultural Research and Technology*: *Open Access Journal* 13(1), 14-15. DOI: 10.19080/ARTOAJ.2018.13.555870.

Islam, M.R., Hu, Y., Mao, S., Jia, P., Eneji, A.E. and Xue, X. (2011). Effects of water saving superabsorbent polymer on antioxidant enzyme activities and lipid per oxidation in corn (*Zea mays* L.) under drought stress. *Journal of the Science of Food and Agriculture* 91, 813–819.

Joshi, P.K., Birthal, P.S. and Bourai, V.A. (2002). Socioeconomic onstraints and Opportunities in rainfed Rabi Cropping in Rice Fallow Areas of India. National Centre for Agricultural Economics and Policy Research, New Delhi.

Kalhapure, A., Kumar, R., Singh, V.P. and Pandey, D.S. (2016). Hydrogels: a boon for increasing agricultural productivity in water-stressed environment. Current Science 111(11), 1773-1779.

Kar, G. and Kumar, A. (2009) Evaluation of post-rainy season crops with residual soil moisture and different tillage methods in rice–fallow of eastern India. *Agricultural Water Management* 96, 931–8.

Khadem, S.A., Galavi, M., Ramrodi, M., Mousavi, S.R., Rousta, M.J. and Rezvani-Moghadam, P. (2010). Effect of animal manure and super absorbent polymer on corn leaf relative water content, cell membrane stability and leaf chlorophyll content under dry condition. *Australian Journal of Crop Science* 4, 642-647.

Kumar, R., Mishra, J.S. and Hans, H. (2018). Enhancing productivity of rice–fallows of eastern India through inclusion of pulses and oilseeds. *Indian Farming* 68(08), 7–10.

Mangold, J.M. and Sheley, R.L. (2007). Effects of soil texture, watering frequency, and a hydrogel on the emergence and survival of coated and uncoated crested wheatgrass seeds. *Ecological Restoration* 25, 6-11.

Martinez, F.X., Contreras, F. and Lopez, N. (2004). Influence of polyacrylamide on physical properties of a *Petperlite mixand* on the growth of *Argynthemum coronopifollium. Ishs Acta Horticulture.* pp. 554.

Hedayetullah, M. and Sadhukhan, R. (2019). Effect of Hydro Gel on Nutrient Acquisition in Chickpea (*Cicer arietinum* L.) Under Rice Fallow of West Bengal. Paper presented in International Conference on "Pulses as the Climate Smart Crop: Challenges & Opportunities" organized by ISPRD & ICAR-IIPR at Minto Hall, Bhopal, M.P., India.

Narjary, B., Aggarwal, P., Kumar, S. and Meena, M.D. (2013). Significance of hydrogel and its application in agriculture. *Indian Farming* 62(10), 15-17.

Narjary, B., Aggarwal, P., Singh, A., Chakraborty, D. and Singh, R. (2012). Water availability in different soils in relation to hydrogel application. *Geoderma* 187, 94-101.

Neetu, T.M., Dubey, P.K. and Kaswala, A.R. (2018). Prospects and Applications of Hydrogel Technology in Agriculture. *International Journal of Current Microbiology and Applied Sciences* 7(5), 3155-3162.

Pande, S., Sharma, M., Ghosh, R., Rao, S.K., Sharma, R.N. and Jha, A.K. (2012). Opportunities for chickpea production in rainfed rice-fallows of India–Baseline survey report. International Crops Research Institute for the Semi-Arid Tropics Patancheru 502 324, Andhra Pradesh, India. pp. 56.

Singh, H. (2012). Effect of hydrogel on growth, yield and water use efficiency in pearl millet (*Pennisetum glaucum*) production. *Forage Research* 38(1), 27–28.

Wichterle, O. and *Lim*, D. (1960). Hydrophilic Gels for Biological Use. Nature 185, 117–118. https://doi.org/10.1038/185117a0.

Yangyuoru, M., Boateng, E., Adiku, S.G.K., Acquah, D., Adjadeh, T.A. and Mawunya, F. (2006). Effects of natural and synthetic soil conditioners on soil moisture retention and maize yield. *Journal of Applied Ecology* 9, 1-8.

Yazdani, F., Allahdadi, I. and Akbari, G.A. (2007). Impact of superabsorbent polymer on yield and growth analysis of soybean (*Glycine max* L.) under drought stress condition. *Pakistan Journal of Biological Science* 10, 4190-4196.

CHAPTER 11

Deploying contingency measures in drought situation for sustaining of pulses productivity in central India

S.R.K. Singh[1], Anupam Mishra[1], A.A. Raut[1], Anita Deshmukh[1], A.K. Singh[2] and Narendra Kumar[3]

1. *ICAR-Agricultural Technology Application Research Institute (ATARI), Jabalpur, Madhya Pradesh*

2. *Krishi Vigyan Kendra, Jabalpur, Madhya Pradesh*

3. *ICAR-Indian Institute of Pulses Research, Kanpur, Uttar Pradesh*

**Corresponding author email: singhsrk@yahoo.co.in*

Abstract

Pulses are an integral part of Indian agricultural economy next to cereals and oilseeds in terms of acreage, production and economic values. Drought, a natural hazard, through its multi-facet characteristics causes damage to ecology and economy. The impact of drought on rainfed crops like pulses is much higher compared to irrigated crops. Drought proofing through contingency plan implementation consists of land treatments, cover crops with pre-monsoon rains, tank silt application, normal/contingency seed arrangements with needed inputs and crop management strategies in the case of delay in onset of monsoon (drought tolerant & short duration cultivars, alternate crops, intercrops, changes in plant population, etc.), mid-season droughts (thinning, mulching, foliar sprays, lifesaving irrigation of harvested rain water) and terminal drought (lifesaving irrigation, harvesting for fodder) with suitable post rainy rabi action plan depending upon rains received at the end of the season or in the off-season. If implemented properly, the contingency plans ensure stable food production and sustained livelihood in the rainfed dryland and contribute to country food security.

Keywords: Intercropping, Irrigation, Mitigation, Monsoon, Nutrient, Planting techniques, Summer ploughing

1. INTRODUCTION

India is one of the major pulse producing countries contributing about 25% to the global pulse production. Pulses are also an important component of Indian agricultural economy next to foodgrains and oilseeds. Pulses are an integral part of vegetarian diet of a large population in India. Besides being a rich source of proteins and essential amino acids; they also maintain soil fertility through biological nitrogen fixation in symbiotic association with *Rhizobium* bacteria present in their root nodules. Thus, pulses play a vital role as nitrogen fixing mini-factories, which help in sustaining crop productivity and soil health.

Pulses are rich source of protein and energy, but in India, these are largely cultivated under energy starved conditions, mostly on marginal and sub-marginal land and more than three-fourth of the area under pulses is still rainfed resulting in poor crop productivity. India is the largest producer of pulses in the world with 25% share in the global production. The important pulse crops are chickpea, pigeonpea, greengram, blackgram, lentil and fieldpea. The pulses production in the country reached to 25.23 Mt from an area of 29.99 M ha with productivity of 841 kg/ha (2017-18). A large share in area (28%) and production (35%) of pulses is contributed by central Indian states of Madhya Pradesh and Chhattisgarh. Madhya Pradesh alone contributes 25% in area and 32% in production of pulses in the country. The average productivity of pulses in the country is still less than 1.0 t/ha which is far below the productivity of rice, wheat and maize. The poor productivity of pulses is mainly because of its cultivation on marginal and sub-marginal lands under rainfed where drought occurrence is a common phenomenon.

India is reeling under an intense drought situation which has affected close to 330 million people from state *viz*., Madhya Pradesh, Chhattisgarh, Maharashtra, Bihar, Gujarat, Andhra Pradesh, Uttar Pradesh and Jharkhand. Bad monsoons and weak policies around water security have further compounded the problem and precipitated a crisis. Rising temperature and acute water shortage are adversely affecting human health as well as the economy which is primarily reliant on agriculture. Due to its varying geographical features, climate conditions and vast area, the Indian state of Madhya Pradesh has been consistently prone to drought like conditions for years. Every year, many districts of Madhya Pradesh have been repeatedly facing the drought situation. To fight with the drought situation in the central Indian states not only focusing on weather forecasting, early warning, vulnerability and preparedness measures for the information of community, but also following of best practices for rainwater harvesting and *in-situ* soil moisture management. Focus should also be given for development of contingency crop planning and measures as well for the mid-season corrections based on the weather condition and the forecasting reports.

2. DROUGHTS IN CENTRAL INDIA

Rainfall deficiency during crop growing season results into meteorological as well as agricultural drought causing decrease in crop yields. The impact of drought on rainfed crops is much higher as compared to crops grown in irrigated regions. According to Goswami *et al*. (2006) the frequency of more intense rainfall events in many parts of Asia has increased whereas the number of rainy days and total annual precipitation has decreased. Due to its varying geographical features, climate conditions and vast area, the Indian states in central India have been consistently prone to drought like conditions for years (Anonymous, 2016). Many districts of Madhya Pradesh and Chhattisgarh have been repeatedly facing the drought situation. Uprety *et al.* (1996) reported that with the type of climate change in India (variation in temperatures and CO_2 concentration), the production of crop is likely to be shifted in relatively drier areas than where it is grown presently. Variability in the rainfall conditions directly affecting the yield performance of the crops in the region. In year 2015-16, Balaghat, Jhabua and Guna districts of Madhya Pradesh have faced early season drought, whereas Morena, Ratlam, Chhattarpur and Satna of Madhya Pradesh and Bhatapara of Chhattisgarh faced mid-season drought. The terminal drought was observed in Datia and Bilaspur of Madhya Pradesh and Chhattisgarh, respectively. In year 2016-17, the mid-season drought occurred in Balaghat, Morena, Datia and Chhatarpur of Madhya Pradesh and Bhatapara of Chhattisgarh state, while the terminal drought occurred in Ratlam and Bilaspur of Madhya Pradesh and Chhattisgarh, respectively. Similarly in the year 2017-18, the mid-season drought occurred in Guna and Morena, and the terminal drought in Satna districts of Madhya Pradesh. Similar kind of variations in occurrence of drought was observed in the year 2018-19. This showed that there is no definite pattern of occurring drought in the different regions of central India. The rainfall patterns of some of the representative districts of central Indian states are depicted in Fig 1.

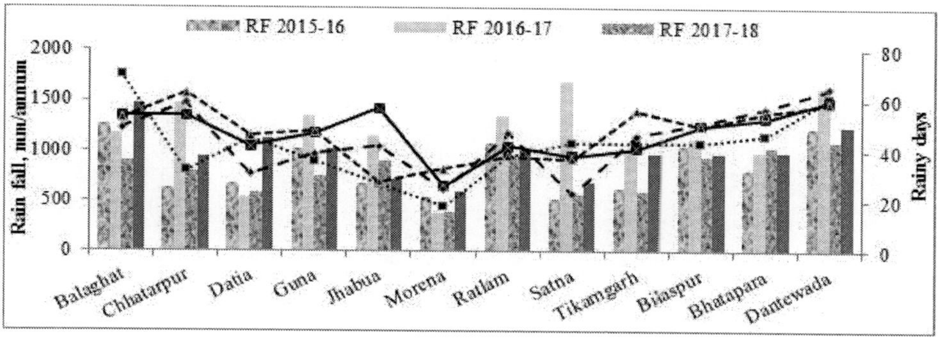

Fig. 1. Total rainy days and rainfall in some of the pulses growing districts of central India

3. AGRONOMIC PRACTICES FOR DROUGHT MITIGATION

Drought management strategies in the past were adopted generally during or after the onset of drought which lacked pro-active preventive interventions during normal or excessive rainfall years (Anonymous, 2009). Hence, watershed developments and development of percolation tanks, pond and check dams at the village level can be feasible approaches for the drought mitigation strategies within the village. It helps in conserving water, thereby recharging the ground water and at the same time providing employment to the villagers within their surroundings (Anonymous, 2016). To fight with the drought situation in the state of Madhya Pradesh, many research and development initiatives have been taking place in the recent past. The state is focusing towards best practices for rainwater and soil management through linking on station and on farm research activities and also focusing on contingency crop planning as well as mid-season corrections based on the weather conditions and the forecasted reports. Some of the practices recommended for rainfed cultivation of pulses in central India are explained below.

3.1 Land and seed bed preparation

Seed bed preparation for crop is an important factor which allows direct contact of seed to soil, even planting depth and moisture absorption during drought situation to a uniform germination. Dwivedi *et al.* (2002) reported that adoption of proper land preparation will help to increase income of the farmers due to higher crop yield besides preventing land degradation due to runoff and erosion. All the growth parameters were significantly improved when the seed bed prepared with only one ploughing (Tomar and Singh, 1991). In dryland areas, deep ploughing resulted in better moisture conservation, root proliferation and higher productivity over shallow cultivation (Vadi *et al*., 2006). If land is deep ploughed during summer, then only shallow tillage with the onset of monsoon is sufficient for seed bed preparation.

3.2 Selection of suitable varieties

Varietal improvement programme in pulses in India was initiated in 1966 with start of All India Coordinated Project on Pulses Improvement Programme (AICRP on PIP). Till date more than 400 varieties of different pulses have been developed for different agro-ecological regions of the country. The early maturing high yielding varieties of pulses resistant to diseases and pests, and tolerant to heat and drought are suitable for the rainfed cultivation in central India. The early maturity varieties during *rabi* season can save an irrigation requirement of pulse crops and thus crop can be successfully grown under rainfed. The some of the varieties of blackgram (IPU-94-1, Shekhar, PU 40, Pratap Urad-1, etc), greengram (PDM 139, TJM-3,Virat, HUM 16, etc), pigeonpea (TT-401, Pusa-991, Pusa-992, TJT-501, ICPL 88039, Pusa-16), chickpea (JG 12, JG 14, PA-291, RVG-202) are already popular among farmers of this region.

3.3 Sowing time and planting geometry

Time of sowing is the most important vital non-monetary input to achieve the potential yield of crop. Sowing time is governed by onset of monsoon, atmospheric temperature and soil moisture. The delayed sowing restricts vegetative growth and pod bearing branches, decreases biological-nitrogen fixation and also leads to forced maturity. At the same time, delay sowing of *rabi* pulses like chickpea resulted in higher incidence of pests, especially pod borer (Ali *et al*., 1998). Padhi (1995) reported that sowing in first week of June recorded highest grain yield under both narrow and wider spacing in different varieties of pigeonpea and sowing beyond that date reduced the yield. Optimum sowing depth depends on type of crop/cultivar, growing season, soil moisture, soil texture, and more importantly on seed size of the respective pulse crops (Dass *et al.*, 1997). Planting geometry depends on type of crop and cultivar, growing season and planting system. Most of short duration pulse varieties need narrow spacing, while long duration varieties perform well under wider spacing. An appropriate planting density of crop lead to better harness of the solar radiation to translate into higher crop yields (Choudhary *et al.*, 2014b). In general, *kharif* sown crop requires wider spacing and less plant population compared to summer sowing due to fairly warm temperature, favourable soil moisture, prolonged vegetative growth and profuse branching (Prasad, 2012).

3.4 Optimum seed rate

The seed rate is the most important factor to maintain optimum plant population of pulse crops. It has been observed that the yield of crop is greatly affected by seed rate. Singh *et al*. (1992a) reported that seed rate of 20, 30 or 40 kg seeds/ha in blackgram produced seed yields of 1.22, 1.39 and 1.26 t/ha, respectively. Sekhon *et al*. (2006) reported that optimum seed rate for higher yield of greengram is 37.5 kg/ha during summer and 30 kg/ha during rainy season in Punjab.

3.5 Summer ploughing

Summer ploughing is defined as the ploughing the field across the slope during hot summer with the help of specialized tools with the primary objective of opening of the soil crust accompanied by deep ploughing and simultaneously overturning of the soil underneath to disinfect it with the help of piercing sun rays. Deep summer ploughing (off season tillage) with pre-monsoon showers (during May) helps to recharge the soil profile, increases soil water content and reduces runoff. It also reduces insect-pests, diseases and pest and weed infestation (Kumar *et al.*, 2016). The number and depth of ploughing depends on weed intensity. Two times summer ploughing at an interval of 15-20 days prior to advent of monsoon gives better results. Kumar *et al.* (2014 and 2016) reported reduced density and biomass of purple nutsedge (*Cyperus rotundus*) and also reduced infestation of *Orobanche*. In dryland areas, deep ploughing resulted in better moisture conservation, root proliferation and higher productivity over shallow cultivation (Vadi *et al.*, 2006; Bahera *et al.*, 2016). Under drought stress, deep tillage could facilitate uptake of sub-soil water and thus stabilise crop yields (Doty *et al.*, 1975). In farmer's field demonstrations at Tikamgarh conducted by KVK recorded 60% higher yield of blackgram under deep summer ploughing. Furthermore, increased soil organic carbon sequestration is also reported under deep summer ploughing (Alcantara *et al.*, 2016). Thus, deep summer tillage might be a tool to make crops more resilient to climate change and to minimise the yield loss caused by drought.

Deep summer ploughing

3.6 Planting techniques

Several land configurations are recommended for *in-situ* soil moisture conservation and rainwater management. The commonly used practices for pulses cultivation are as follows:

3.6.1 Ridge & furrow

Ridge and furrow practice found beneficial in *in-situ* soil moisture conservation and to prevent water logging condition during heavy rains because of proper drainage system. The wider row crops like pigeonpea are fit to be planted on ridges. Mishra *et al.* (2009) has reported 27% higher yield of pigeonpea under ridge planting over conventional flat bed planting. Desai *et al.* (2000) reported significant effect of different land configurations on grain yield of pigeonpea and recorded significantly highest grain yield under ridge and furrow planting. Ridge and furrow method of planting also reduced the incidence of *Phytophthora* blight in pigeonpea (Chauhan *et al.*, 2005). KVK Tikamgarh recorded 27.4% higher grain yield of blackgram and KVK Satna recorded 29.8% higher yield of pigeonpea on ridge and furrow method of planting during 2015-16 and 2016-17, respectively over farmer's practice of flat bed sowing.

3.6.2 Broad bed and furrow

The broad bed and furrow (BBF) system has been developed by the International Crops Research Institute for the Semi-arid Tropics (ICRISAT) in India. The recommended ICRISAT system consists of broad beds of about 100 cm wide separated by sunken furrows of about 50 cm wide in a slope of 0.4 and 0.8 per cent on Vertisols. Depending on crop or variety, 2-4 rows can be grown on the broad beds. The technique worked best on deep black soils in areas with average rainfall of 750 mm or more. BBF system is an improved *in-situ* soil and water conservation, and drainage technology for the Vertisols. Higher infiltration rate and lower bulk density under BBF system was also reported. The BBF system has many other advantages such as irrigation water saving, mechanical weeding, fertilizer placement, less lodging, better crop standand increased yield (Astatke *et al.*, 2002). Verma *et al.* (2018) reported higher number of pods per plant, pod length and seed index of soybean under BBF system in Chhattisgarh. Shivkumar *et al.* (1997) reported that grain yield of greengram was higher when sown in BBF system compared to flat bed sowing. Similarly, Kantawa *et al.* (2006) reported higher biomass and grain yield of pigeonpea under broad bed and furrow system of planting. In farmers fields demonstrations conducted by KVK, Tikamgarh have shown 60% yield advantage in blackgram sown in BBF over farmer's practice of flat bed sowing.

Sowing of crop under BBF system

Blackgram crop performance under BBF

3.6.3 Furrow-irrigated raised bed

Furrow-irrigated raised bed (FIRB) system of planting is a drought proofing technology for cultivation of pulses in rainfed agro-ecosystem of central India. Benefits of FIRB system also include lesser weeds, facilitates seeding into relatively aerated soils, vigorous and better crop stands, less seed, reduced crop-lodging and, seed and fertilizer contact, better drainage, improved *in-situ* rainwater conservation, higher crop productivity, reduced wilt infestation and avoid temporary water logging problems (Kumar *et al.*, 2012, 2015b). Additionally, furrows can also be used for drainage when there is excess moisture or rainfall (Sayre and Moreno Ramos, 1997) and can also be managed to grow certain leafy vegetables. In a study conducted under farmer's participatory trial in India showed increased yields of field crops such as maize (37.4%), urdbean (33.6%), mungbean (21.8%), green peas (14.5%), wheat (6.4%), rice (6.2%), pigeonpea (46.7%) and chickpea (37.0%) under FIRB system of planting compared to flat bed planting (Connor *et al.*, 2003). The results of farmer's field demonstrations conducted KVK Morena showed 38.6% increased yield of blackgram under FIRB planting. Similarly, demonstrations of greengram variety 'Virat' showed 40.9% higher yield. Further, the average yield of 2002 kg/ha of pigeonpea varieties Pusa-991, UPAS-120, PA-291, ICPL-88039, PUSA-992 and PUSA-2001 was recorded under FIRB in comparison to 1637 kg/ha under farmer's practices. According to Pramanik and Singh (2000) blackgram planted on raised bed during *kharif* season recorded significantly better growth than that planted on flat beds and significantly increased the branching, nodulation and root growth. There was also saving in seeds and fertilizer to the extent of 25-30% following raised bed system of planting (Kumar *et al.*, 2012).

FIRB planting system of pulse crop

3.7 Intercropping for drought proofing

In general, growing of 2 or more crops in proximity is called intercropping. The most common goal of intercropping is to produce a greater yield on a given piece of land by making efficient use of available resources that would otherwise not be utilized by a single crop. Traditionally, pulses are grown with cereals (mostly coarse cereals) and oilseeds under rainfed for ensuring a definite yield. The slow initial growth of pulses like pigeonpea is most suitable for intercrop with fast growing short duration crops like blackgram, greengram, sorghum, pearl millet, soybean, etc (Table 1). Apart from increasing profitability and resource use efficiency, pulses as intercrop act as safeguard under unprecedented moisture stress condition. Under rainfed conditions where chances of crop failures are greater, an intercropping is more stable and dependable than growing a sole crop (Willay and Rao, 1980; Kumar *et al.*, 2008). Intercropping of rainfed crops such as maize + cowpea and maize + pigeonpea is an important risk minimizing strategy for drought proofing in the scarce rainfall zones (Bahera *et al.*, 2016).

Table 1. Some of the intercropping systems for central India

Intercropping	Reference
Soybean + pigeonpea (3:2)	Lakhena and Maurya, 2009
Gram + safflower (3:1 and 6:2) Chickpea + linseed (3:1 and 6:2)	Thakur *et al.*, 2000
Blackgram + pigeonpea (6:2)	Upadhyay *et al.*, 1998
Blackgram + pigeonpea (1:1)	Katre, 2018
Soybean + pigeonpea (2:1)	Ghosh *et al.*, 2006; Kumar *et al.*, 2015c

Pigeonpea + mungbean/ soybean/ urdbean (1:3)	Dudhade *et al.*, 2009
Banana + cowpea	Singh *et al.*, 2012
Pigeonpea + pearl millet	Rajput and Bhadouriya, 2019
Caster/wheat + lentil	Singh *et al.*, 2011
Finger millet + pigeonpea (4:1)	Pradhan *et al.*, 2014

3.8 Nutrient management

In rainfed agro-ecological regions of central India, nutrient imbalance is one of the major abiotic constraints limiting pulses productivity. Thus, balanced supply of essential plant nutrients is important key for successful production of pulses in these regions. The management strategies involved addition of right amount of nutrients at right time through an appropriate method and sources so as to it minimizes the nutrient losses, thereby, making efficient nutrient-use for enhancing crop productivity and maintaining soil fertility (Dass *et al.*, 2014). The inbuilt mechanism of biological N_2 fixation enables pulse crops to meet 80-90% of their N requirements, hence a small dose of 15-25 kg N/ha is sufficient to meet the requirement of most of the pulse crops. In a cropping system, 25% higher dose of nitrogen (30-40 kg N/ha) is recommended to pulse crops grown after rice, sorghum and maize than those after urdbean, mungbean or fallow (Ali and Mishra, 2000). However, pulses need adequate phosphorus and sulphur for their root proliferation and synthesis of sulphur containing amino acids (Choudhary, 2009). Phosphorus deficiency is wide spread and most of the pulse crops have shown good response to 20-60 kg P_2O_5/ha depending upon nutrient status of soil, cropping system and moisture availability. Response to K application (20-40 kg K_2O) is also commonly observed in pulses grown under rainfed. In the recent years, use of 20-30 kg S/ha and some of the micronutrients such as Zn, B, Mo and Fe have improved productivity of pulse crops. Some micronutrients like molybdenum and iron required in higher amount by pulses because they are integral constituents of nitrogenase enzyme, essentially required for nitrogen fixation (Choudhary *et al.*, 2014a). The beneficial effects of foliar application of 2% urea and micronutrients during reproductive stage were also reported by many researchers under rainfed. Foliar application of 2% urea at 75 DAS increased the grain yield of chickpea by 11% under rainfed conditions (Venkatesh and Basu, 2011). Growth and yield attributes were also increased significantly due to foliar spray of urea. To improve nutrient-use efficiency and sustainability of pulses production under rainfed regions, biofertilizers can be incorporated in nutrient management strategies as a supplement of fertilizers. The biofertilizers are low-cost inputs compared to the chemical fertilizers. Some of the important biofertilizers include Rhizobium, Azotobacter, Azospirillum, Phosphate-Solubilizing Microorganisms (PSMs), Arbuscular Mycorrhiza (AM) and Plant Growth Promoting Rhizobacteria (PGPR) can be effectively used in pulses for increasing yield. The inoculation of biofertilizers also improves tolerance against abiotic stresses like drought.

3.9 Water management

Pulses are somehow tolerant to soil moisture deficit stress due to deep root system, but very sensitive to water logging. Since soil moisture deficit stress or drought is more commonly observed in rainfed pulses growing regions of central India, the adoption of soil moisture conservation is pre-requisite for maximizing pulses productivity grown in this region. Thus, it is desirable to ensure maximum water harvesting as *in-situ* rainwater conservation by integrating conservation and production technologies. A deep summer ploughing coupled with levelling and land modification (bunding, raised bed, FIRB, etc.) are essential for moisture conservation and drainage of excessive water during heavy rainfall to avoid water logging. Rainwater harvesting structures like on farm ponds and check dams may be constructed for village or *taluka* level runoff water harvesting to give supplemental or lifesaving irrigation at critical crop growth stages of pulses under drought condition. In general, all the *kharif* pulses are grown under rainfed conditions without irrigation in India. However, *kharif* pulses respond to irrigation at critical growth stages of flowering and pod formation, if there is dry-spell (Prasad, 2012). Rabi pulses in this region are mostly grown on residual soil moisture after harvest of *kharif* crops. The mid-season and terminal drought is more common in *rabi* pulses due to non-occurrence of winter rainfall in this region. Thus, supplemental lifesaving irrigation is necessary with micro-irrigation systems at critical crop growth stages for successful cultivation of *rabi* pulses.

3.10 Weed management

Weeds are a major biotic stress in pulses production under rainfed which cause about 30-40% yield reduction. Weeds compete with crop plants mainly for nutrients, moisture, light and space and also secrete toxic allelochemicals which affect the growth and development of crops adversely. As a result of the severe crop-weed competition, the quantity and quality of the produce are substantially reduced (Kumar *et al.*, 2014). Efficient weed control can be difficult to achieve because pulses have a relatively slow growth habit and open canopy early in the growing season. Weed management takes away nearly one third of total cost of production of field crops. In India, the manual method of weed control is quite popular and effective. However, due to higher labour cost, herbicides use is increasing in pulses. Kumar *et al.* (2017) reported the yield loss due to weed competition in *kharif* season greengram to the tune of 62%. They further reported highest weed control efficiency in pendimethalin @ 1.25 kg/ha + imazethapyr @ 100 g/ha as pre- and post-emergence herbicides combination in *kharif* pulses. Whereas in rabi pulses, aapplication of pendimethalin @ 1.0-1.25 kg/ha *fb* quizalofop-ethyl @ 50-100 g/ha and pendimethalin @ 1.0-1.25 kg/ha *fb* manual weeding recorded highest weed control efficiency and yield (Kumar *et al.*, 2015a).

4. CONTINGENCY MEASURES FOR DROUGHT MITIGATION

Contingency crop management refers to implementing a plan for making alternate crop or cultivar choices in tune with actual rainfall situation and soils in a location. Information on alternate crop varieties/ crops to be chosen in case of delay in onset of monsoon or early season drought and also agronomic measures for mid- and terminal- season droughts is necessary to develop contingency crop management strategies (Umate *et al.*, 2011; Prasad *et al.*, 2012; Prasad *et al.*, 2013). Extreme variability in the quantity, time and duration of rains expose the crop to soil moisture stress either on account of delayed monsoon, longer dry spells or early withdrawal of monsoon, has been identified as one of the major limiting factor for poor performance of pulse crops (Tiwari, 2014; Gupta *et. al.*, 2018). Thus, droughts have multiplier effect on agricultural production. Its impact is felt during the subsequent years also due to deficit in ground water recharge, non-availability of quality seeds, and reduced draught power for agricultural operations due to distress sale of cattle and land degradation. Contingency crop management strategies for pulse crops under different situations have been described in Fig 2.

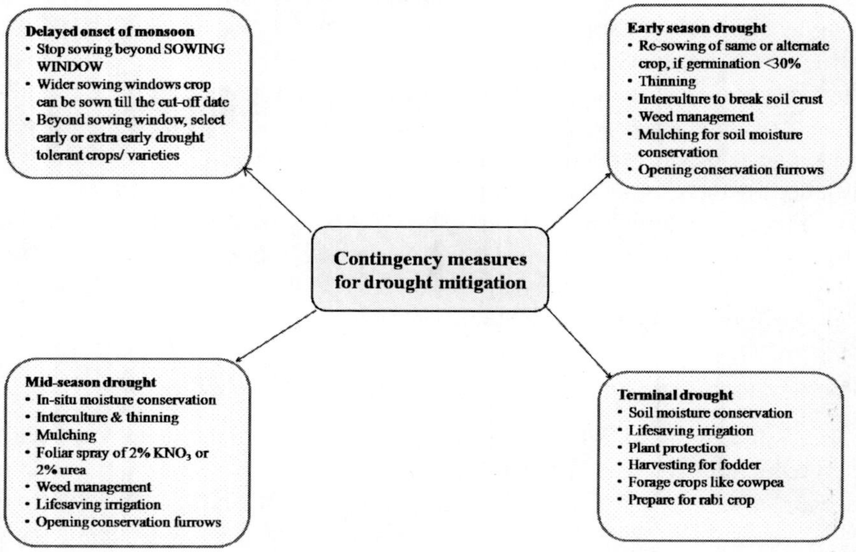

Fig. 2. Contingency measures for drought mitigation in different situations

4.1 Delayed onset of monsoon

In general, sowing of crops in rainfed with start the onset of monsoon is considered as best practice to harvest potential yield. The sowing window of crop is major decisive factor for selection of crops and varieties with delay in monsoon. The crop with narrow sowing window cannot be taken up if the delay is beyond this cut-off date. However, crop with wider sowing window can be taken till the cut-off date. Beyond the sowing window, choice of alternate crops or cultivars depends on the farming situation, soil, rainfall and cropping pattern in the location and extent of delay in the onset of monsoon (DACFW, 2016). During 2018-19 under delay monsoon, sowing of short duration and YMV tolerant blackgram variety 'RBU-38' on ridge & furrow gave 37.8% higher yield over farmer's practice in Ratlam district.

4.2 Early season drought

Early season drought occurs due to delay in commencement of sowing rains. Sometimes, early rains may occur tempting the farmers to sow the crops followed by a long dry spell leading to withering of seedlings and poor crop establishment. The germination of seed will become difficult which results in poor crop establishment and need for partial or sometimes complete re-sowing. The management options to cope up with early season drought are re-sowing of the same crop/variety or alternate crop (if germination is <30%) within a week to 10 days with subsequent rains and under favourable soil moisture conditions, thinning to maintain minimum required population, interculture operations to break soil crust and remove weeds, soil mulch for conserving soil moisture, opening conservation furrows, foliar spray of 2% urea, etc.

4.3 Mid-season drought

Mid-season droughts occur in association with long gaps between two successive rain events if the moisture stored in the soil falls short of water requirement of the crop during the dry period. During certain times, the mid-season droughts may be associated with low and inadequate rainfall in the growing season to meet the crop water needs as per its phonological stage. Therefore, the strategies to cope up with mid-season drought aims at increase the total soil water available to the crop either through short term measures like weed control and mulching or through long term measures like land configurations. It results in stunted growth if occurs at vegetative stage and adversely affects the yield if occurs at flowering or early reproductive stage. The contingency measures include *in-situ* soil moisture conservation, plant protection measures, avoid top dressing of fertilizer until favourable soil moisture condition, intercultural operation to create dust mulch for soil moisture conservation, weed management, surface mulching of crop residues or weed biomass to conserve soil moisture, opening conservation furrows, thinning, foliar spray of 2% KNO_3 or 2% urea solution or 1% water soluble fertilizers like 19-19-19, 20-20-20, 21-21-21 to supplement nutrition during dry spells, providing supplemental irrigation, if available.

4.4 Terminal drought

Early cessation of rainfall or terminal drought under rainfed conditions affects grain filling, fewer productive branches, reduced pod size and seed weight and ultimately poor yield. The varieties suitable for growing under terminal drought and heat tolerant should be promoted in rainfed region of central India. The contingency measures in this situation include application of lifesaving or supplemental irrigation (if available) from harvested pond water or other sources, use of organic mulches, keep the field weed free, harvesting crop at physiological maturity with some realizable yield or harvest the crop for fodder purpose under severe condition and prepare for winter (*rabi*) sowing in double- cropped areas.

5. SUCCESS STORY

5.1 System of Pigeonpea Intensification

Convergence model has been implemented successfully in Madhya Pradesh by Agricultural Technology Application Research Institute, Zone-IX, Jabalpur, State and ICAR KVKs, Jawahralal Nehru Krishi Vishwa Vidyalaya, Jabalpur and Government of Madhya Pradesh (Mishra *et al.*, 2013) for dissemination of system of pigeonpea intensification (SPI). It is an agro-ecological based innovation for quick and healthy plant establishment of pigeonpea for improving productivity and resilient to climate change.

Technology details: Pigeonpea nursery is raised for transplanting in polybags. The polybags are filled with soil and organic manure (compost or FYM) mixture. In each polybags 1-2 pigeonpea seeds are sown and watered regularly. Out of 2 seedlings in a polybag, the poor one is removed by cutting at ground level. The pigeonpea seedlings will be ready for transplanting at 25-30 days after sowing. Pigeonpea seedlings are planted at wider spacing (100-120 cm x 40-50 cm) than recommended spacing to get better results. By raising pigeonpea seedlings, delay in monsoon and early season drought situation can be avoided. Secondly, the establishment and growth of seedlings after transplanting is faster that helps in sustaining plant during later season drought.

***Benefit of technology*:** The benefits of SPI system are

- Maintenance of optimum plant population
- Saving of seed because of low seed rate
- Early and fast growth and development of crop
- Fast development of root systems even to deeper layer which help to survive plants under drought better than farmer's practice
- Shortened the crop period in field condition and thus escape from delay onset of monsoon or early season drought stress
- Higher input use efficiency and yield
- The results of SPI demonstrations in Madhya Pradesh showed more no of pods/ plant, higher yield, net return and B: C ratio in SPI as compared to flat bed planting (Table 2).

Table 2. Comparison of SPI and sowing in furrows at flat bed

Observation	SPI	Sowing in furrows at flat bed
No of pod/plant	115.1	61.5
Yield (kg/ha)	1780	1010
Net Return (Rs/ha)	68553	33961
B:C Ratio	2.07	1.56

Source: Annual Progress Report, ICAR-ATARI, Jabalpur (2015-16)

Impact: In Madhya Pradesh, pigeonpea is mostly grown under rainfed condition in every district of the state. With the efforts of ICAR, SAUs and state department in the state, SPI technology spread to many districts of the state. KVKs took the lead in dissemination of technology through farmer's field demonstrations (Table 3). At present, the area under SPI technology had been spread up to 14203 ha in Madhya Pradesh.

Table 3. Horizontal spread of technology by KVKs

ACZs & KVKS	No. of village covered	No. of farmers	Area in ha	Mean yield (kg/ha)
Umaria	250	1500	200	2750
Shahdol	315	4750	1900	2750
Balaghat	65	265	95	1400
Betul	30	120	500	2000
Chhatarpur	09	160	46	1500
Narsinghpur	22	500	500	2700
Panna	12	450	200	2150
Seoni	39	146	214	1250
Sidhi	20	85	200	1300
Rewa	719	4300	5800	1900
Mandla	145	490	48	1400
Sagar	200	5650	4500	2100
Total	**1826**	**18416**	**14203**	**1933**

(a) Sowing of pigeonpea seed in polybags

(b) Transplanting of pigeonpea seedlings

(c) Crop performance at reproductive stage

5.2 FIRB planting method in Jhabua district of Madhya Pradesh

Jhabua district in Jhabua hills of Madhya Pradesh is recognized as drought prone area due to its climatic features like uncertain and erratic rainfall, long dry spells, late arrival and early withdrawal monsoon, etc. Climate in district varies from dry sub-humid to moist sub-humid. The rainfall is erratic, uni-modal and with spatio-temporally variability. The average annual rainfall of the district is 975.8 mm. The details of rainfall received and the dry spells observed in last 5 years are given in following Table 4.

Table 4. Rainfall pattern in last 5 years in Jhabua district of Madhya Pradesh

Year	Rainfall (mm)	No. of rainy days	No. of dry spells	Remarks
2015	671.6	27	2	Deficit rainfall, Early season drought
2016	725.2	57	Nil	Deficit rainfall
2017	1153.6	42	1	Excess rainfall
2018	1418.6	28	2	Excess rainfall, terminal drought
2019	909.8	48	Nil	Excess rainfall

Major crops grown in the district include maize, blackgram, cotton, soybean, pigeonpea and greengram in *kharif* season, and wheat and chickpea in *rabi* season. Blackgram is the second most major crop after maize being grown in about 65.35 thousand ha area in the district. The frequent dry spells and mid season droughts adversely affect the crop yield in this region. Sometimes due to excessive rains during crop maturity deteriorate the seed quality of blackgram.

***Technological intervention*:** KVK, Jhabua after identification of the problem in blackgram production in the district has started promoting furrow irrigated raisedbed planting (FIRB) since 2015. Training cum awareness programmes were conducted in collaboration with State Agriculture Department, ATMA and KVK scientists. Front Line Demonstrations (FLDs) and method demonstrations are regularly being conducted since 2015. Technology is also being scaled up in KVK adopted villages including NICRA villages in the district since last 5 years.

***Technology details*:** Blackgram was sown under FIRB system (67.5-75 cm bed width). The furrows act as both, drainage of excess soil moisture and *in-situ* soil moisture conservation in case of deficit rainfall. During making of raised bed, most of the applied fertilizers concentrated on raised bed by tractor drawn raised bed planters. The lifesaving irrigation can be applied in the furrows during drought which will save time as well as irrigation water requirement.

***Benefit of technology*:** The benefits of technology during dry spells are as follows:

- Crop stand improved by 70-75%
- Time saving (25-30%) in irrigation (1.5 h/ha/irrigation)
- Input saving: 20-25% seeds and fertilizers can be saved
- Irrigation water saving was up to 25-30%
- Better crop management (intercultural operations and insect-pests and disease management)
- Blackgram yield increased by 50-55%.

***Impact*:** The FIRB technology was found suitable for drought prone area of Madhya Pradesh. The results of FLDs reported an average yield increase of 52.2% over the farmer's practice. This technology was spread to the adjoining districts of Jhabua like Ratlam, Dhar, Alirajpur, etc. During the last 5 years, there was no report of crop failure of blackgram planted under this technology in the region.

FIRB planted blackgram at farmer's field in Jhabua

6. CONCLUSION

During last few decades, growth in pulses production has increased significantly. India achieved a record in pulses production of 25.23 Mt in 2017-18 with all-time high production achieved in chickpea (11.23 Mt), greengram (2.01 Mt) and blackgram (3.56 Mt). Although, sustaining this growth rate is a big challenge for researchers, extension agencies and policy makers. Central India has major share in country pulses production. But, the pulses area in this region is prone to drought at different crop growth stages. The review of above text suggests that following of drought proofing contingency measures like deep summer ploughing, land configuration for soil moisture conservation, drought tolerant & short duration cultivars, alternate crops, intercrops, plant population management, mulching, foliar sprays of nutrients and life saving irrigation from harvested rainwater with suitable post-rainy *rabi* action plan depending upon rains received at the end of the season or in the off-season can certainly increase the pulses production and improve the livelihood of resource poor farmers of this region.

REFERENCES

Alcantara, V., Don, A., Well, R., and Nieder, R. (2016). Deep ploughing increases agricultural soil organic matter stocks. *Global Change Biology* 22, 2939–2956.

Ali, M. and Mishra, J.P. (2000). Nutrient management in pulses and pulse-based cropping systems. *Fertiliser News* 45(4), 57-69.

Ali, M., Mishra, J.P., Ahlawat, I.P.S., Kumar, R. and Chauhan, Y.S. (1998). *Effective management of legumes for maximizing biological nitrogen fixation and other benefits.* In: Residual effects of legumes in rice and wheat cropping systems of the Indo-Gangetic plain: proceedings of the Workshop, 26-28 August 1998, Patancheru, India.pp.127–8.

Anonymous. (2009). Drought Management Strategies – National Rainfed Area Authority, Ministry of Agriculture Government of India, New Delhi.

Anonymous. (2016). *southasiadisasters.net,* towards drought free India, Issue No. 147 Pp. 7.

Astatke, A., Jabbar, M., Mohamed, M.A. and Erkossa, T. (2002). Technical and economical performance of animal drawn implements for minimum tillage-experience on vertisols in ethopia. *Experimental Agriculture* 38(2), 185-196.

Bahera, S.K., Bastia, D.K. and Mishra, A. (2016). Effect of in-situ moisture conservation methods on productivity in North Eastern Ghat zone of Odisha. *Indian Journal of Dryland Agricultural Research and Development* 31(2), 27-30.

Chauhan, V.B., Singh, V.B. and Singh, O.N. (2005). Integrated management on Phytophthora blight of pigeonpea. *Indian Journal of Pulses Research* 18(1), 46–49.

Choudhary, A.K. (2009). Role of phosphorus in pulses and its management. *Indian Farmers' Digest* 42(9), 32–34.

Choudhary, A.K. and Suri, V.K. (2014a). Frontline demonstration programme – An effective technology transfer tool for adoption of oilseeds production technology in Himachal Pradesh, India. *Communications in Soil Science and Plant Analysis* 45(11), 1480–98.

Choudhary, A.K. and Suri, V.K. (2014b). Scaling up of pulses production under frontline demonstrations technology programme in Himachal Himalayas, India. *Communication in Soil Science and Plant Analysis* 45(14), 1934–48.

Connor, D.J., Gupta, R.K., Hobbs, P.R. and Sayre, K.D. (2003). Bed planting in rice-wheat systems. In: Addressing Resource Conservation Issues in Rice-Wheat Systems of South Asia: A Resource Book. Rice-Wheat Consortium for the Indo-

Gangetic Plains. Intl. Maize and Wheat Impr. Cent., New Delhi, India. pp. 103-108.

DACFW. (2016). Manual for drought management. Department of Agriculture, Cooperation & Farmers Welfare, Ministry of Agriculture & Farmers Welfare, Government of India New Delhi. 154p.

Dass, A., Kharwara, P.C. and Rana, S.S. (1997). Response of gram varieties to sowing dates and phosphorus level under on-farm conditions. Himachal. *Journal of Agricultural Research* 23(1 & 2), 112–115.

Dass, A., Suri, V.K. and Choudhary, A.K. (2014). Site-specific nutrient management approaches for enhanced nutrient-use efficiency in agricultural crops. Research and Reviews: *Journal of Crop Science and Technology* 3(3), 1–6.

Desai, N.C., Adreshna, R.B. and lntwala, C.G. (2000). Increasing pigeonpea productivity by providing land configuration in vertisols. *International Chickpea and Pigeonpea Newsletter* 7, 66-67.

Doty, C.W., Campbell, R.B. and Reicosky, D.C. (1975). Crop response to chiseling and irrigation in soils with a compact A2 horizon. *Transactions - American Society of Agricultural Engineers* 18, 668–672.

Dudhade, D.D., Deshmukh, G.P., Harer, P.N. and Patil, J.V. (2009). Studies on intercropping of pulse crops with pigeonpea under rainfed condition. *Legume Research* 32(3), 215-217.

Dwivedi, S.P., Ramana R.S., Vadivelu, K.V., Navalgund, A. and Pande, A.B. (2002). Spatial distribution of rainy season fallows in Madhya Pradesh. Potential for increasing productivity and minimizing land degradation. Global Theme 3: Water, Soil and Agro diversity Management for Ecosystem Health. Report No. 3. Patancheru 502324, Andhra Pradesh, India: International Crops Research Institute for Semi Arid Tropics. 40p.

Ghosh, P.K., Mohanty, M., Bandyopadhyay, K.K., Painuli, D.K. and Mishra, A.K. (2006). Growth, competition, yield advantage and economics in soybean/pigeonpea intercropping system in semi-arid tropics of India: I. Effect of sub-soiling. *Field Crops Research* 96(1), 80-89.

Goswami, B.N., Venugopal, V., Sengupta, D., Madhusoodanan, M.S. and Xavier, P.K. (2006). Increasing trend of extreme rain events over India in a Warming Environment. *Science* 314, 1442-1445.

Gupta, R., Kulmi, G.S., Basediya, A.L. and Jadav, M. (2018). Influence of furrow irrigated raised bed seed drill on growth characteristics and yield of soybean in Mandsaur district of Madhya Pradesh, India. *Plant Archives* 18(1), 320-324.

Kantawa, S.R., Ahlawat, I.P.S. and Gangaiah, B. (2006). Performance of sole and intercropped pigeonpea (*Cajanus cajan*) as influenced by land configurations, post-monsoon irrigation and phosphorus fertilization. *Indian Journal of Agricultural Sciences* 76(10), 635-637.

Katre, Y.K. (2018). Production potential and profitability of blackgram based intercropping system for crop diversification in rainfed upland of Chhattisgarh plains. M.Sc.(Ag) Thesis. IGKVV, Raipur, Chhattisgarh. 132 pp.

Kumar, N., Hazra, K.K. and Nadarajan, N. (2017). Efficacy of pre- and post- emergence herbicides in rainy season mungbean (*Vigna radiate* L.). *Indian Journal of Agricultural Sciences* 87(9), 1219–24.

Kumar, N., Hazra, K.K. and Yadav, S.K. (2014). Issues related to weed management in RCT. In- Ghosh, P.K., Narendra Kumar, Venkatesh, M.S., Hazra, K.K. and Nadarajan, N. (Ed.) Resource Conservation Technology in Pulses. Scientific Publisher, Jodhpur. Pp. 567-575.

Kumar, N., Hazra, K.K., Yadav, S.L. and Singh, S.S. (2015a). Weed dynamics and productivity of chickpea (*Cicer arietinum*) under pre- and post-emergence application of herbicides. *Indian Journal of Agronomy* 60(4), 570-575.

Kumar, N., Nath, C.P., Hazra, K.K. and Sharma, A.R. (2016). Efficient weed management in pulses for higher productivity and profitability. *Indian Journal of Agronomy* 61(4th IAC Special Issue), S93-105.

Kumar, N., Prakash, V., Meena, B.L., Gopinath, K.A. and Srivastva, A.K. (2008). Evaluation of*toria* (*Brassica compestris*) and lentil (*Lens culinaris*) varieties in intercropping system with wheat (*Triticum aestivum*) under rainfed conditions. *Indian Journal Agronomy* 53(1), 47-50.

Kumar, N., Singh, M.K., Praharaj, C.S., Singh, U. and Singh, S.S. (2015b). Performance of chickpea under different planting method, seed rate and irrigation level in Indo-Gangetic Plains of India. *Journal of Food Legumes* 28(1), 40-44.

Kumar, R., Paslawar, A.N., Singh, U. and Nagar, R.K. (2015c). Effect of conservation tillage on growth, production and energetics of pigeonpea based intercropping system under rainfed area. *The Ecoscan* 9(1&2), 153-157.

Lakhena, K.K. and Maurya, B.M. (2009). Intercropping of soybean and pigeonpea under rainfed condition. *Mysore Journal of Agricultural Sciences* 43(2), 369-373.

Mishra, A., Chahal, V.P., Singh, S.R.K., Chand, P. and Athare, T. (2013). Strategy for Common Mission for sustaining double digit growth in Madhya Pradesh: KVK-ATMA Interface. Zonal Project Directorate, Zone VII, ICAR Jabalpur. M.P. 32p.

Mishra, A., Prasad, K., Swarnkar, G.B., Rai, G. and Khan, N. (2009). Studies on grain yield and yield components of long duration pigeonpea (*Cajanus cajan* L. Millsp) as influenced by planting methods and plant populations. *Legume Research* 32(3), 230-232.

Padhi, A.K. (1995). Effect of sowing date and planting geometry on yield of redgram (*Cajanus cajan*) genotypes. *Indian Journal of Agronomy* 40(1), 72–76.

Pradhan, A., Thakur, A., Sao, A. and Patel, D.P. (2014). Biological efficiency of intercropping in finger millet (*Eleusine coracana* L. Gaertn) under rainfed condition. *International Journal of Current Microbiology and Applied Sciences* 3(1), 719-723.

Pramanik, S.C. and Singh, N.B. (2000). Boost pulse production through new planting techniques. *Indian Farming* 58(1), 4–6.

Prasad, R. (2012). Textbook of Field Crops Production-Food Grain Crops, Vol I, pp. 248–319.

Prasad, R., Chary, G.R., Rao, K.V., Prasad, Y.G., Rao, C.S., Ramana, D.B.V., Sharma, N.K., Rao, V.U.M. and Venkateswarlu, B. (2013). District Level Contingency Plans for Weather Aberrations in Himachal Pradesh. CSK HPKV, Palampur, HP and Central Research Institute for Dryland Agriculture. Hyderabad-500 059, India. 222p.

Prasad, Y.G., Venkateswarlu, B., Chary, R.G., Ch. Srinivasarao, Rao, K.V., Ramana, D.B.V., Rao, V.U.M., Reddy, S.G. and Singh, A.K. (2012). Contingency Crop Planning for 100 Districts in Peninsular India. Central Research Institute for Dryland Agriculture, Hyderabad-500 059, India. 302p.

Rajput, R.L. and Bhadouriya, N.S. (2019). Effect of moisture conservation practices on growth and yield of pearl millet + pigeonpea intercropping system. *Legume Research* 42(4), 547-549.

Sayre, K.D. and Moreno Ramos, O.H. (1997). Application of raised bed planting system to wheat. In: Wheat Program Special Report, Number 31, CIMMYT, Mexico.

Sekhon, H.S., Singh, G., Sharma, P. and Sharma, P. (2006). Agronomic management of mungbean grown under different environments. Improving income and nutrition by incorporating mungbean in cereal fallows in the Indo-Gangetic Plains of South Asia, DFID Mungbean Project for 2002–2004. (In) *Proceedings of the Final Workshop and Planning Meeting*, Punjab Agricultural University, Ludhiana.82–103.

Shivkumar, B.G., Mishra, B.N. and Gautam, R.C. (1997). Nutreint economy through land configuration and residue management in a greengram (*Phaseolus radiatus* L.) wheat (*Triticum aestivum*) cropping sequence with limited water supplies. In: Division of Agronomy, I.A.R.I., New Delhi, India.

Singh, M.K., Chitale, S., Sarawgi, S.K., Tiwari, A. and Jaiswal, V. (2011). Productivity, economics, nutrient status and weed studies in intercropping in *rabi* cereal, legume, oilseeds and spices on *inceptisols* in Chhattisgarh Plain. *Journal of Soils and Crops* 21(2), 165-169.

Singh, S., Yadav, D.S. and Singh, M. (1992). Effect of row spacing and seed rate on nutrient uptake and grain protein of summer blackgram (*Vigna mungo L. Hepper*). *Narendra Deva Journal of Agricultural Research* 7(1), 127-130.

Singh, A., Pathak, J., Mustafa, M.M. and Gautam, U.S. (2012). Evaluation of different intercropping options with banana in Nimar region of M.P. *Progressive Horticulture* 44(2), 342-343.

Thakur, N.S., Pannase, S.K. and Sharma, R.S. (2000). Production potential of gram (*Cicer arietinum*) - based intercropping systems under rainfed condition. *Indian Journal of Agronomy* 45(3), 534-539.

Tiwari, S.P. (2014). Raising the yield ceiling in soybean-An Indian overview. *Soybean Research* 12(2), 1-43.

Tomar, S.P. and Singh, R.R. (1991). Effect of tillage, seed rates and irrigation on the growth, yield and quality of lentil. *Indian Journal of Agronomy* 36(2), 143–147.

Umate, M.G., More, G.R., Prasad, Y.G., Chary, G.R., Mandal, D.K., Ramana, D.B.V., Sarkar, D. and Venkateswarlu, B. (2011). District Level Contingency plans for Weather Aberrations in Marathwada Region, Maharashtra. Marathwada Krishi Vidhyapeeth, Parbhani-431 402 and Central Research Institute for Dryland Agriculture, Hyderabad- 500 059, India. 272p.

Upadhyay, V.B., Vishwakarma, S.K. and Kewat, M.L. (1998). Performance of blackgram (*Phaseolus mungo*) - based intercropping in Malwa zone of Madhya Pradesh. *Indian Journal of Agronomy* 43(3), 393-395.

Uprety, D.C., Chakravarty, N.V.K., Katiyal, R.K. and Abroal, I.P. (1996). Climate variability and Brassica, Climate variability and Agriculture Eds. Abrol, YP, Sulochana Gadgil, Pant, G B. 264-280, Narosa Publishing House, New Delhi, India.

Vadi, H.D., Kachot, N.A., Polara, J.V., Sekh, M.A. and Kikani, V.L. (2006). Effect of tillage and mulching on yield and yield attributing characters of pigeonpea. *Advances in Plant Sciences* 19(2), 497–499.

Venkatesh, M.S. and Basu, P.S. (2011). Effect of foliar application of urea on growth, yield and quality of chickpea under rainfed conditions. *Journal of Food Legumes* 24(2), 110-112.

Verma, P.D., Jain, P.V. and Rajput, A.S. (2018). Evaluation of Sowing Methods of Soybean in Bhatapara District of Chhattisgarh. *Journal of Krishi Vigyan* 6(2), 109-112.

Willey, R.W. and Rao, M.R. (1980). A competitive ratio for quantifying competition between intercrops. *Experimental Agriculture* 16(2), 117-125.

CHAPTER 12

Foliar nutrition of pulses under rainfed condition

Anit Das[1], Asik Dutta[2*], Narendra Kumar[2], C.P. Nath[2], K.K. Hazra[2], C.S. Praharaj[2]

1. *ICAR-Indian Agricultural Research Institute, New Delhi*

2. *ICAR- Indian Institute of Pulses Research, Kanpur*

*Corresponding author email: asikdutta975@gmail.com

Abstract

Pulses are integral component of daily diet as it is rich in protein and other nutritional components. The role of pulses in restoring soil health is immense. Rainfed agro-ecosystems support more than half of world food productions despite of its constraints like degraded soil, lack of soil moisture, poor soil fertility and organic carbon etc. In India, majority of pulse growing areas are in rainfed regions resulting very poor utilization of applied nutrients. Foliar nutrition in the rainfed regions can be an important strategy to increase efficacy of applied nutrients. Experimental results showed significant increase in yield and yield attributes with foliar application of major nutrients like nitrogen (N), phosphorus (P), potassium (K). Interestingly, foliar nutrition improves the drought tolerance and stabilize osmotic pressure. In case of secondary nutrients the result is also prominent with profound increase in photosynthesis and carbohydrate level. Foliar nutrition with micronutrients (zinc, iron, boron and molybdenum) has significant impact on photosynthesis, biological nitrogen fixation, nodulation and enzymatic activity etc. hence higher yield compared to soil application. In the end, it can be said that foliar nutrition in the rainfed regions is very critical for achieving the targeted yield goal.

Keywords: Foliar nutrition, Nutrients, Pulses, Rainfed agro-ecosystem, Soil fertility

1. INTRODUCTION

Pulses are integral part of the human diet, popularly known as "poor man's meat" as pulses are good source of vegetable protein, ranged from 15-28%. Beside protein, pulse also supplies the essential minerals, vitamins and fibre to the human being (Mudryj *et al.*, 2014). Apart from human, pulse also plays significant roles in soil heath and soil fertility due to its leguminous in nature. Inclusion of legume in the cropping system alone or in combination i.e. either as cover crop or as intercrop have numerous benefits like improvement of soil organic carbon (SOC) and prevent rising carbon-di-oxide (CO_2) via C-sequestration, enrich soil nitrogen content through biological nitrogen fixation (BNF), augmenting soil biodiversity, maintain soil water balance and secure financial and nutritional security to the farmers (Srinivasarao *et al.*, 2017, Kumar *et al*, 2019). Beside this, legume also has especial capacity to uptake nutrients from lower depth through their deep root system and helps in better recycling of nutrients by bring them in surface layer from sub-surface layer. In term of production, India is the world leader with 22% global pulse production but productivity in India (550-625 kg/ha) is far behind from most the advance nation like, USA (1600 kg/ha) and China (1400 kg/ha) (Shashikumar *et al.*, 2013).

The cultivation of pulses in rainfed situation with poor agronomic practices is the foremost factor leading to poor productivity of pulses. Crop production in rainfed region faces severe challenges from water scarcity, wind and soil erosion, soil health deterioration and fertility loss, poor input use efficiency, emerging issues like soil salinity and alkalinity and these situations potentially be aggravated under climate change scenario (Srinivasarao *et al.*, 2009; 2015). Beside these, another major issue that most commonly encountered in dryland farming is poor utilisation of applied nutrient resources due to lack of soil moisture as the nutrient ions requires water for movement in soil. The easiest option for addressing the problem is foliar nutrition in periodic interval or during critical periods (spraying 0.5% of Zn with 0.25% lime at 4-8 weeks after sowing). The positive sides of foliar fertilisations are better and quick response over conventional methods, preventing fertiliser for further any reaction with soil constituents, no chance of leaching, uniform application and requirement of lower amount of fertiliser compared to blanket method of recommendations (Sarma *et al.*, 2015; Srinivasarao *et al.*, 2019). Nonetheless, the benefit of foliar nutrition can be duly achieved with soluble fertilizers, application time and method and places where ground water contains less content of salt (carbonates, bicarbonates of calcium or magnesium etc.) (Srinivasarao *et al.*, 2015a). Here in this chapter we will discuss in details the response of foliar fertilisation in pulses with special emphasis to rainfed situation.

2. STATUS OF RAINFED FARMING

Rainfed agriculture is one of most primitive type of farming which solely rely on natural rainfall to meet the water requirement of crops, has great significance on world agriculture as rainfed farming alone fulfils the 60% of the global food demands (Biradar *et al.*, 2009). According to recent estimate, 1.13 billion ha of crop land come under rainfed farming which account for 74.4% of global cultivated land (Table 1). Therefore, rainfed agriculture plays more crucial role in livelihood of marginal or subsistence farmers of Sub-Saharan Africa (95.5%) and South American continent (99.9%). Among the countries, USA is the global leader in terms of area rainfed farming (10.8% of global rainfed area) followed by Russia (10.1%), China (8.1%), Brazil (7.72%), India (4.31%), Australia (3.25%), Canada (3.09%), and Argentina (3.03%). These eight countries account for 51.4% of world rainfed cultivated area (Biradar *et al.*, 2009).

Table 1. Distribution of the rainfed area among the different continent of the world

Continents	Rainfed area (M ha)	% of total cultivated land
Africa	189.05	95.6
Asia	327.29	52.9
Europe	227.89	87.1
North America	190.67	84.3
South America	158.44	99.9
Australia	36.76	75.6
Oceania	1.46	7.56
Total	1131.56	74.3

Source: Biradar *et al.* (2009)

In India, around 57% of the net arable land (80 M ha) is under rainfed farming which account for 40% of foodgrains (cereal and pulse) production. This rainfed agriculture is mainly distributed over arid, semi-arid and sub-humid region of India, providing foods to 40% of the country's population. Geographically, rainfed eco-system covers states like desert regions of Rajasthan, central plateau of Madhya Pradesh, Gujarat and Maharashtra, Bundelkhand areas of Uttar Pradesh and even some areas of Tamil Nadu also (Fig. 1) (Singh *et al.*, 2008; Srinivasarao *et al.*, 2015). Besides this, soils of rainfed region are also highly diversified in nature with respect to its origin and physico-chemical properties, depending upon the climate and physiography. For example, soils of peninsular India are predominantly belonging to Alfisols and Vertisols while Aridisols along with Entisols and Inceptisols are dominant soil order in dry regions of India. However, majority of soils in rainfed regions are belonging to order of Inceptisols followed by Entisols, Alfisols and Vertisols.

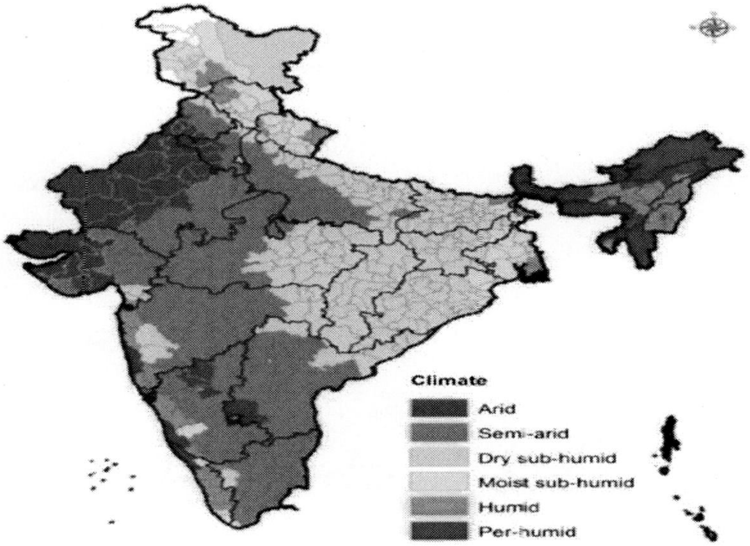

Fig. 1. Different rainfed ecosystems of India (Arid, Semi-Arid, Dry sub-humid) (Source: Srinivasarao *et al.*, 2015)

3. SCOPE OF ENHANCING PULSE PRODUCTION IN RAINFED REGION

The population of India is expected to reach 1.81 billion in 2050 that will create the projected foodgrains demand of 377 Mt by 2050. On the other hand, proportion of arable land slowly decreased over the time due to rapid urbanization and industrialization. Thus, this increased foodgrains demand will be achieved through enhancing the productivity of cereals and pulses. In this context, importance of the rainfed farming significantly increased as majority of coarse cereals (85%) and pulses (83%) is cultivated under rainfed condition by utilizing the residual soil moisture (Venkateswarlu and Prasad, 2012). The pulses are highly recommended for healthy and balanced human diet because nutritional quality of pulses is higher compared to majority of the cereals. In addition, pulses also acts as major source of vegetable protein and essential minerals in human diet. In India, commonly grown pulses are chickpea (41%) followed by pigeonpea (15%) but much of area under pigeonpea (96%) and nearly 67% chickpea area is rainfed. One of the most important limitations of the rainfed farming is lower productivity of the pulses. It has been well established fact that on-farm yield (actual yields obtained from farmers' fields) of the major pulses grown in dryland region are lower as compared to potential yields (maximum yield obtained when all growth condition is optimum), resulted in significant yield gaps (Table 2).

Table 2. Yield gap of important pulse crops under rainfed farming

Soil order	Crop	Climate	Average yields (kg/ha)		Yield gap (kg/ha)	Reference
			Potential	On-farm		
Vertisols	Chickpea	Semiarid	2200	900	1300	Reddy (2009); Neenu *et al*. (2014); Bhatia *et al*. (2006)
	Pigeonpea	Semiarid	2400	700	1700	Rathod *et al*., (2004) and Bhatia *et al*., (2006)
	Soybean	Semiarid	3100	1300	1800	Meghvansi and Mahna (2009); Ghosh *et al*., (2006); Sharma *et al*., (2006); Bhatia *et al*., (2006)
Inceptisols	Chickpea	Semiarid / Sub-humid	4000	1000	3000	Yadav *et al*. (2007)
	Soybean	Semiarid / Sub-humid	5500	600	4900	AICRPDA (2009)
Entisols and Aridisols	Chickpea	Semiarid	1400	800	600	Aziz *et al*., (2013); Sangwan and Raj (2004); AICRPDA (2011)
	Cluster bean	Semiarid andarid	1100	500	600	Srinivasarao *et al*., (2013a) and Sankar *et al*., (2012)

The yield gap (difference between potential and on-farm yield) of the major pulses such as chickpea and soybean are ranged between 1300-3000 and 1800-4900 kg/ha, respectively depending upon the nature of soil and climatic condition (Table 2) which offers great potential for improvement on-farm productivity of the pulses with suitable management practices. This lower on-farm yield under rainfed agriculture is directly related to poor soil fertility, scarcity of water, poor crop management and loss of the top soil through erosion. Therefore, adoption of appropriate nutrient management options in synergy with residual soil moisture provides huge scope to unleash the untapped potential of rainfed agriculture with respect to pulse production.

4. FERTILITY STATUS IN SOILS OF RAINFED REGIONS OF INDIA

It has been well known fact that soils of rainfed region are deficient in various nutrients which is the most limiting factor to achieving the higher crop yield. A details and systematic study was conducted by ICRISAT after collecting large number of soil samples (28270 soil samples) from farmers' fields in the rainfed region of Andhra Pradesh, Karnataka, Madhya Pradesh and Rajasthan to prepare a detail fertility map of rainfed India and results highlighted that 82% of collected samples were deficient in $CaCl_2$ extractable S followed by hot water extractable B (68%) and DTPA extractable Zn (62%). The results also pointed out the occurrence of large scale deficiencies of primary nutrients, specifically N and P but K content in collected samples shows the promising result with only 16% deficiency (Sahrawat *et al.*, 2011). Deficiency of the secondary nutrients is mainly confined in intensively cultivated area because rapid depletion secondary nutrients from soil reserves by nutrients responsive high yield varieties of the crops. Recent report indicated that sandy soil of red and laterite region shows the widespread deficiency of Ca and Mg (Rao *et al.*, 2015). Therefore, majority of rainfed crops, particularly pulses suffers more yield loss due to multinutrient deficiencies rather than the water scarcity. In recent time, deficiencies of the trace elements, mainly B and Zn are poses serious threat to the sustainable pulse production under rainfed condition (Table 3). The possible reason behind this deficiency of the trace elements is excess use of micronutrient free high analysis chemical fertilizers that causes nutrients imbalance in soil and also accelerate the depletion of trace elements from soil through enhancing biomass production of the crops (Srinivasarao and Rani, 2013).

Table 3. Deficient nutrients in rainfed soils of India under different pulse based cropping system

State	Location	Soil order	Cropping system	Limiting nutrients
Uttar Pradesh	Varanasi	Inceptisol	Rice –Lentil	N, Zn, B
Odisha	Phulbani	Oxisol	Rice-horsegram	N, Ca, Mg, Zn, B
Madhya Pradesh	Indore	Vertisol	Soybean	-
	Rewa	Inceptisol	Soybean	N, Zn
Chhattisgarh	Kawardha	-	Rice-chickpea	N, P, K, Mn, Fe, Zn
	Sahaspur Lohara	-	Rice-chickpea	N, P, K, Mn, Fe

Source: Srinivasarao and Vittal (2007); Kumar *et al.* (2017)

5. FACTORS RESPONSIBLE FOR LOWER FERTILITY IN SOILS OF RAINFED REGION

5.1 Low organic matter content

Lower soil organic carbon (SOC) in soils of rainfed region is the principle factor responsible for poor fertility because concentration of N, P, S and majority of cationic micronutrients in soil directly co-related with soil organic matter content. Beside this, soil organic matter also considers key to the soil fertility because of its central roles in nutrient recycling through controlling the microbial activity and diversity. Details analysis of soils collected from various rainfed region of India revealed that most the soils fall under category of low organic matter content (Fig 2) (Srinivasarao and Vittal, 2007). However, it is very difficult to maintain high level of organic matter in soils of rainfed conditions as rate of carbon mineralization is very rapid due to high temperatures in dry part of India. Beside this, low residue addition and erosion of top soils are the few other factors contributing lower SOC in dryland regions.

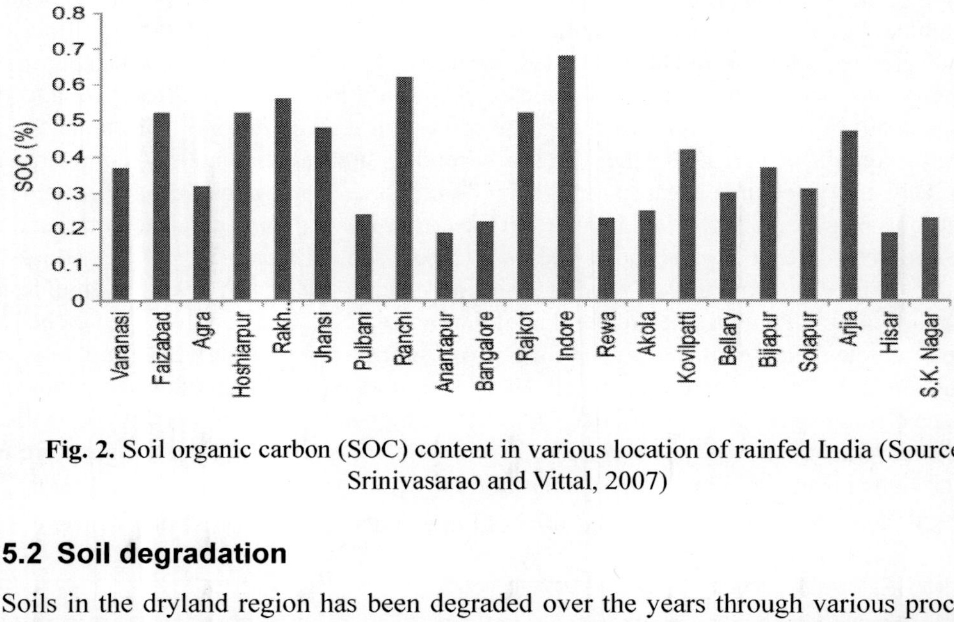

Fig. 2. Soil organic carbon (SOC) content in various location of rainfed India (Source: Srinivasarao and Vittal, 2007)

5.2 Soil degradation

Soils in the dryland region has been degraded over the years through various processes such as accelerate erosion by wind and water, acidification and salinization that resulted in loss of soil fertility and SOC content. The area and extent of degradation of lands in various states of India that predominately follow the rainfed agriculture are presented in Table 4. Among various degradation processes, accelerate erosion severely impacted the soil fertility. Erosion declines the soil fertility through removing the top soils that are highly rich in various essential nutrients while acidification or salinization is responsible for nutrients imbalance in soil. For example, soil salinization is dealing with toxicity of Na^{+}, Cl^{-} and deficiency of Zn, Cu, Fe, Mn whereas acidification is responsible for Al^{3+}, Fe^{3+} toxicity and Ca^{2+}, Mg^{2+} deficiency.

Table 4. Statistics of degraded land in major rainfed states of India

States	Geographical area (M ha)	Degraded land (M ha)	% of degradation
Assam	7.8	4.60	59
Jammu & Kashmir	22.2	20.9	94
Jharkhand	8.0	4.30	54
Karnataka	19.2	8.70	46
Madhya Pradesh	30.8	15.3	50
Rajasthan	34.2	21.0	61
Uttar Pradesh	24.1	15.3	63
Uttarakhand	5.3	3.2	60

Source: Maji *et al.* (2010)

5.3 Poor nutrient management

Before 1970, application of fertilizer inputs for crop production was very less but the situation was drastically changed for both irrigated and rainfed agriculture during the green revolution era with the introduction of high analysis chemical fertilizers. Although consumption of fertilizers input in rainfed farming is relatively lower compared to irrigated agriculture, balanced fertilization in rainfed crops is rarely achieved. Thus, continuous use of chemical sources of plant nutrition in fertilizers schedules of rainfed crops without any organic manure resulted in deficiency of secondary and micronutrients particularly S, Zn and B in various pulse based cropping system as discuss earlier.

5.4 Scarcity of water

Crops responses to the applied nutrients are greatly influenced by the soil moisture content. The excess or deficiency of the soil moisture may negatively impact the uptake of various nutrients by the crops. Occurrence of prolong dry spell during critical crop growth stage is a common phenomena in dryland agriculture due to uneven distribution of rainfall that resulted in lower nutrient use efficiency of applied fertilizers (Rayan, 2011).

6. FOLIAR SUPPLY OF NUTRIENTS TO PULSE CROPS IN RAINFED AGRICULTURE

Foliar supply of nutrients is gaining importance in the situations where plant roots are unable to absorbed essential nutrients from the soil. Such situations are mainly found when roots are submersed under water (waterlog soils), heavy metals contaminated soils where toxic metals inhibits the nutrients uptake and dryland soils where deficit of water hinders the nutrients movement in soil. Therefore, spraying of nutrients solution directly to plant shoots is accepted solution under such situations. Hence, effect of foliar application of major and micronutrients on growth of various pulse crops are discussed below:

6.1 Nitrogen

In dryland agriculture, nitrogen demand of pulses can easily meet through foliar fertilization because N requirement of pulses is significantly lower as compared to other crops. Pulse crops have special capacity to fix atmospheric N_2 gas in their root nodules which can be sufficient for their growth. Urea is the most commonly used N source for foliar fertilization because of it high solubility, rapid absorption and low toxicity. Numerous experiments have been conducted to examine the growth characteristics of pulse crops under foliar application of N and they found that foliar feeding of N may be a best alternative to soil application. It was found that spraying of 2% urea solution at pre-flower stage enhanced the initiation of flower in mungbean that resulted in higher grain yield (Kandagal *et al.*, 1990). Nitrogen taken up through leaves seems to be translocated to inflorescence and since increases the number of flowers (Singh *et al.*, 2013). In another experiment, foliar supplement of N in the form of 2% urea was accounted for 14.2%, 13.5%, 26.9%, 24.6% and 32.4% enhancement of grain yields in greengram, blackgram, lathyrus, lentil and chickpea, respectively over the soil fertilization (20 kg N/ha) (Das and Jana, 2015). Foliar applied urea enhances the seed survival under water stress condition during physiological maturity of pulses that increases the pods containing more number of seeds and hence improves the crop yield (Palta *et al.*, 2005). Likewise, foliar feeding of 0.5% urea at pre-flower and 50% flowering stages resulted in 25% and 19% increasement of total N, respectively at maturity of the chickpea and also significantly increased the seed protein contents in chickpea (Palta *et al.*, 2005).

6.2 Phosphorus

Phosphorus (P) in soil exists in both organic and inorganic form. The bioavailability is depend on release of P from this two pools which greatly influenced by soil pH and microbial activities. Beside this, when soluble phosphatic fertilizers applied to soil to meet the immediate crop requirement, major portion of applied P is fixed with Al, Fe and Ca compounds resulted in poor fertilizer use efficiency (15-20%). This poor utilization of applied P is even higher under dryland condition due to its lower diffusion coefficient of P (10^{-9} cm^2/sec) among the all macronutrients. In this context, importance of foliar P nutrition came into the picture to enhance the crop growth, especially for pulses in rainfed farming. It has been observed that foliar feeding of DAP (2%) along with chelated micronutrient (0.5%) and growth regulator (40 ppm NAA) increases the uptake of various nutrients from soil which can help to improve the various growth attributes such as the branching, pod formation and percentage fruit setting and grain yield in blackgram (*Vigna mungo*) (Shashikumar *et al.*, 2013). As requirement of macronutrient especially P is higher in case of pulses, it is more economical to supplement the P at critical growth stage through foliar mode along with basal soil application rather than the individual soil or foliar application. This statement is well supported by finding of Pandian *et al.* (2001). They found that foliar supplement of 2% DAP in greengram twice in growing season along with soil application of recommended dose fertilizers provided higher net return to per rupee of investment. In addition, P has particular importance in rainfed farming because of its critical role in drought resistance. Recent investigation on greengram suggested that foliar addition of phosphate either alone or in combination with salicylic acid stimulated the accumulation of sugar in plant which helps better absorption of water during dry period of crop growth through maintaining higher osmotic pressure inside the cells (Kuttimani and Velayutham, 2011).

6.3 Potassium (K)

Potassium requirement is significantly higher in dryland farming because of occurrences of frequent water stress condition. This deficit of water limits the utilization of soil applied K. Therefore, interest on foliar application of K is greatly increased in recent time with an especial emphasis in enhancing yield of pulse crops. Govindan and Thirumurugan (2000) found that poor pod setting, lower partitioning of photosynthates to pods and poor photosynthetic efficiency were the major factors responsible for lower productivity of greengram in rainfed region of Tamil Nadu which was easily alleviated through foliar application of 1% KNO_3 and KCl solution either individually or in combination. Similar finding also reported in case of pigeonpea where combined application of 0.5% KNO_3 and growth regulator (20 ppm NAA) brought significant improvement in yields and biomass production (Reddy *et al.*, 2004). Aside from greengram and pigeonpea, blackgram also showed good responses to foliar spraying of macronutrients. An improvement in plant height, pod bearing branches, seeds per plant and harvest index were observed in blackgram with the spraying of DAP and K solution and the values were increased as concentration of nutrients in spray solution increased (Tahir *et al.*, 2014).

6.4 Secondary nutrients

Calcium, magnesium and sulphur are commonly known as secondary nutrients. Most of the Indian soils except highly weathered soils, contained sufficient amount of Ca and Mg in bioavailable form which is enough for meeting crop requirement. Therefore, foliar application of Ca and Mg in pulses only recommended when deficiency symptoms seem to be appeared in crops. On other hand, deficiency of sulphur is frequently encountered in rainfed farming because of low organic matter content that causes severe yield loss. Hence, incorporation of foliar spraying of sulphur in the fertilizers schedule provides huge potential to improve productivity of pulses. Blackgram cultivated in rainfed of Andhra Pradesh and sprayed with nutrients solution containing 1% $Ca(NO_3)_2$, 1% $Mg(NO_3)_2$ and 1% elemental S, had significantly higher leaf area index, plant height and harvesting index compared to plants grown in control treatment (Lakshmi *et al.*, 2017). Similar type of finding also reported in case of lentil where foliar application of 50 ppm $MgSO_4$ solution at 30 and 50 DAS (days after sowing) resulted in higher plant height, branching, leaf area, branching and pod formation (Thalooth *et al.*, 2006). These enhancements were associated with higher chlorophyll production in plants with foliar applied Mg and hence improve the photosynthesis and net accumulation of sugar (Thalooth *et al.*, 2006).

6.5 Zinc

Pulse in rainfed farming responds well to foliar application of Zn. Yield responses of various pulse crops under differential dose of $ZnSO_4$ is represent in Table 5. Beside this, response of mungbean with respect to yield was also examined under graded dose Zn (0.5%, 1% and 1.5% Zn) and the significant increase in grain yield was observed with increasing the Zn concentration in spray solution upto 0.5% but further increased (1% and 1.5%), had a negative impact on biomass production (Haider *et al.*, 2018). Report also suggested that foliar Zn has positive effect on biological nitrogen fixation and grain protein content (Ved *et al.*, 2002). A comparative study of soil versus foliar use of Zn was carried out on chickpea to examine the impact on nodulation and it was observed that average number of nodules per plant increased from 18.4 under soil application of recommended dose NPK (25:50:25 kg/ha) + 7.5 t FYM/ha + biofertilizer (Rhizobium and phosphate solubilizing bacteria) to 26.4 in the plants receiving foliar feeding of 0.5% $ZnSO_4$ solution along with the soil addition of recommended dose NPK, FYM and biofertilizers (Rathod *et al.*, 2020). This is good for enhancing biological nitrogen fixation and hence improves protein content in plants. An improvement of 21.4% seed protein content was observed in cowpea due to repeated application of 2 ppm Zn at an interval of 15 days (Salih *et al.*, 2013).

Table 5. Yield response of pulses to foliar application of Zn

Crop	Rate foliar application	Yield responses	References
Chickpea	0.2% $ZnSO_4$	37% higher yield	Habbasha *et al.* (2013)
Black gram	0.5% $ZnSO_4$	18% higher yield	Pandey and Gupta (2012)
Cowpea	0.1% $ZnSO_4$	15% higher yield	Salehin and Rahman (2012)
Cowpea	0.5% $ZnSO_4$	43% higher yield	Anitha *et al.* (2005)
Soybean	0.5% $ZnSO_4$	29% higher yield	Gowthami and Rao (2014)

6.6 Iron

Iron is essential for growth and development of plants, important component of chloroplast, ferredoxins, nitratereductase and nitrogenase that involved in metabolism of protein, nucleic acid and lipid (Singh *et al.*, 2013). Despite fourth most abundant element in earth crust, plant frequently suffered with deficiency of iron due to strong association with soil components particularly carbonate. Thus foliar application of $FeSO_4$ to the standing crop is most commonly followed practices to control the carbonate induced Fe chlorosis. In a recent experiment, Fe in form of 0.5% $FeSO_4$ applied to chickpea at various critical growth stages and the responses with respect to grain yield and grain Fe content were recorded (Pal *et al.*, 2019). They concluded that there were 4.97% and 10.4% increment in grain yield and grain Fe content, respectively when $FeSO_4$ was sprayed at only 50% flowering stage but the value being slightly higher (8.81% and 13.73%) when Fe was applied at both 50% flowering stage and pod formation stage. Apart from yields and other growth attributes, Fe has particular significance in pulse production because of its active role in synthesis chlorophyll and biological nitrogen fixation. Several studies have been carried out on various pulses to investigate the effectiveness of foliar Fe on synthesis of photosynthetic pigment. One of such experiment found that foliar addition 0.5% $FeSO_4$ to the mungbean twice in growing season (25 and 40 DAS) resulted in 7.14% increment in leaf chlorophyll content coupled with higher number of nodules formation (16.27%) over the control treatment (Meena *et al.*, 2013). These enhancements associated with higher uptake of Fe in Fe applied plants. Beside this, mungbean plant also exhibited 11.4% higher uptake of nitrogen over the control. This is predominantly related with higher infection and colonization of Rhizobium and other biological nitrogen fixers under enhanced nodulation (Meena *et al.*, 2013).

6.7 Boron

Crops in dryland areas most commonly encountered the deficiency of boron which severely impacted the cell elongation, fruit setting and pod formation. Therefore, foliar spraying of boron to the standing crops proved beneficial for quick recovery of deficiency in many crops especially pulses. Rathod *et al.*, (2020) studied the effect of foliar boron nutrition (0.2% solubor solution) on growth of chickpea and they found that all the growth attributes namely, uptakes of micronutrients, plant heights, number of bunch per plants and biomass production significantly increased over the untreated plants. Other than the growth attributes, boron is also essential for protein and nucleic acid metabolism (Debnath and Ghosh, 2013). Previous study indicated that foliar applied boron (2%) significantly improved the protein content in cowpea seed (Salih *et al.*, 2013). Beside this, numerous studies have been conducted to examine the effectiveness of foliar B on nodulation and nitrogen fixation but the results are not consistent. In one study, it was found that foliar spraying of B significantly increased the nodules number (16% higher) in soybean over untreated plants (Bellaloui *et al.*, 2010) whereas in another experiment, nodule number of cowpea did not showed any significant changes with foliar spraying 1.5 g/l B (Chatterjee and Bandyopadhyay *et al.*, 2017).

6.8 Molybdenum

Molybdenum is the least required essential nutrient to plants, make it best choice for the foliar application. Mo is very much essential for pulse crops as Mo is structural component of nitrogenase and nitrate reductase (Singh *et al.*, 2013). Therefore, knowledge regarding the crop responses to foliar Mo is very important in order to sustained higher productivity of pulse in rainfed region. Ali and Mishra (2001) concluded that combine spraying of Mo and B caused significant improvement of plant height in chickpea as compared to their individual application. In addition, Patra and Bhattacharya (2009) conducted an experiment on rainfed mungbean at red and laterite region of West Bengal with foliar spraying of graded dose of Mo (0.03% and 0.05% ammonium molybdate) and B (0.1, 0.2% and 0.3% borax solution) and the results revealed that both B and Mo and their combined use had significant positive impact on all growth and yield attributes of mungbean. The enhance growth is mainly associated with higher Rhizobium infection and more nitrogen fixation under elevated concentration of Mo and B (Singh, 1993). Report also suggested that there was higher activity of nitrate reductase and glutamine synthetase in shoot of common pea when Mo was sprayed twice in week (Hristozkova *et al.*, 2006) that helps in better metabolism of nitrogenous compound.

7. CONCLUSION

Rainfed ecologies have tremendous potential to cater the need of food, fuel and fibre demand if it is managed properly and pulses on the other hand due to less input intensive nature are well adapted to rainfed regions. Poor nutrient utilization due to lack of soil moisture is the foremost reason for poor crop performance in the arid regions. So, foliar fertigation appears to be a win-win strategy for saving water and increasing nutrient use efficiency. Future research strategies like multi-locational trials in the rainfed regions considering all the factors, integrated approaches like co-application with any growth hormones or bio-chemical compounds and breeding of pulses with higher response towards fertigation are the need of the hour.

REFERENCES

AICRPDA. (2009). All India Coordinated Research Project on Dryland Agriculture. Annual report, 2008e2009.Central Research Institute for Dryland Agriculture, Hyderabad.

AICRPDA. (2011). All India Coordinated Research Project on Dryland Agriculture. Annual reports, 2010-2011. Central Research Institute for Dryland Agriculture, Hyderabad.

Ali, M. and Mishra, J.P. (2001). Effect of foliar nutrition of boron and molybdenum on chickpea. *Indian Journal of Pulses Research* 14(1), 41-43.

Anitha, S., Sreenivasan, E. and Purushothaman, S.M. (2005). Response of cowpea {*Vigna unguiculata* (L.)Walp.} to foliar nutrition of zinc and iron in the oxisols of Kerala. *Legume Research-An International Journal* 28(4), 294-296.

Aziz, M.A., Amees, T.A., Raina, B., Pharande, A.L. and Ali, T. (2013). Influence of potassium application in conjunction with FYM on growth, yield and quality of chickpea in shrink swell soil series. *Progressive Agriculture 1*3(2), 151-155.

Basavarajappa, R., Salakinkop, S R., Hebbar, M., Basavarajappa, M.P. and Patil, H.Y. (2013). Influence of foliar nutrition on performance of blackgram (*Vigna mungo* L.), nutrient uptake and economics under dry land ecosystems. *Legume Research-An International Journal* 36(5), 422-428.

Bellaloui, N., Reddy, K.N., Gillen, A.M. and Abel, C.A. (2010). Nitrogen metabolism and seed composition as influenced by foliar boron application in soybean. *Plant and Soil* 336(1-2), 143-155.

Bhatia, V.S., Singh, P., Wani, S.P., Rao, A.K. and Srinivas, K. (2006). Yield Gap Analysis of Soybean, Groundnut, Pigeonpea and Chickpea in India Using Simulation Modeling: Global Theme on Agroecosystems Report no. 31.

Biradar, C.M., Thenkabail, P.S., Noojipady, P., Li, Y., Dheeravath, V., Turral, H. and Xiao, X. (2009). A global map of rainfed cropland areas (GMRCA) at the end of last millennium using remote sensing. *International Journal of Applied Earth Observation and Geoinformation* 11(2), 114-129.

Chatterjee, R. and Bandyopadhyay, S. (2017). Effect of boron, molybdenum and biofertilizers on growth and yield of cowpea (*Vigna unguiculata* L. Walp.) in acid soil of eastern Himalayan region. *Journal of the Saudi Society of Agricultural Sciences* 16(4), 332-336.

Das, S.K. and Jana, K. (2015). Effect of foliar spray of water soluble fertilizer at pre flowering stage on yield of pulses. *Agricultural Science Digest-A Research Journal* 35(4), 275-279.

Debnath, P. and Ghosh, S.K. (2011). Determination of critical limit of available boron for rice in terai zone soils of West Bengal. *Journal of the Indian Society of Soil Science* 59(1), 82-86.

El Habbasha, S., Mohamed, M.H., El-Lateef, E.A., Mekki, B. and Ibrahim, M. (2013). Effect of combined zinc and nitrogen on yield, chemical constituents and nitrogen use efficiency of some chickpea cultivars under sandy soil conditions. *World Journal of Agricultural Sciences* 9(4), 354-360.

Ghosh, P.K., Mohanty, M., Bandyopadhyay, K.K., Painuli, D.K. and Misra, A.K. (2006). Growth, competition, yield advantage and economics in soybean/pigeonpea intercropping system in semi-arid tropics of India: I. Effect of subsoiling. *Field Crops Research* 96(1), 80-89.

Govindan, K. and Thirumurugan, V. (2000). Response of greengram to foliar nutrition of potassium. *Journal of Maharashtra Agricultural Universities* 25(3), 302-303.

Gowthami, P. and Rao, R.G. (2014). Effect of foliar application of potassium, boron and zinc of growth analysis and seed yield in soybean. *International journal of Food, Agriculture and Veterinary Sciences* 4(3), 73-80.

Haider, M.U., Farooq, M., Nawaz, A. and Hussain, M. (2018). Foliage applied zinc ensures better growth, yield and grain biofortification of mungbean. *International Journal of Agriculture and Biology* 20(12), 2817-2822.

Hristozkova, M., Geneva, M. and Stancheva, I. (2006). Response of inoculated pea plants (*Pisum sativum* L.) to root and foliar fertilizer application with reduced molybdenum concentration. *Applied Plant Physioogy* (Special Issue), 73-79.

Jayarani Reddy, P.K., Narasimha Rao, L., Narasimha Rao, C.L. and Mahalakshmi, B.K. (2004). Effect of different chemicals on growth, yield and yield attributes of pigeonpea in vertisol. *Annals of Plant Physiology* 17(2), 120-124.

Kandagal, S.S., Panchal, Y.C. and Manjunath, S. (1990). Effects of growth regulators and nutrients on yield components of mung bean genotypes. *Journal of Maharashtra Agricultural Universities* 15(2), 199-200.

Kumar, A., Srivastava, L.K., Mishra, V.N. and Banwasi, R. (2017). Major and micro nutrient status of rice-chickpea grown in soils of Chhattisgarh plain region of India. *Indian Journal of Agricultural Research* 51(1), 1-8.

Kumar, N., Nath, C.P., Hazra, K.K., Das, K., Venkatesh, M.S., Singh, M.K., Singh, S.S., Praharaj, C.S. and Singh, N.P. (2019). Impact of zero-till residue management and crop diversification with legumes on soil aggregation and carbon sequestration. *Soil & Tillage Research* 189(6), 158–167.

Kuttimani, R. and Velayutham, A. (2011). Foliar application of nutrients and growth regulators on yield and economics of greengram. *Madras Agricultural Journal* 98(4), 141-143.

Lakshmi, E.J., Rameshbabu, P.V., Reddy, G.P., Umamaheswari, P. and Reddy, A.P.K. (2017). Effect of foliar application of secondary nutrients and zinc on growth and yield of blackgram. *International journal of chemical Studies* 5(6), 944-947.

Maji, A.K. (2007). Assessment of degraded and wastelands of India. *Journal of the Indian Society of Soil Science* 55(4), 427-435.

Maji, A.K., Reddy, G.P.O. and Sarkar, D. (2010). Degraded and wastelands of India: status and spatial distribution. Directorate of Information and Publications of Agriculture, IndianCouncil of Agricultural Research, New Delhi and National Academy of Agricultural Sciences, India.

Meena, K.K., Meena, R.S. and Kumawat, S.M. (2013). Effect of sulphur and iron fertilization on yield attributes, yield and nutrient uptake of mungbean (*Vigna radiata*). *Indian Journal of Agricultural Science* 83(4), 472-476.

Meghvansi, M.K., and Mahna, S.K. (2009). Evaluating the symbiotic potential of *Glomus intraradices* and *Bradyrhizobium japonicum* in vertisol with two soybean cultivars. *American-Eurasian Journal of Agronomy* 2(1), 21-25.

Mudryj, A.N., Yu, N. and Aukema, H.M. (2014). Nutritional and health benefits of pulses. *Applied Physiology, Nutrition, and Metabolism* 39(11), 1197-1204.

Pal, V., Singh, G., and Dhaliwal, S.S. (2019). Yield enhancement and biofortification of chickpea (*Cicer arietinum* L.) grain with iron and zinc through foliar application of ferrous sulfate and urea. *Journal of Plant Nutrition* 42(15), 1789-1802.

Palta, J.A., Nandwal, A.S., Kumari, S. and Turner, N.C. (2005). Foliar nitrogen applications increase the seed yield and protein content in chickpea (*Cicer arietinum* L.) subject to terminal drought. *Australian Journal of Agricultural Research* 56(2), 105-112

Pandey, N. and Gupta, B. (2012). Improving seed yield of blackgram (*Vigna mungo* L. var. Dpu-88-31) through foliar fertilization of zinc during the reproductive phase. *Journal of Plant Nutrition* 35(11), 1683-1692.

Pandian, B.J., Kumar, A.A., Veerabadran, V. and Ravichandran, V.K. (2001). Growth and yield of rice fallow green gram as influenced by methods of sowing, stubble management and nutrient application in Tambiraparani command area. *Madras Agricultural Journal* 88(7/9), 406-408.

Patra, P. K. and Bhattacharya, C. (2009). Effect of different levels of boron and molybdenum on growth and yield of mungbean [*Vigna radiata* (L.)Wilczek (cv. Baisakhi Mung)] in Red and Laterite Zone of West Bengal. *Journal of Crop and Weed* 5(1), 111-114.

Rao, C.S., Gopinath, K.A., Rao, C.R., Raju, B.M.K., Rejani, R., Venkatesh, G. and Kumari, V.V. (2016). Dryland agriculture in South Asia: Experiences, challenges and opportunities. In: Farooq M., Siddique K. (eds) Innovations in Dryland Agriculture; Springer: Cham, Switzerland. 345–392.

Rao, C.S., Lal, R., Prasad, J.V., Gopinath, K.A., Singh, R., Jakkula, V.S., Sahrawat, K.L., Venkateswarlu, B., Sikka, A.K. and Virmani, S.M. (2015). Potential and challenges of rainfed farming in India. *Advances in Agronomy* 133, 113-181.

Rathod, P.S., Halikatti, S.I., Hiremath, S.M. and Kajjidoni, S.T. (2004). Comparative performance of pigeonpea based intercropping systems in northern transitional zone of Karnataka. *Karnataka Journal of Agricultural Sciences* 17(2), 203-206.

Rathod, S., Channakeshava, S., Basavaraja, B. and Shashidhara, K.S. (2020). Effect of soil and foliar application of zinc and Boron on growth, yield and micro nutrient uptake of Chickpea. *Journal of Pharmacognosy and Phytochemistry* 9(4), 3356-3360.

Ryan, J. (2011). Plant Nutrient Management in Rainfed Farming Systems. In: Tow P., Cooper I., Partridge I., Birch C. (eds) *Rainfed Farming Systems.* Springer, Dordrecht. Pp. 133-147.

Sahrawat, K.L., Wani, S.P., Rao, A.S. and Pardhasaradhi, G. (2011). Management of emerging multinutrient deficiencies: A prerequisite for sustainable enhancement of rainfed agricultural productivity. In: Wani, S.P., Rockstrom, J., Sahrawat, K.L.

(Eds.), Integrated Watershed Management in Rainfed Agriculture, CRC Press/ Balkema, 2300 AK Leiden, The Netherlands. pp. 281–313.

Salehin, F. and Rahman, S. (2012). Effects of zinc and nitrogen fertilizer and their application method on yield and yield components of *Phaseolus vulgaris* L. *Agricultural Sciences* 3(1), 9-13.

Salih, H.O. (2013). Effect of Foliar Fertilization of Fe, B and Zn on nutrient concentration and seed protein of Cowpea '*Vigna unguiculata*'. *Journal of Agriculture and Veterinary Science* 6(3), 42-46.

Sangwan, P.S. and Raj, M. (2004). Effect of zinc nutrition on yield of chickpea (*Cicer arietinum* L.) under dryland conditions. *Indian Journal of Dryland Agricultural research and development* 19(1), 1-3.

Sankar, G.M., Sharma, K.L., Padmalatha, Y., Bhargavi. K., Babu, M.V.S., Sravani, P.N., Ramachandrappa, B.K., Dhanapal, G., Sharma, S, Thakur, H.S., Devi, A.R., Jawahar, D., Ghabane, V.V., Abrol, V., Singh, B., Sharma, P., Kumar, N.A., Girija, A., Ravi, P., Venkateswarlu, B. and Singh, A.K. (2012). Efficient Nutrient Management Practices for Sustainable Crop Productivity and Soil Fertility Maintenance Based on Permanent Manorial Experiments in Different Soil and Agro-Climatic Conditions. In: Abrol, V. and Sharma, P. (eds), *Resource Management for Sustainable Agriculture*. IntechOpen. DOI: 10.5772/50572.

Sarma, P.K., Hazarika, M., Sarma, D., Saikia, P., Neog, P., Rajbongshi, R., Kakati, N., Bhattacharjee, M. and Rao, C.S. (2015). Effect of foliar application of potassium on yield, drought tolerance and rain water use efficiency of toria under rainfed upland situation of Assam. *Indian Journal of Dryland Agricultural Research and Development* 30, 55-59.

Sharma, R.A., Sharma, S.K. and Thakur, H.S. (2006). Sustainability of improved management practices of dryland agriculture in western Madhya Pradesh. *Indian Journal of Dryland Agricultural Research and Development* 21(2), 157-162.

Singh, J., Singh, M., Jain, A., Bhardwaj, S., Singh, A., Singh, D.K. and Dubey, S.K. (2013). An introduction of plant nutrients and foliar fertilization: a review. In: T. Ram, S.K. Lohan, R. Singh, P. Singh (eds), *Precision Farming: A New Approach.* Daya Publishing House, New Delhi. pp. 252-320.

Singh, K.K., Srinivasarao, C. and Ali, M. (2008). Phosphorous and mung bean residue incorporation improve soil fertility and crop productivity in sorghum and mungbean-lentil cropping system. *Journal of Plant Nutrition* 31(3), 459-471.

Singh, K.P. (1993). Response of boron and molybdenum in soil of Singbhum and Santal Parganas. *Annals of Agricultural Research* 14, 100-102.

Srinivasarao, C. and Rani, Y.S. (2013). Zinc deficiency: a productivity constraint in rainfed crop production systems of India. *Journal of SAT Agricultural Research, 11.*

Srinivasarao, C. and Vittal, K.P.R. (2007). Emerging Nutrient Dificiencies in Different Soil Types under Rainfed Production Systems of India. *Indian Journal of Fertilisers* 3(5), 37.

Srinivasarao, C., Vittal, K.P.R., Venkateswarlu, B., Wani, S.P., Sahrawat, K.L., Marimuthu, S. and Kundu, S. (2009). Carbon stocks in different soil types under diverse rainfed production systems in tropical India. *Communications in Soil Science and Plant Analysis* 40(15-16), 2338-2356.

Srinivasarao, Ch., Subha Lakshmi, C. and Kundu, S. (2019). Critical role of nutrient management in climate adaptive dryland farming systems in India. *Indian Journal of Fertilizers* 15(11), 12-26.

Srinivasarao, Ch., Veeraiah, R., Rammohan, S., Dixit, S., Sanjeeva Reddy, B., Rao, K.V., Ramana, D.B.V., Nagasree, K., Dastagiri, S., Kundu, Sumanta, Jakkula, Vijay, S., Anuradha, B. and Venkateswarlu, B. (2013a). Livelihood Improvement in Tribal Rainfed Region: Experiences from Participatory On-farm Interventions in Nalgonda District, Andhra Pradesh. Central Research Institute for Dryland Agriculture, Hyderabad, Andhra Pradesh, pp.142.

Tahir, M., Maqbool, R., Majeed, A., Rehman, A. and Zafar, M.A. (2014). Potential of foliar applied diammonium phosphate (DAP) and potassium (K) in achieving maximum productivity and quality of mash bean (*Vigna mungo* L.). *Scientia Agriculturae* 7(3), 147-149.

Thalooth, A.T., Tawfik, M.M. and Mohamed, H. M. (2006). A comparative study on the effect of foliar application of zinc, potassium and magnesium on growth, yield and some chemical constituents of mungbean plants grown under water stress conditions. *World Journal of Agricultural Science* 2(1), 37-46.

Venkateswarlu, B. and Prasad, J.V.N.S. (2012). Carrying capacity of Indian agriculture: issues related to rainfed agriculture. *Current Science*, 882-888.

Yadav, S.S., Redden, R.J., Chen, W. and Sharma, B. (2007). Chickpea Breeding and Management. CAB International, Oxford, UK.

CHAPTER 13

Emerging technologies for sustainable pulses production in Madhya Pradesh

A.A. Raut, S.R.K. Singh, Shrishti Bilaiya* and Anupam Mishra
ICAR-Agricultural Technology Application Research Institute (ATARI), Jabalpur, Madhya Pradesh
*Corresponding author email: bshrishti1000@gmail.com

Abstract

Pulses are the crops of nutritive, biological and economic importance. Because of their high protein content, low price and wide spread access to the poor, they are considered as poor man's meat. Although, India is the largest pulse producing country in the world, pulses share to total foodgrains production is only 6-7% in the country and has not been able to keep pace with their domestic demand resulting in imports of 2.5 Mt of pulses in 2018-19. As per an estimate of IIPR, Kanpur the country would require 39 Mt of total pulses by 2050. To fulfill the growing requirement, the country has to produce enough pulses as well as remain competitive to protect the domestic production. It is imperative to develop and adopt more efficient crop production technologies along with favourable policies and market support to encourage farmers to bring more area under pulses. The small change through land configuration in flat bed and adopting *in-situ* conservation technologies like broad-bed furrow, ridge and furrow, raised bed, etc. for soil moisture conservation, minimum runoff and soil erosion, proper drainage and aeration in the root zone may help in minimizing the risk of crop failure and substantially improving the productivity of pulses.

Keywords: Pulses, Productivity, *In-situ* moisture conservation, Broad-bed and furrow, Ridge & Furrow, Raised Bed

1. INTRODUCTION

Pulse crops also referred as "grain legumes" are the most important food crops after cereals. They are edible dry seeds of plants belonging to the Leguminosae family. They are consumed in the form of whole seed, split grain, dehulled split grain and flour. Many types of pulses are grown all over the world. Of these, the major ones, in terms of global production and consumption quantities are chickpea, fieldpea, pigeonpea, blackgram, greengram, lentil, horsegram, lathyrus, etc. They are important and inexpensive source of plant-based proteins, minerals, fibre, amino acids and various vitamins; thiamine (B1), riboflavin (B2) and E, calcium, potassium, iron and phosphorus. Because of their high protein content, low price and its wide spread access to the poor, pulses are also known as "poor man's meat" (Venkidasamy *et al*, 2019). If we consider some of the major sources of proteins, pulses turn out to be one of the most economical sources of protein for human consumption. Pulses, supplemented with cereals, provide a perfect mix of vegetarian protein of high biological value.

Besides serving as an important source of protein for a large portion of the global population, pulses contribute to healthy soils and climate change mitigation through the mechanism of fixing atmospheric nitrogen in their root nodules and thus meet their nitrogen requirements to a great extent. They tend to fix 72 to 350 kg N/ha/year (Nulik *et al*, 2013). They improve the physical condition of the soil like soil aeration, water holding capacity, improving microbial population, breaking of hard pans and moisture retention. Pulses can be use as catch crop, cover crop, intercrop and in crop rotation.

India is the largest producer (28% of global production), consumer (27% of world consumption) and importer (14%) of pulses in the world (FAOSTAT, 2016). About 80% of global pigeonpea (*Cajanus cajan*), 65% of chickpea (*Cicer arietinum*), 37% of lentil (*Lens culinaris*) and more than 65% of mungbean (*Vigna radiata*) and urdbean (*Vigna mungo*) are produced in India. During 2017-18, pulses were cultivated in 29.30 M ha area and recorded the highest ever production of 25.23 Mt at a productivity level of 841 kg/ha and ranked first in area and production in the world (DPD Annual report, 2017-18). In India, 10 states are the major producers of pulses contributing more than 90% of the total production. These are Madhya Pradesh (> 8 Mt), Rajasthan (>3 Mt), Maharashtra (>3 Mt) Uttar Pradesh (>2 Mt), Karnataka (2 Mt) and Andhra Pradesh (>1 Mt) followed by Gujarat, Jharkhand, Tamil Nadu, and Chhattisgarh producing < 1 Mt each (Anonymous, 2018).

State Madhya Pradesh contributes remarkable 25% of the country's pulses area with 32% (8.11 Mt) production, thereby ranking first both in area and production followed by Rajasthan and Maharashtra with 13% each and Uttar Pradesh at 9%. Out of the total pulses area (29.30 M ha) more than 7.40 M ha is confined to Madhya Pradesh alone, earning a prime status in pulses production. Compared to normal production, the estimated production during 2017-18 is 30% higher in case of total pulses, 32% chickpea, 27% pigeonpea, 58% urdbean, 18% mungbean and 40% higher lentil production (DPD Annual report, 2017-18).

2. REASON FOR LOW PRODUCTIVITY OF PULSES IN MADHYA PRADESH

As we know, India is the highest pulse producing country in the world, though the productivity of pulses in India is less than half of the productivity levels in the USA and Canada, as the pulses are mainly grown under rainfed condition. Similar is true in case of Madhya Pradesh. The pulses experiences moisture stress during the dry spell ranging from 15 to 21 days at any growth stage under rainfed conditions resulting significant reduction in the yield. Extreme variability in the quantity, time and duration of rains exposed the pulse crops to soil moisture deficit as well as excess moisture, either on account of delayed monsoon, longer dry spells or early withdrawal of monsoon or excessive rainfall events (Tiwari, 2014; Gupta *et al*, 2018). Currently, rainfall-use efficiency for crop production ranges between 30-45% as sizable part of the incipient rainwater is lost as surface runoff. Besides reducing fertility of soil it also causes erosion to the tune of 2.2 to 6.4 t/ha annually (Ahlawat *et al*., 2016). The problems of soil moisture related stress get aggravated due to the inadequate adoption of production technology risk and low level of irrigation which are responsible for stagnation in the productivity of pulses in Madhya Pradesh.

3. NEED FOR PARADIGM SHIFT IN SOWING METHOD

The country would require 39 Mt of total pulses by 2050 which will require pulses production to grow at an annual rate of 2.2% (IIPR, 2015). To meet the increasing demands, the country has to produce enough pulses as well as remain competitive to protect domestic production. It is imperative to develop and adopt more efficient crop production techniques along with market support to boost the morale of farmers to bring more area under pulses. Among all legumes crops, pulses are most sensitive to soil moisture. Excess and continuous rains leads to water logging problems due to improper drainage that restrict aeration which results in non-availability of plant nutrients and poor microbial activities, root proliferation and final yield (Tiwari, 2014; Gupta *et al.*, 2018). The traditional flat bed cultivation system, which is popular among farmers, resulted in more problems of water logging and poor aeration. Water logging adversely affects the growth of crops primarily due to reduced oxygen supply to the roots. The loss in yield can be avoided or minimized if good amount of water is stored in the soil during rainy days and utilized by the crop during moisture stress or dry spell and same time, there must be provision for the drainage of the excess rainwater.

The small change through land configuration in flat field conditions may help in improving the productivity. The pulse crop planted on ridges/ broad bed/ raised bed has shown significantly higher yield than planted on flat bed. Studies on soil management for increasing crop production revealed that use of various tillage methods and modification in land configuration such as broad bed furrow, ridges and furrow for pulses were superior over flat bed and recommended in watershed development for moisture conservation as well as for safe removal of excess rainwater (Raut and Taware, 1997). It also helps in soil moisture conservation, minimum runoff and soil erosion, proper drainage and proper aeration in the root zone and thus, helps in minimizing risk of crop failure and substantially improving the productivity of pulses (Singh *et al.*, 1999; Nagavallemma *et al.*, 2005).

4. IMPROVED SOWING TECHNOLOGIES IN PULSES FOR MADHYA PRADESH

4.1 Ridge & furrow method

Ridge and furrow (R&F) system involves preparation of a ridge of 20-30 cm, furrow of 15-20 cm and sowing of crop at a row spacing of 40-45 cm (Fig. 1). When seeds are sown using R&F system, they are either sown at the bottom of the furrow or on the side or top of the ridge (Fig 2). Planting on the side or top of the ride is practiced in heavy rainfall areas and under irrigated conditions.

4.1.1 Planting techniques

The location of plants in a furrow system is not fixed. It depends on the natural circumstances. Following are the few examples:

(A) In areas with heavy rainfall, the plants should be planted on top of the ridge in order to prevent damage as a result of water logging (Fig 2A).

(B) If water is scarce, the plants may he put in the furrow itself, to get more benefit from the limited water (Fig 2B).

(C) As salts tend to accumulate at the highest point, a crop on saline soils should be planted away from the top of the ridge. Usually it is planted in two rows at the sides (Fig 2C). However, it is important to make sure there is no danger of water logging.

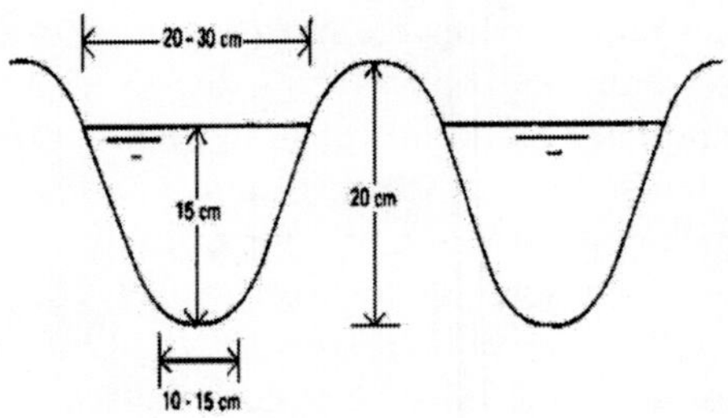

Fig. 1. Diagrammatic view of Ridge & Furrow system

(*Source:* http://www.fao.org/3/S8684E/s8684e04.htm)

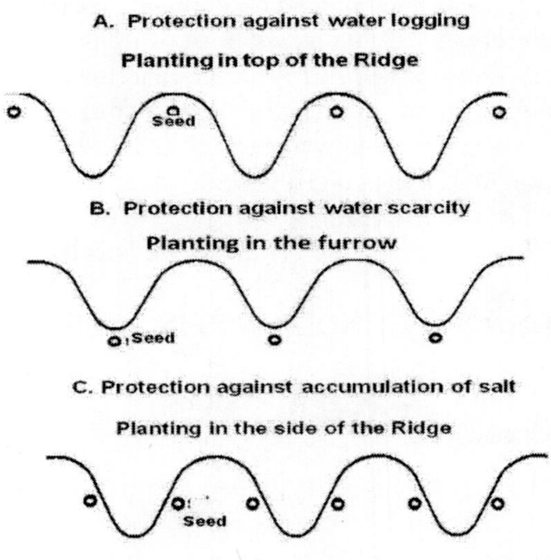

Fig. 2. Planting technique in Ridge and Furrow sowing

(Source:http://www.fao.org/)

4.1.2 Advantage of R&F method

This method controls over irrigation and drainage as well as its impact on transport and transformations of nutrients, and rainwater management during the monsoon season. It saves the irrigation water of about 30-35% over flat method. This rainwater harvesting system is an effective and simple method for increasing the soil water content (SWC) and improving the rainwater use efficiency in dryland farming (Ren *et al.*, 2008; Li *et al.*, 2016).

In this system, the plastic mulched ridges provide a runoff surface that directs rainwater to the furrows so that it can penetrate deeper into the soil (Li and Gong, 2002; Tian *et al.*, 2003), as well as reducing the loss of soil water by evaporation from the ridges and extending the period of moisture availability for crops (Zhang *et al.*, 2007; Liu *et al.*, 2014). Therefore, ridge and furrow system has been applied in increasing the crop water availability, improving soil productivity, and enhancing food security (Gan *et al.*, 2013).

4.1.3 Crops performance under R&F

The demonstrations conducted by KVK's Tikamgarh, Jhabua, Damoh, Guna, Agar Malwa, Ashoknagar and Dhar under ATARI, Zone-IX, Jabalpur (MP) revealed that blackgram, chickpea, pigeonpea and greengram crops sown under R&F method resulted in 36.39%, 41.46%, 40.67% and 16.30% increase in yield, respectively, in comparison to flat bed sowing (Table 1).

Table 1. Performance of different crops under ridge & furrow (R&F) method

Crop	Season	Area (ha)	No. of Farmers	Yield (kg/ha)		% Change	Net return (Rs/ha)		B:C	
				Flat bed	R&F		Flat bed	R&F	Flat bed	R & F
Blackgram	*Kharif*	150	355	709	967	36.39	19368.03	30691.14	2.19	2.85
Pigeonpea	*Kharif*	72.4	181	890	1252	40.67	30405.00	38639.25	2.39	2.66
Greengram	*Kharif*	30	75	1515	1762	16.30	34041.00	41233.00	3.13	3.45
Chickpea	*Rabi*	119.20	250	1235	1747	41.46	33178.17	53257.67	2.51	3.17
Total		371.60	861							

(Source : KVK demonstrations under ATARI, Zone-IX, Jabalpur (MP) in 2018-19)

Further, it was depicted that, ridge planting helps farmers in saving approximately 30-35% water and obtaining ease in irrigation management in comparison with farmers practice. Farmers with sizable land holding and tube well were the initial adopter of technology for its simplicity and availability of ridger at farm level, water saving, and better yield. In addition to yield enhancement, R&F method helps farmers of Madhya Pradesh in increasing seed germination percentage, maintaining plant population, increased plant growth and development and higher income (Table 2).

Table 2. Performance of blackgram (var. PU1) under ridge & furrow

Important Parameters	Germination (%)	Plant population (per m^2)	Pest infestation (per m)	Plant height (cm)	Pods/ plant (No.)
Ridge & Furrow	85	35	Semilooper- 2/m^2 Anthracnose - 3%	38	22
Flat bed (local practice)	79	32	Semilooper- 4/m^2 Anthracnose - 6%	35	18

Source: KVK demonstrations under ATARI, Zone-IX, Jabalpur (MP) during 2018-19

SUCCESS STORY OF RIDGE & FURROW METHOD OF SOWING

Shri Antar Singh of Village Baijapur, district Khargone was earlier cultivating old variety of greengram with traditional practices of flat bed planting, hence was obtaining poor productivity. After contacting KVK scientists, demonstration of improved greengram variety 'IPM 2-14' was conducted in his field under CFLD pulses program. This high yielding and YMV resistant variety, IPM 2-14' was sown using ridge and furrow method of sowing after undergoing seed treatment with Carboxin 2.0 gm/kg + Trichoderma 5 gm + Rhizobium culture 5 gm/kg. Further, proper nutrient management along with insect and pest management was given to the crop greengram. It was found that the yield increased by 35.15% with 7.59% increment in germination, 14.14% in plant height, 13.51% in plant population and 21.56% in No. pods/plant. The average yield obtained by adopting R&F method in greengram was 1115 kg/ha with B:C ratio of 5.61.

Greengram IPM 2-14 under ridge and furrow method

Practice	Yield (kg/ha)	Cost of Cultivation	Gross income	Net income	B:C	% increase in gross income
		(Rs/ha)				
Flat bed	825	12250	57543	48293	4.69	**35.15**
Ridge & furrow	1115	13805	77771	68121	5.61	

4.2 Broad bed & furrow

The broad bed furrow (BBF) based sowing of crop is an emerging practice in rainfed farming for sustainable yield. Broad bed furrow system is a precision method for *in-situ* soil moisture conservation which makes the moisture availability for longer period to the crops. This system was developed at ICRISAT, India (Krantz, 1981; Pathak *et al.*, 1985). This method of sowing includes ploughing to improve effectiveness and efficiency of traditional drainage practices for *in-situ* moisture conservation in case of dryland zone and helps to overcome water logging under excessive rainfall and also improves soil structure, along with enhanced crop yield. It is an improved *in-situ* soil and water conservation and drainage technology best suitable for the Vertisols with slope upto 1%. The system consists of relatively flat bed approximately 90 cm to 180 cm wide and shallow furrow about 30-50 cm wide (Fig 3).

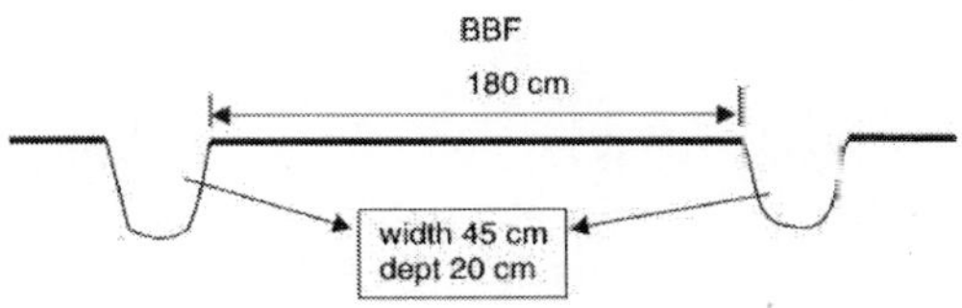

Fig. 3. Broad bed furrow method of sowing

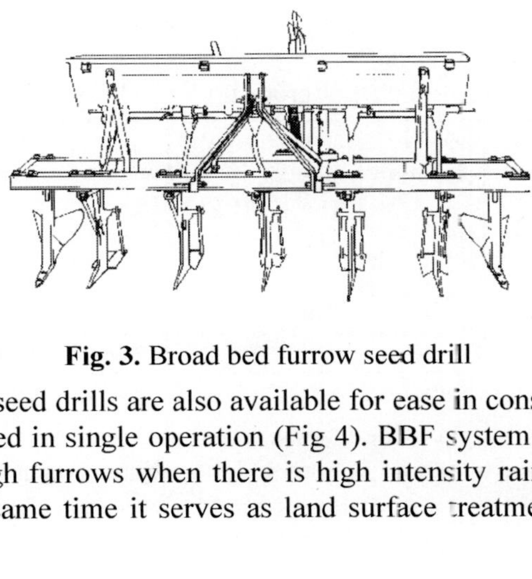

Fig. 3. Broad bed furrow seed drill

Tractor operated BBF seed drills are also available for ease in constructing broad bed and furrow, and sowing seed in single operation (Fig 4). BBF system helps for safe disposal of excess water through furrows when there is high intensity rainfall with minimal soil erosion, while at the same time it serves as land surface treatment for in-situ moisture conservation.

4.2.1 Objectives of BBF (Source: Hudson, 2018):

(i) To encourage moisture storage in the soil profile which supports growth even during the subsequent dry but cooler post-rainy season.

(ii) To dispose safely of surplus surface run-off without causing erosion.

(iii) To provide a better drained and more easily cultivated soil in the beds.

(iv) The possibility of the re-use of run-off stored in small farm tanks for giving life-saving irrigation during dry spells, particularly on soils with lower storage capacity than the deep vertisols.

4.2.2 Advantage of BBF planting

The main function of BBF planting method of sowing is to control erosion and to conserve moisture in the soil during rainy days. In a BBF system, runoff water is diverted into field furrows (30 cm wide and 30 cm deep). The field furrows are blocked at the lower end. When one furrow is full, the water backs up into the head furrow and flows into the next field furrow. Between the field furrows are broad beds about 120-180 cm wide, where crops are grown. This method of sowing has various advantages over flat bed method of sowing. In addition to increasing crop productivity by 5-10%, it helps in increasing water use efficiency (WUE) and saves irrigation water up to 25-30%. Also, BBF helps in reducing irrigation time by 25-30%, seed rate by 20-25% and crop lodging. Further, BBF also helps in better weed management and during non- rainy days have less moisture stress.

4.2.3 Crop performance under BBF

The farm level performance of BBF was demonstrated through frontline demonstrations (FLD) for blackgram, greengram and pigeonpea in KVK's of Madhya Pradesh under ATARI, Zone-IX, Jabalpur. FLD results revealed that following of BBF enhanced the yield of blackgram by 74%, greengram by 19.90% and pigeonpea by 12.66% compared to conventional flat bed system used by the farmers. The BCR under BBF increased by 44.00, 9.09 and 13.61%, respectively in blackgram, greengram and pigeonpea over conventional flat bed method (Table 3).

Table 3. Performance of crops under BBF planting system

Crop	Season	Area (ha)	No. of Farmers	Yield (Kg/ha)		% Change	Net return (Rs/ ha)		B:C	
				Flat bed	BBF		Flat bed	BBF	Flat bed	BBF
Blackgram	Kharif	2	5	770	1340	74.02	26950	46900	2.5	3.6
Greengram	*Kharif*	10	25	730	870	19.17	25294	31340	3.3	3.6
Pigeonpea	*Kharif*	15	40	1390	1566	12.66	37850	44340	2.13	2.42

Source : KVK demonstrations under ATARI, Zone-IX, Jabalpur (MP) in 2018-19)

Frontline demonstrations recorded consistent improvements in yield and reductions in irrigation inputs, together with cost savings in labour, land preparation, fertilizer, and seed inputs, on raised beds. The participating farmers were enthusiastic about BBF because the practice improved livelihoods and food security for their families. Also, BBF helped farmers in increasing the germination percentage of crop, along with increase in plant population, plant height, plants/pod and thus, helping farmers in fetching higher net return (Table 4). Interest in the BBF technology expanded beyond the initial group to farmers in the surrounding areas which convinced farmers in different districts of Madhya Pradesh to adopt BBF technology in pulses.

Table 4. Performance of blackgram (Pratap Urd 1) under improved planting technique

Planting method	Germination %	Plant population (per m²)	Weed count (per m²)	Pest infestation (per m²)	Plant height (per cm)	Pods/ plant (No.)
BBF	98	28	10	1	26	60
Flat bed	68	15	35	35	23	34

Source : KVK demonstrations under ATARI, Zone-IX, Jabalpur (MP) in 2018-19)

SUCCESS STORY OF BROAD BED FURROW METHOD OF SOWING

Shri Govind Singh Thakur is farmer of Village Charpuwan, district Tikamgarh, practicing farming from last 10 years in his 1 acre land, but his income was low due to unavailability of water and practicing of traditional farming system. After coming in contact with KVK, he started adopting technologies suggested by the scientists. So, in spite of using traditional practice of flat bed planting for cultivation of blackgram, he adopted BBF method of sowing along with improved variety 'Pratap Urd-1'. As a result, he obtained 1340 kg/ha yield of blackgram which is more than 100% higher over conventional flat bed. Further, he obtained a net profit of Rs 34500/ha with B:C ratio of 3.7. Along with increase in yield, there was increase in germination percentage, plant height and pods/ plant.

Blackgram (Pratap Urd-1) under Broad bed & furrow

Practice	Yield (kg/ha)	Cost of Cultivation (Rs/ha)	Gross income (Rs/ha)	Net income (Rs/ha)	B:C ratio	% of income increase
Flat bed sowing	640	10500	22400	11900	2.1	109.3
BBF sowing	1340	12400	46900	34500	3.7	

4.3 Raised bed planting

This system is often considered more appropriate for growing high value crops that are more sensitive to temporary water logging stress. Raised bed planting is a method of planting a crop on the top of raised beds with a definite number of rows (Fig 4). Raised beds are prepared by moving soil from the furrows to the area of the bed by raising its surface level using raised bed planter. The furrows serve as irrigation channels, drains and passage for agricultural operations. The irrigation water moves laterally from the furrow into the bed, and is driven upwards towards the bed surface by evaporation and capillarity, and downwards largely by gravity. Raised bed planting will be good strategy for risk reduction due to water logging under high rainfall by facilitating drainage for sustainable pulses production during *kharif* season (Ram and Kler, 2007).

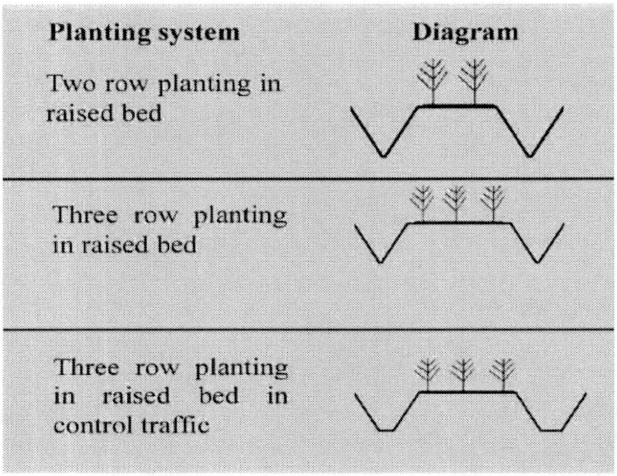

Fig. 4. Planting System in Raised bed planting

Changing from flat bed to raised bed layout alters the hydrology of the system and allows greater control of irrigation, better surface drainage and possibly better capture and use of rainfall. Recent works showed that the system of raised bed planting of crops may be particularly advantageous in areas where groundwater level are falling and herbicide resistant weeds are becoming a problem. Planting on raised bed accelerate easy germination and development of nodules due to better aeration in the root zone. The fertilizers can be placed conveniently and nutrient use efficiency is higher in this method. These raised beds facilitates better crop stand by reducing crop lodging (Tripathi *et al.*, 2005) and proper interception of solar radiation. The furrow made between raised beds facilitates safe drainage of excess rainwater, percolation through field and conserve moisture which may be utilized by crop during dry spell and also check soil erosion (Motwani and Ashish, 2018). In addition to these, through raised bed planting method of sowing, yield potential is enhanced over conventional method through higher nutrient and water availability (Singh *et al.*, 2012). The advantage of this system is reduced cost, soil conservation, less soil compaction and improvement in soil physiological properties (Sayre *et al.*, 2005).

4.3.1 Raised bed planter

The sowing of the crops is done with the help of raised bed planter (Fig 5). The machine has adjustable blades for making raised beds of different widths and heights that can be adjusted by the shifting of the blades on the frame and roller on the rear. This planter can make two rows at a time and has seed-cum-fertilizer drilling mechanism for sowing in rows. The width of bed can be adjusted from 65-75 cm. Raised bed planter forms two beds and three furrow in single operation. These furrows have 25 cm top width and 15 to 20 cm depth for medium to heavy soils (Singh *et al.*, 2020).

Fig 5. Raised bed planter

4.3.2 Advantage of raised bed planting

Raised beds can improve production as well as save space, time, and money. It is also perfect solution for dealing with difficult soils such as heavy clay. Raised beds are effective in increasing inputs use efficiency and yields. The crops produce better as the soil in the beds is deep, loose, and fertile. Weeds can be easily uprooted. The seed rate in this planting method can be reduced by 25-30% (Ram and Kler, 2007). Planting on raised bed accelerate easy germination and development of nodules due to better aeration in the root zone and facilitates in application of fertilizer and management of weeds (Jha *et al.*, 2014). These raised beds facilitates better crop stand by reducing crop lodging (Tripathi *et al.*, 2005) and proper interception of solar radiation. The furrow made between raised beds facilitates safe drainage of excess rain water, percolation through field and conserve moisture which may be utilized by crop during dry spell and also check soil erosion (Motwani and Ashish, 2018). The conserved soil moisture is available in soil profile in later crop stage after withdrawal of monsoon (Kumar *et al.*, 2002; Hati *et al.*, 2013). Yield potential is enhanced in raised planting over conventional method through higher nutrient and water availability (Singh *et al.*, 2012). An additional advantage is apparent when beds are permanent that it is maintain over medium term and no tillage operations are performed for sowing of next crop. These permanent beds give opportunity for resource conservation and saving of input cost (Sayre and Ramos, 1997). The system of reduced tillage is followed in next crop of soybean can directly sown in raised bed without field preparation. The advantage of this system is reduced cost, soil conservation, less soil compaction and improvement in soil physiological properties (Sayre *et al.*, 2005). This technology is more suitable in medium to heavy soils in plain topography.

4.3.3 Crop performance

Cluster frontline demonstrations (CFLD) on pulses were conducted in different areas of Madhya Pradesh under ICAR-ATARI, Zone-IX, Jabalpur in 2018-19. Improved sowing technology viz. raised bed planting was demonstrated in blackgram, chickpea, greengram and pigeonpea. The farm level performance under FLD results revealed that this system saved 25-30% irrigation water, 30-45% planting cost, and ensured higher crop yield of 24-52% as compared to conventional system of flat bed planting. Further, it was found that, there was increase in yield of blackgram, chickpea, greengram and pigeonpea by 52.43, 25.02, 24.63 and 26.09%, respectively with increase in BCR by 41.40, 21.63, 15.67 and 20.09% (Table 5).

Table 5. On farm performance of pulses under raised bed planting

Crop	Area (ha)	No. of Farmers	Yield (kg/ha)		% Change	Net return (Rs/ha)		B:C ratio	
			Flat bed	Raised bed		Flat bed	Raised bed	Flat bed	Raised bed
Blackgram	50	124	719	1096	52.43	21648.00	41050.50	2.15	3.04
Chickpea	55.2	136	1527	1909	25.02	45649.62	63047.05	2.79	3.39
Greengram	20	49	934	1164	24.63	44729.00	59180.00	3.19	3.69
Pigeonpea	30	75	920	1160	26.09	25087.00	35836.00	2.19	2.63

Source : KVK demonstrations under ATARI, Zone-IX, Jabalpur (MP) in 2018-19)

It was also observed that planting on raised bed accelerate easy germination and development of nodules due to better aeration in the root zone. Weed management operations can be easily performed due to space between beds provide opportunity for mechanical weeding in early stage of crop life-cycle and the fertilizers can be placed conveniently and nutrient use efficiency is higher in this method. Under agro-climatic region of Madhya Pradesh, sowing seeds using raised bed method helps in conserving moisture by 25-30%, along with increasing plant height, No. of branches/plant, No. of pods /plant and grains per pod, while decreasing insect-pest and weed infestation (Table 6).

Table 6. Performace of pigeonpea under raised bed system of planting

Important Parameters	Germination (%)	Plant population (per m^2)	Weed count (per m^2)	Pest infestation (per m^2)	Plant height (per cm)	Pods/ plant (No.)
Raised bed	82	27-30	20-22	1-2 larvae	170-190	365
Flat bed	73	16-18	25-28	2-3 larvae	200-225	205

Source : KVK demonstrations under ATARI, Zone-IX, Jabalpur (MP) in 2018-19)

SUCCESS STORY OF RAISED BED PLANTING OF BLACKGRAM

Shri Santosh Singh is farmer of village Padiya of district Rewa. He along with the other farmers of village Padiya were using old variety of blackgram with injudicious use of fertilizer, which resulted into low yield of blackgram with increase in cost of production. Further, due to broadcast sowing on flat bed, weeding and other interculture operations were difficult. Due to poor yield and high cost of production, Shri Singh and other farmer's of this village were facing financial problems and it was very hard to run their family satisfactorily.

With the suggestions of scientist of KVK Rewa, he started cultivating blackgram using raised bed technology. Improved variety UH 1 with seed rate of @ 15 kg/ha was sown using raised bed method. Further, STV based nutrient management (20:50:20:20 kg NPKS/ ha) was given at the time of sowing with proper insect, pest and weed management. This resulted in enhancement of yield of blackgram by 80.20% along with increase in income by 44.90%.

Blackgram under raised bed planting

Practice	Yield (Kg/ha)	Cost of Cultivation	Gross income	Net income	B:C ratio	% of income increase
		(Rs/ha)				
Raised bed	1098	18550	61488	42938	3.31	44.90
Flat bed	609	16750	34104	17354	2.03	

5. CONCLUSION

It can be summarized that in order to provide the nutritional security to the poor masses relying on vegetarian diet, making pulses affordable through boosting domestic production is the best alternative. Thus, to augment the supply of pulses to poor masses under the climate changing scenario, adoption of selective mechanization and improved practices to sustain productivity and soil quality is need of an hour. This can be achieved by adopting in-situ moisture conservation technologies like broad-bed furrow, ridge and furrow and raised bed planting. These technologies will help to maximize use of rainfall by increasing infiltration and moisture retention, encouraging surface drainage and reducing runoff and soil erosion.

REFERENCES

Ahlawat, I.P.S., Sharma, P. and Singh, U. (2016). Production, demand and import of pulses in India. *Indian Journal of Agronomy* 61(4th IAC Special issue), S33-S41

Anonymous. (2018). Pulses Revolution from Food to Nutritional Security. Ministry of Agriculture and Farmers Welfare (DAC&FW), GOI. Pp. 17.

APEDA Report. (2019). Agricultural and Processed Food Products Export Development Authority, Ministry of Commerce and Industry, GOI.

Confederation of Indian Industry (CII) Report. (2010). Overcoming the Pulses Crisis: Key Interventions Recommended.

DPD Annual Report. (2017-18). Directorate of Pulses Development, Department of Agriculture, Cooperation and Farmers Welfare, MINISTRY OF Agriculture & Farmers welfare, Government of India. DPD/Pub./TR/29/2017-18.

FAOSTAT (2016) (http://faostat3.fao.org/home/E)]

Gan, Y., Siddique, K., Turner, N., Li, X., Niu, J. and Yang C. (2013). Ridge-furrow mulching systems-an innovative technique for boosting crop productivity in semiarid rain-fed environments. *Advances in Agronomy* 118, 429–476.

Global Nutrition Report. (2018). Shining a light to spur action on nutrition, November, 2018.

Gupta, R., Kulmi, G.S, Basediya, A.L. and Jadav, M. (2018). Influence of furrow irrigated raised bed seed drill on growth characteristics and yield of soybean in Mandsaur district of Madhya Pradesh, India. *Plant Archives* 18(1), 320-324.

Hati, K.M., Chaudhary, R.S., Manda, l K.G., Mishra, A.K., Singh, R.K., Wani, S.P. and Pathak, P. (2013). Effect of land management and cropping systems on runoff, soil water dynamics and crop yield in vertisol of central India. *Journal of the Indian Society of Soil Science* 61, 79-88.

Hudson, N.W. (2018). Soil and Water Conservation in Semi-arid Areas. Chapter 5. Water Conservation. FAO, Pg 61. (http://www.fao.org/ 3/T0321E/ t03 21e-11.htm)

IIPR. (2015). Vision 2050. Indian Institute of Pulses Research Kanpur, Uttar Pradesh, India. 50 pages.

Jha, A.K., Dwivedi, B.S., Shrivastava, A.K., Soni, M., Taniwaki, K., Kokuryu, T., Kitagawa, I., Tateishi, K., Tsuji, K. and Kobayashi, S. (2014). Effect of Drainage, Tillage and Land Configurations on Root, Nodules and Yield of Soybean in Vertisols. *Soybean Research (Special issue)*, 194-198.

Kumar, S., Singh, M. and Singh, B.R. (2013). Feasibility and economic viability of raised bed planter in western plane zone of Uttar Pradesh, India. Elsevier. *Soil and Tillage Research* 128, 37-43.

Kumar, V., Sharma, D.N. and Sardana, P.K. (2002) Techno-economic evaluation of bed seeding techniques in wheat crop. *Journal of Agricultural Engineering* 39(2), 32-43.

Li, R., Hou, X., Wang, X., Jia, Z. and Han, Q. (2016) Research progress on the dual-mulching of ridge and furrow technology in dry farming regions of northern China. *Chinese Journal of Applied Ecology* 27, 1314–1322.

Li, X.Y. and Gong, J.D. (2002). Effects of different ridge: furrow ratios and supplemental irrigation on crop production in ridge and furrow rainfall harvesting system with mulches. *Agricultural Water Management* 54, 243-254.

Liu, E., He, W. and Yan, C. (2014). 'White Revolution'to 'White Pollution'—agricultural plastic film mulch in China. *Environmental Research Letters* 9, 091001.

Mohanty, N., Sinha, N.K., Mcdermid, S.P., Chaudhary, R.S., Reddy, K.S., Hati, K.M., Somasundaram, J., Lenka, S., Patidar, R.K., Prabhakar, M., Cherukumalli, S.R. and

Patra, A.K. (2017). Climate change impacts vis-à-vis productivity of soybean in vertisol of Madhya Pradesh. *Journal of Agrometeorology* 19(1), 10-16.

Motwani, S. and Ashish. (2018). Performance of Soybean Planted on Broad Bed Furrow System under Real farm Conditions. *Soybean Research* 16(1&2), 113-117.

Mothapo, N., Grossman, J. and Maul, J. (2011). Hairy vetch cultivation history affects nodulation and biological nitrogen fixation across host genotypes. ASA-CSSA-SSSA International Annual Meetings, Oct 16–19, 2011, San Antonio, TX. (Available online at: http://a-c-s.confex.com/crops/2011am/webprogram/Paper66799.html) (verified 20 April 2012).

Nagavallemma, K.P., Wani, S.P., Reddy, M.S. and Pathak, P. (2005). Effect of landform and soil depth on productivity of soybean based cropping systems and erosion losses in Vertic Inceptisols. *Indian journal of Soil Conservation* 33(2), 132–136.

Nulik, J., Dalgliesh, N., Cox, K. and Gabb, S. (2013) Integrating herbaceous legumes into crop and live-stock systems in eastern Indonesia. Australian Centre for International Agricultural Research (ACIAR), Canberra.

Pathak, P., Miranda, S.M. and El-Swaify, S.A., (1985) Improved rainfed farming for semi-arid tropics- Implications for soil and water conservation. In: El-Swaify, S.A., Moldenhauer,W.C., Andrew, L. (Eds.), Soil Erosion and conservation . Soil conservation Society of America, Ankeny, Lowa, USA, 338-35

Ram, H. and Kler, D.S. (2007) Growth analysis of soybean [*Glycine Max* (L.) Merill] and wheat (*Triticum Aestivum* L. emend Fiori and Paol) in sequence under no tillage and permanent raised bed planting. *Indian Journal of Ecology*. 34, 154-157.

Ramesh, K., Patra, A.K. and Biswas, A.K. (2017) Best Management Practices for Soybean under Soybean-Wheat System to minimize the Impact of Climate Change. *Indian Journal of Fertilizers*. 13(2), 42-55

Ramteke, R. and Gupta, P. (2011) Tractor-drawn broad bed furrow seed drill machine to overcome moisture stress for soybean (*Glycine max*) in Vertisols. *Indian Journal of Agricultural sciences*. 81(10), 941-944.

Raut, V.M. and Taware, S.P. (1997) Comparison of different sowing methods in soybean. *Journal of Maharashtra Agriculture University*. 25 (2), 218-219.

Ren, X., Chen, X. and Jia, Z. (2010) Effect of rainfall collecting with ridge and furrow on soil moisture and root growth of corn in Semiarid Northwest China. Journal of *Agronomy Crop Science* 196, 109–122.

Sayre, K.D. and Ramos, O.H.M. (1997) Applications of Raised Bed Planting Systems to Wheat. Wheat Special Report No.31. Mexico, D.F.: CIMMYT

Sayre, K.D., Limon A. and Goaverts, V. (2005) Experiences with permanent bed planting system CIMMYT, Mexico. Evaluation and performance of permanent raised bed cropping systems in Asia, Australia and Mexico. ACIAR Proceeding No.121.

Singh, D., Ramteke, R., Vyas, A.K., Billore, S.D. and Khan, I.R. (2012) Tractor Operated Furrow Irrigated Raised Bed Systems (FIRBS) Seed Drill for Rainfed Soybean in Vertisols. *Soybean Research*.10, 88-92.

Singh, P., Alagarswamy, G., Pathak, P., Wani, S. P., Hoogenboom, G. and Virmani, S.M. (1999) Soybean-chickpea rotation on Vertic Inceptisol: 1.Effect of soil depth and landform on light interception, water balance and crop yields. *Field Crops Research* 63:211-224. II. Long-term simulation of water balance and crop yields. *Field Crops Research*. 63(3): 225-236.

Singh, S.R.K., Mishra A. and Raut, A.A. (2020) Policy Paper on TAD Modified Sowing Techniques in Soybean for Risk Minimization. TAD Modified Sowing Techniques in Soybean for Risk Minimization. Pg : 07.

Technical Bulletin (2017) Raised Bed Planter for *Kharif* Pulse Cultivation in Assam. Directorate of Research (Agri), Assam Agricultural University. Technical Bulletin no. FIM /5 /2017. Pg : 7

Tripathi, S.C., Sayre, K.D. and Kaul, J.N. (2005) Planting Systems on lodging behavior, yield components and yield of irrigated spring bread wheat. *Crop Science* 45, 1448-1455.

Krantz, BA (1981) In: Advances in Food Producing Systems for Arid and Semiarid Lands. Part A, Univ. of California. 339-378.

Tian Y., Su D., Li F., Li X. (2003) Effect of rainwater harvesting with ridge and furrow on yield of potato in semiarid areas. *Field Crop Research*. 84, 385–391.

Tiwari, S.P. (2014). Raising the yield ceiling in soybean.An Indian overview. *Soybean Research* 12(2), 1-43.

Varma, P. (2019) Production, Markets and Trade: A Detailed Analysis of Factors Affecting Pulses Production in India. Pg 13.

Venkidasamy, B., Selvaraj, D., Nile, A.S., Ramalingam, S., Kai, G. and Nile, S.H. (2019) Indian pulses: A review on nutritional, functional and biochemical properties with future perspectives. *Trends in Food Science & Technology* 88, 228–242.

Zhang J., Sun J., Duan A., Wang J., Shen X. and Liu X. (2007). Effects of different planting patterns on water use and yield performance of winter wheat in the Huang-Huai-Hai Plain of China. *Agriculture Water Management* 92, 41–47.

CHAPTER 14

Current pulses production trends, major constrains and management strategies for extension of pulses cropping area in North-Western Himalayas

Mahipal Choudhary*, Ram Prakash Yadav, Manoj Parihar, Sher Singh, Jaideep Kumar Bisht*, Vijay Singh Meena, Suresh Chandra Panday, Lakshmi Kant

ICAR-Vivekananda Parvatiya Krishi Anusandhan Sansthan, Almora-263 601, Uttarakhand, India

E-mail address: Jaideep.Bisht@icar.gov.in; Mahipal.Choudhary@icar.gov.in

Abstract

In spite of impressive growth of Indian agriculture food grain production from 50 Mt in 1950 to 292 Mt in 2020, ensuring household food and nourishing security is yet a challenge due to imbalanced development in agricultural production mainly in terms of one-sided wheat and rice only. In spite of the fact that production of pulses in the new decade has expanded but isn't in speed with the expansion in population. The pulses are such a class of food that has the power to eradicate malnutrition. In hills of North Western Himalaya (NWH) such as rajmash, blackgram, lentil, field pea, horsegram, rice bean, moth bean, pigeonpea, etc. are major pulse crops. In this context, pulse production scenario in India as well as NWH has been represented. Further, prominence has been given on pulses from food and nutritional security concern in hills of NWH and a strategy for enhancing productivity and potential of pulse crops in hills has also been explored.

Keywords: Pulses, Hills, Productivity, Biotic and abiotic constraints

1. INTRODUCTION

India is the largest producer (25% of global production), consumer (27% of world consumption) and importer (14%) of pulses in the world (Yadav *et al.*, 2019b). In India, pulses were cultivated over more than 29.8 million ha (M ha) of area and recorded the highest ever production of 25.2 million tonnes (Mt) with an average productivity of 841 kg/ha in 2017-18 (Choudhary *et al.*, 2017). The six states *viz.* Madhya Pradesh (7.81 Mt) > Rajasthan (3.68 Mt) > Maharashtra (3.0 Mt) > Uttar Pradesh (2.40 Mt) > Karnataka (2.0 Mt) > Andhra Pradesh (1.0 Mt) were the major pulse producers contributing around more than 75% pulses production (DES, 2020). In order to maintain self-sufficiency, it is expected that more than 39.0 Mt of total pulses will be required for the escalating population of India, which will grow to 1.69 billion by 2050. To achieve up to this level of pulses requirement, a yearly development pace of 2.2% is required (Singh *et al.*, 2017; Yadav *et al.*, 2019b). The pulses play a vital role in providing economical as well as nutritional food security because these are a significant wellspring of proteins, minerals, iron, zinc, folate, and magnesium for the entire veggie lover in the world. The pulses contain 22-26% protein, which is almost double of protein content in wheat and triple that of rice. Other than the huge wholesome and health benefits; pulses contain phytochemicals, saponins, and tannins possess antioxidant and anti-carcinogenic effects which are helpful to reduce some illnesses, *i.e.,* cardiovascular sicknesses and colon malignant growth (Mudryj *et al*., 2014).

The United Nations General Assembly had declared year 2016 as the "International Year of Pulses"in recognition of the importance of pulses for human well-being (Mann *et al*., 2020). Major pulses grown in India are chickpea, pigeonpea, blackgram and mungbean. Pulses can be cultivated along an extent of land and climatic conditions and assume significant role in crop diversification, mixed and inter-cropping, keeping up soil richness through biological nitrogen fixation, release of soil-bound phosphorus, and consequently contribute critically to the sustainability of the farming systems (Mukherjee *et al*., 2018).

Northern-Western (NW) Himalayas comprising of Jammu and Kashmir, Himachal Pradesh and Uttarakhand spreads to 33.0 M ha, and is habited by 25.0 million people and 19.0 million domesticated animals. Although, the entire economy of the locals depend on agriculture but cultivable land is just 3.20 M ha, which is absolute undulating topographical territory of the regions (Srivastva *et al*., 2009). NW Himalayas are equipped for supporting the development of an assortment of yields as a result of varied agro-climatic conditions. Notwithstanding, little and dissipated land holdings, and remoteness of villages, tainted watershed and scarcity of resources are some serious challenges to the food security of hill populations (Rajkhowa *et al*., 2015). Irrigation facilities are very limited and so agriculture of these regions is totally depended on rainfall and moreover, most of the rainfall occurs only in rainy season (Choudhary *et al*., 2020).

Pulses are mainly cultivated in rainfed areas that accounted for around more than 75% on marginal and sub-marginal land thus resulting in poor crop productivity (Choudhary, 2013). Furthermore, low profitability in numerous rainfed agricultural crops is frequently because of debased soil health and restricted water and nutrient supply at the time of different development stages (Panday *et al*., 2018). Other than this, absence of specialized information, inaccessibility of value seed and not following plant protection measures further exacerbate the issue of poor productivity and profitability in this region (Paul *et al*., 2011). The change in area from pulses cultivation to different harvests has been very noteworthy in these regions replaced by other grain crops chiefly rice and wheat. This decrease in pulses area has been generally ascribed to comparatively higher return from rice and wheat as compared to pulse crops (Singh *et al*., 2017). In this context, current pulses production trend, important constrains and management strategies for extension of pulses growing area in North-Western Himalayas biomes more important in present scenario.

2. PULSE PRODUCTION TREND IN INDIA

India is leading in the production of pulses. In 2017-18, 25.2 Mt of pulses were produced on 29.8 M ha land with productivity of 853 kg ha^{-1} (Agricultural Statistics at a glance, 2019) (Table 1). In spite of such a huge volume of pulse production, India still import pulses around 2.6 to 6.0 Mt every year for domestic consumption of growing population (DES, 2020; DoC, 2020) (Table 2).

Table 1. Area, production and productivity of total pulse crops in India

Year	Area (M ha)	Production (Mt)	Yield (kg/ha)	Area (%) under irrigation
1950-51	19.09	8.41	441	9.40
1960-61	23.5 (2.37)	12.7 (4.69	539 (2.25)	8.01 (-1.78)
1970-71	22.5 (-0.49)	11.8 (-0.79)	524 (-0.31)	8.80 (1.06)
1980-81	22.4 (-0.04)	10.6 (-1.17)	473 (-1.13)	9.01 (0.25)
1990-91	24.6 (0.04)	14.2 (3.32)	578 (2.25)	10.5 (1.73)
2000-01	20.3 (1.04)	11.1 (-2.76)	544 (-0.67)	12.5 (1.96)
2010-11	26.4 (2.93)	18.2 (5.69)	691 (2.69)	14.8 (1.89)
2017-18	29.8	25.4	853	-

Figures in parentheses indicates decadal growth rate. Source: Agricultural Statistics at a Glance (2019).

Table 2. Demand, supply and availability status of pulses in India

Year	Import (Mt)	Export (Mt)	Total availability for domestic consumption (Mt)
2016-17	6.61	0.14	29.6
2017-18	5.61	0.20	30.6
2018-19	2.23	0.26	25.2

Sources: DES, 2020; DoC, 2020

It is serious concern that India is not self-sufficient in pulses production over after 55 years of green revolution. Pulses in India recorded under 3% yearly development rate underway in the previous 63 years while its per-capita availability declined from 60 g/day in 1951 to less than 53 g/day in 2017, which is less than the recommendation of Indian Council of Medical Research *i.e.*,65 g/day/capita (Sinha *et al.*, 2019). It has been a drowsy yearly growth rate in the pulses cultivation area (0.52%) and productivity (1.2%) throughout the previous 63 years. Among the different pulses, chickpea alone contributes more than 40% share of the total production, followed by pigeonpea (15-21%), blackgram (10-15%), greengram (7.0-10%) and lentil (5.0-6.0%) (Table 3).

Table 3. Production trend of different pulses in India

Name of pulses	2015-16 (000 t)	% share of total production	2016-17 (000 t)	% share of total production	2017-18 (000 t)	% share of total production
Chickpea (Gram)	7060	43.2	9380	40.6	11230	44.5
Pigeonpea (Tur)	2560	15.6	4870	21.1	4250	16.8
Blackgram (Urad)	1950	11.9	2830	12.2	3560	14.1
Greengram (Moong)	1590	9.72	2170	9.38	2010	7.96
Lentil (masoor)	980	6.00	1220	5.30	1610	6.38
Others pulses	2210	19.5	2660	16.7	2570	10.2
Total pulses	16350		23170		25230	

Source: DES, 2020

3. PULSE PRODUCTION TRENDS IN NW HIMALAYAS

In NW Himalayas major pulse crops grown include rajmash, blackgram, horsegram, lentil, field pea, rice bean, moth bean, pigeonpea on nearly 109 thousand ha with the production of 122 thousand tonnes (FAI, 2018).

Table 4. Area, production and productivity of total pulse crop in North Western Himalayas in 2017-18

State	Area (000 ha)	Production (000 t)	Productivity (kg/ha)	Requirement (000 t)	Surplus/ deficit (000 t)
Uttarakhand	62.1	54.2	874	155	-101
Himachal Pradesh	28.5	57.5	2019	111	-53.5
Jammu and Kashmir	19.2	10.6	553	184	-173
NW Himalaya	109	122	1148	449	-327.5

Source: FAI, 2018

Average pulses productivity (1158 kg/ha) of NW Himalayas is far better from national average. Among the NWH states, area under total pulses in Uttarakhand (6,21,000 ha) is near more than double of Himachal Pradesh (2,85,000 ha) and more than triple of Jammu & Kashmir (1,92,000 ha) (Table 4) due to encouraging pulses production programmes conducted by the Uttarakhand state government in recent years. However, productivity of pulses is higher in Himachal Pradesh as compared to Uttarakhand and Jammu & Kashmir. Pulse productivity of Jammu & Kashmir is far low from national average. Neither India nor NWH states are excess in pulse production (Table 4). There is deficit in pulse production and requirement which is highest in Jammu & Kashmir (1,73,000 tonnes), trailed by Uttarakhand (1,01,000 tonnes) and Himachal Pradesh (5,35,000 tonnes).

4. PRODUCTION AND SUSTAINABILITY CONSTRAINTS OF PULSES CULTIVATION IN NW HIMALAYAS

There are so many pulses production constrains but biotic constrains have much more impact on reduction in productivity of pulses. Classification of various constraints is presented in Figure 1.

4.1 Infrastructural constraints

4.1.1 Non availability and acceptability of HYV of seed

A number of varieties of pulses are released in the country every year, however timely accessibility of seeds of released varieties to hills farmers due to very remote locations is a big constraint. Besides these, only a few varieties are well accepted by the farmers for cultivation (Mahajan and Shukla, 2006; Gowda *et al.*, 2013). The seed replacement rate is very low; it is about 2.0 to 6.5% against the target of 10%. The farmers are thus compelled to grow land races, which are poor yielder. Plant breeder primarily select the genotype for high economic yield (to increase economic benefit), resistant to diseases and pests (to minimize the inputs) and early maturity (to fit into the cropping system).

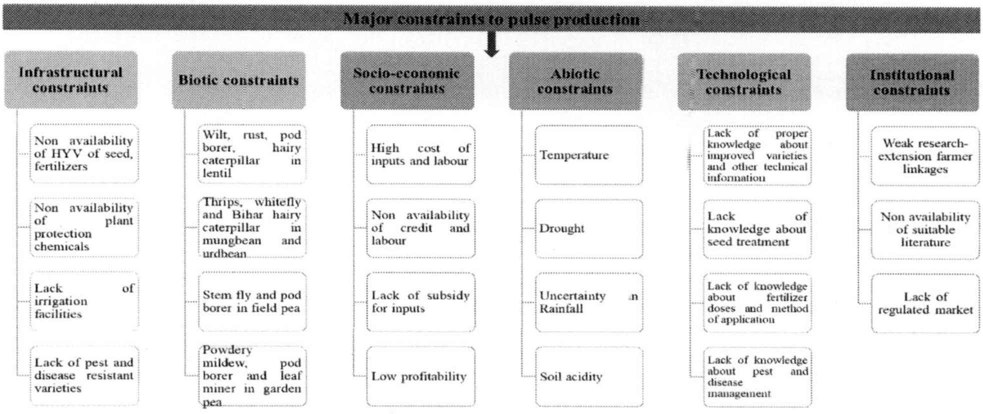

Fig. 1. Outline diagram of major constraints to pulses production

4.1.2 Non availability of balanced fertilizers

Although India is the largest consumer and importer of pulses in the world, but farmers in India treat pulses as secondary crop. Until recently, the government also gave less importance to pulses compared to the staple cereals like rice, wheat and maize. In general farmers' access to agriculture inputs is limited, both because of low purchasing power and accessibility to markets to purchase these inputs. Because of this situation, the farmers give first priority to staple cereals and cash crops for allocating inputs and the second priority to pulses. As a result, pulses continue to be grown on poor soils with deficient of several nutrients (Gowda *et al.*, 2013). Low use of organic manure and higher use of imbalanced fertilizers *i.e.,* urea leads to degradation of soil and reduces production. So, there is need to ensure ample supply of balanced fertilizers to farmers for achieving targeted production goals.

4.1.3 Non availability of plant protection chemicals

In the present era of climate change, it is important that region specific agro-advisories should be issued for guiding pulse growers on disease and pest control. However, pulse crops are very sensitive to many insect-pests and diseases; a major cause of concern is its incidence if not under control may demolish the entire crop. So, necessary plant protection chemicals are required. But good quality pesticides are very expensive and not accessible to farmers on time. Under this situation, if diseases and insects are not controlled timely it results into manifold yield loses.

4.1.4 Lack of irrigation facilities

Growing demand of water for raising the agricultural production, decreasing availability of water for irrigation and its low system performance, competing demand of water from industry and other sources and poor productivity from rainfed agriculture including hill farming, present serious challenge to sustainable pulse production as well as food production (Kumar *et al.*, 2010). The mountains are known as earth's water towers, but they face acute water shortages during the lean period (October to June) (Choudhary *et al.*, 2020). Suitable water conservation techniques are not available to supplemental irrigation during off season crop cultivation. If irrigation facilities are provided by the managed rainwater harvesting reservoir it increased research site's cropping intensity from 155 to 300% (Islam *et al.*, 2017).

4.1.5 Lack of pest and disease resistant varieties

Insect-pests and diseases are one of the majors constrains which are responsible for the low productivity of pulses. Pests are one of the major determinants for achieving higher production in agriculture crops. It is mainly because of very limited number of suitable disease and insect pest resistant varieties are available. Therefore, if farmers have pest and disease resistant varieties, it may avoid loss from pest and diseases.

4.2 Biotic constrains

The richness of pulses in N and P makes them attractive for insect pests and diseases (Gowda *et al.*, 2013). Production of major pulse crops is constrained by biotic stresses. For example, pod borers (*Helicoverpa armigera*), fusarium wilt, root rots, ascochyta blight and botrytis graymold are some of the major biotic constraints hindering in the increasing the productivity. The major constraints to productivity in pigeonpea are biotic stresses such as pod borer, pod fly, fusarium wilt, and sterility mosaic disease. Similarly, pod borer, black aphids, cutworm, powdery mildew, rust and wilt are the major pests and diseases affecting lentil production. Anthracnose and stem rot in horsegram; rust, wilt and powdery mildew in field pea are the challenging problems in hill pulses production (Mahajan and Shukla, 2006). It is estimated that harbouring insects consume about 26% of the potential production. It was reported that up to 30% crop losses in India is recording every year due to attacking of various insect-pests and diseases (Kumar and Bourai, 2012; Yadav *et al.*, 2019b).

4.3 Socio-economic constraints

High cost of inputs and labour, non-availability of credit and labour, lack of subsidy for inputs and low profitability are the major socio-economic constraints in pulses production (Subbarao *et al.* 2001).

4.4 Abiotic constrains

More than 87% of the area under pulses is presently rainfed (Reddy, 2009). Although, pulses have ability to withstand moisture stress, but during years with uncertain and erratic rainfall, timely planting and getting desired plant stand of the crop becomes very difficult. Because of high area of pulses under rainfed conditions, temperature and moisture regimes becomes most important factors controlling pulses production. High temperature and moisture stress greatly increase flowers drop and bud abortion causing considerable reduction in yield (Mahajan and Shukla, 2006). Terminal drought and heat stress results in forced maturity with low yields. Drought stress alone may reduce seed yields by 50% in the tropics (Kumar and Bourai, 2012).

Low temperature during night and frost, which are very common in hills, cause severe damage to *rabi* pulses, particularly to chickpea and lentil. Pulse crops are very sensitive to poor soil physical conditions water regime and micro-nutritional regime, which affect germination, plant growth and seed yield. Another side, excessive rainfall or water logging or moisture condition in *kharif* pulses at reproductive phase affects the physically damage of the flowers and developing pods/ seeds and also encourage foliar diseases such as Botrytis grey mould (BGM), rust and stemphylium blight in lentil (Subbarao *et al.*, 2001), ultimately reduces productivity. Soil acidity is a serious constraint to pulses production since Hill soils have pH value less than 6.0. Pulses are more sensitive to soil acidity as compared to cereals and other non-legumes crops (Choudhary and Pandey, 1986; Chong *et al.*, 1987).

4.5 Technological constraints

Inadequate crop management including broadcasting method of sowing leads to uneven plant stand and creates problems for adopting essential agronomic practices such as weeding and inter-culturing (Mahajan and Shukla, 2006). Lack of efficient *Rhizobium* cultures is an area of concern that affects the growth and yield of pulses. Lack of proper knowledge or technical know how about improved varieties, seed treatment, pest and disease management and other technical information is also one of the constraints of pulses production in NW Himalaya regions (Sinha *et al.*, 2019).

4.6 Institutional constraints

4.6.1 Poor researchers-extensioner-farmer linkages

A committed transfer of technology programme is very essential for achieving the goals in pulses production. The extension, training and human resources development components have to be re-energized for achieving self-sufficiency in pulses.

4.6.2 Lack of infrastructure for post-harvest storage

Post-harvest handling, processing, non-availability of proper storage facilities with farmers and the vulnerability of pulses, particularly to storage pests cause severe losses. Non-processed pulses cannot be stored for longer duration. Small scale processing units at the village level can effectively meet the local requirements. Post-harvest losses in pulses at different levels i.e., field, storage, transportation and distribution also reduce total pulses production. Vishwakarma *et al.* (2019) estimated the national-level harvest and post-harvest losses of pulses at INR 38.7 billion per annum at the average annual prices of 2014. The loss ranged between 6.36% (pigeonpea) and 8.41% (chickpea). Chickpea contributed to 63.3% of the total monetary value of pulses lost, and pigeonpea 24.7%.

4.6.3 Lack of regulated market

Prioritization of market linked cultivation of selected pulses is either lacking or insufficient. Assured procurement and adequate price for all pulses will go a long way in promoting pulses and oilseeds farming. It was observed that the linkage between the farmers to research and to the extension personnel was very weak and also found that there was no regulated market nearby or even in State level, where the farmers could sell their produce directly to the consumers (Sinha *et al*., 2019). In the changing market scenario in agriculture, a number of market factors are increasingly getting important, which resulted limited acceptability of the new varieties by the farmers. Thus, there is also a need to explore the agriculture produce to new markets and need to compete with the local produce.

5. STRATEGY FOR ENHANCING PRODUCTIVITY AND PRODUCTION OF PULSE CROPS IN NW HIMALAYAS

5.1 Breeding for stress conditions

Pulses are adapted to harsh environment due to acquisition of such traits, which help in overcoming the stress and completing the life cycle. Some of these traits, which include long maturity duration, indeterminate growth habit, excessive foliage and flower production, flower drop, non-synchronous maturity and pod shattering, are impediments in undertaking crop improvement work in the desired direction due to their varied genetic behaviour. In NWH, agro-climatic variations are quite significant. In such areas, the genotype and environment interactions are very important determinant of crop yields. Most of the breeding programmes for improved yields in pulses depend upon empirical selection for superior seed yield and quality across a wide range of target environments. While, direct selection for seed yield can be effective (White *et al.,* 1994), the approach is difficult and costly, and grains from the selection are often low in stress prone environments. Inheritance or repeatability of seed yield is very low, which is indicative that the observed variation that is attributable to genetic effects (G) is relatively small in comparison to variation observed due to environmental effects (E) and that just as different grain legume species respond differentially depending on environment (G x E) (Blum, 1988; Williams, 1992). Availability of large–scale plant biodiversity in NWH is a great wealth, much of which are yet to be evaluated and exploited. Farmers generally cultivate land races, which are suited to local conditions. The spread of improved and adapted varieties of pulses through aggressive mission-oriented seed replacement coupled with a vigorous quality seed production programme is essential.

5.2 Improved crop management

There is an ample scope for increasing the area under pulses in the hills. This can be achieved by suitable crop intensification, intercropping and modifying the cropping system. The horizontal increase in area is possible through diversion of area from cereals to pulses. Intercropping of pulses with cereals is a better alternative for bringing more area under these crops without much sacrificing the area under cereals. Gan *et al*. (2015) reported that pulse-based crop diversification increased total grain production by 35.5%, improved protein yield by 50.9% and enhanced fertilizer-N use efficiency by 33.0% over the summer fallow system. Further, studies revealed that inclusion of pulses-rich cropping patterns is important for increased yields and efficient input and farm management.

5.3 Diversified agroforestry system

There is a tradition in the Himalaya that farmers plant fruit trees on their agricultural lands (Yadav *et al.*, 2018). In the hilly areas of the Himalaya, the pulses can also be grown under different fruit trees in the form of agroforestry production management (Yadav *et al.*, 2019a). Hence, considering the nutritional importance of the pulse crops these have potential to be cultivated in interspaces of the fruit trees because the inter space under fruit trees is otherwise lying unutilised. Agri-horti systems not only meet the nutritional requirement of the farming families but also serve as a source of food, fuel and fodder. In fruit (17-year-old) based agri-horti system, the soybean in *kharif* and dual-purpose wheat during *rabi* season were grown with four fruit crops, viz., *Citrus limon* (hill lemon), *Pyrus communis* (pear), *Prunus domestica* (plum) and *Prunus armeniaca* (apricot). During initial years, no significant effect on seed yield of soybean was observed with the presence of different fruit trees. However, after 13 years of experimentation the presence of fruit trees affected significantly the seed yield of soybean. The seed yield of soybean was ranged from 1.25 to 1.69 t/ha under different fruit trees and in sole cropping the grain yield was 1.98 t/ha (Fig. 2).

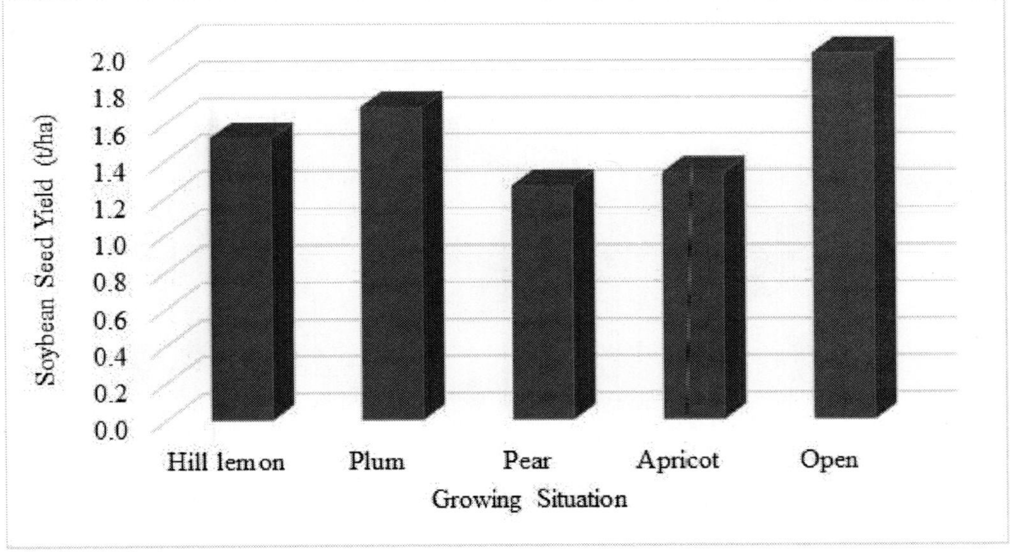

Fig. 2. Performance of soybean under different fruit based agri-horti system

Similarly, four cropping systems, *viz.*, finger millet (*Eluesine coracana*)–lentil (*Lens esculentus*), finger millet (*Eluesine coracana*)–wheat (*Triticum aestivum*), soybean (*Glycine max*)–lentil (*Lens esculentum*) and soybean (*Glycine max*)–wheat (*Triticum aestivum*) were evaluated under pecan and sole cropping (without pecan nut) in rainfed ecosystem (Fig. 3). Among different cropping systems significantly 16.4, 47.2 and 73.9% higher annual grain yield was recorded from soybean-wheat cropping system compared to finger millet-wheat, soybean-lentil and finger millet-lentil cropping systems, respectively.

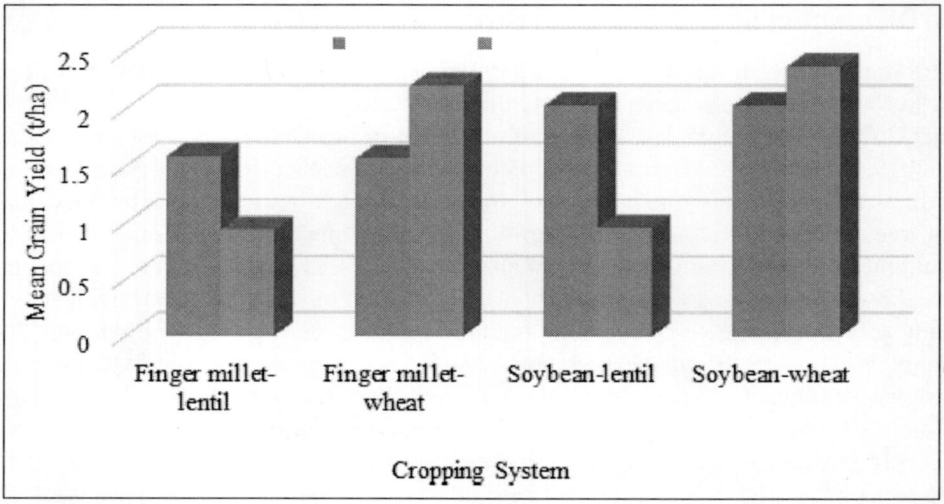

Fig. 3. Performance of different cropping systems under pecan based agri-horti system

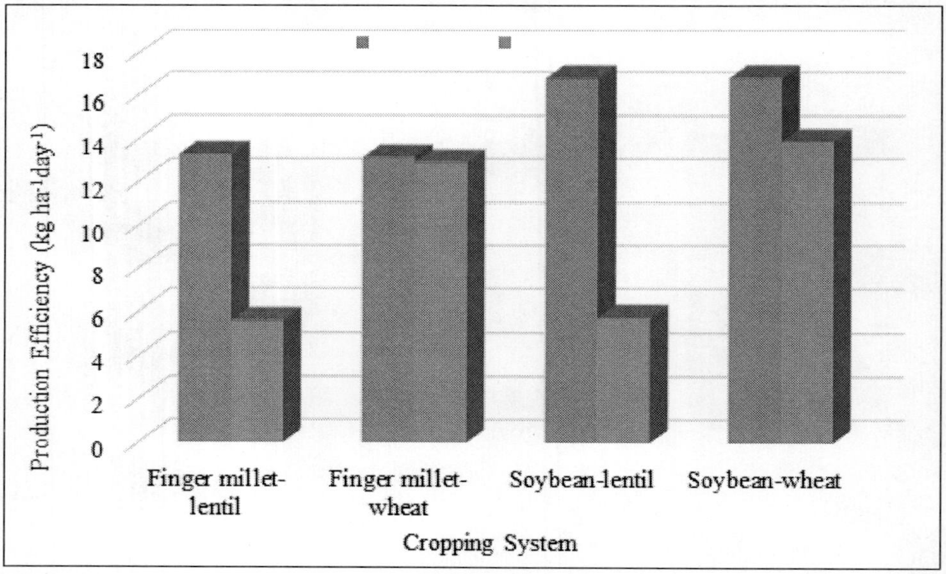

Fig. 4. Production efficiency of different cropping systems under pecan based agri-horti system

Likewise, mean grain production efficiency was recorded significantly higher in soybean-wheat system (16.78 kg/ha/day) compared to finger millet-lentil (13.28 kg/ha/day) and finger millet-wheat (13.13 kg/ha/day). However, it was at par in soybean-lentil cropping system (16.77 kg/ha/day) in *kharif* season (Fig. 4). During *rabi* season, significantly higher mean grain production efficiency was also recorded in soybean-wheat system (13.88 kg/ha/day) compared to finger millet-wheat (12.89 kg/ha/day), soybean-lentil (5.74 kg/ha/day) and least under finger millet-lentil (5.59 kg/ha/day) cropping systems.

5.4 Quality improvement and value addition

The improvement for quality components in pulses is very much required in the present-day context, when the consumers are increasingly becoming quality conscious. In soybean, development of 'green' soybean for table purposes is possible in lines with low lipoxygensase activity. The protein in many pulses cause flatulence and in chickpea, the puffing ability during roasting is variable among varieties. Besides major or common pulses, lesser-known legumes from North-Western Himalaya such as rice bean, horse gram and black seeded soybean are the excellent source of many nutritional compounds with unique medicinal significance. These lesser-known legumes contain lower content of phytic acid and trypsin inhibitor but higher antioxidant properties and unsaturated fatty acid, which make them potential source of human nutrition (Pal *et al.*, 2020). In order to popularize these pulses among users, emphasize must be towards quality, nutraceutical, value addition and health benefits of these traditional foods.

5.5 Physiological approach

Pulses produce high biomass with poor conversion and accumulation of photo-assimilates in the sink. The harvest index in pulses is low as compared to cereals. Improvement in harvest index would lead to increase in productivity. The determinate growth habit of many pulse crops is cause for reduced photosynthesis accumulation in the sink, as in indeterminate plants vegetative and reproductive sinks go hand-in-hand causing competition for photosynthesis. Flower drop and poor pod setting is another reason for low productivity. Various reasons such as limitation of photosynthesis and N availability, reduced light intensity in plant canopy, canopy temperature, hormonal factors, gaseous exchange in canopy, humidity in canopy and soil and water factors are responsible for poor pod set and needs to be addressed.

5.6 Biotechnological options

Use of biotechnological tools in conjunction with conventional breeding approaches is very much needed for development of multiple disease and pest resistant cultivars. In this regard, use of genetic engineering could be a potential tool for the management of insect pests and diseases of pulses. The study of genes for inherited traits such as larger pods, higher protein content, early maturity, resistance to biotic and abiotic stresses through proper genetic manipulation provide an effective alternative for considerable development in productivity and value of pulses. Molecular marker assisted breeding programme is another promising approach to identify key traits for better varietal response towards input response, pest management and pyramid resistance for better sustainability. Molecular marker helps in genomic mapping by identifying and locating genes precisely and being practised in many crops.

5.7 Policy interventions

The lack of market regulation is a major problem for the pulses growing farmers. Due to unorganized market, farmers receive non-remunerative prices for their produces and exploited by middlemen even after better quality of harvest. In this regard, initiatives like public procurement of pulse and coarse cereals as being followed in case of rice and wheat will provide competitive market to the pulse growing farmers and helping them to fetch a higher price. In present scenario, when government has no competent procurement policy for these crops, strict regulatory norms are required to control the market intermediaries. Overall, efficient and competitive agricultural marketing system is necessary for improving the growth of agricultural production and marketed surplus with greater benefits to the poor consumers. To strengthen the agricultural marketing system, investment in market infrastructure is most important which further requires significant investments in technology, management and policy intervention.

5.8 Improved varieties for NW Himalayas

A number of varieties of pulse crops have been developed for cultivation in the states of NWH (Table 5).

Table 5. Varieties of pulses for cultivation in NW Himalaya

Crop	Variety
Pigeonpea	Manak, HPA 1, UPAS 120, Sarita, VL Arhar 1
Horsegram	VL Gahat 1, VL Gahat 8, VL Gahat 15, VL Gahat 19
Urdbean	PantU-19, PantU-35, NarendraU-1, Kula 4, HPU 6, T 9, UG 218, PDU 1, Mash 1-1, UG 218
Moong bean Mungbean	PDM 54, PDM 11, Pant M-4, Narendra M-1, PS 16, PS 7, PS 10, K 857, Shining M-1, T 44
Lentil	VL Masoor 1, VL Masoor 4, VL Masoor 103, VL Masoor 125 (black small seeded), VL Masoor 126, VL Masoor 129, VL Masoor 507 (bold seeded), L 9-12, Vipasha, PL 406, PL 639, PL 5, T 36, VL Masoor 133, VL Masoor 514, VL 150
Rajmash	VL Rajma 63, IPR 96-4, Hans, Jwala, Him 1, HPR 35, VL Rajma 125
Field pea	Pant P 14, VL 64, VL Matar 42, VL Matar 47
Bengal gram Chickpea	C 235, PBG 1 Gaurav, C 235, C 214, H 355, H 208, G 130, K 468, L 550

6. CONCLUSION

Promotion of cultivation of pulses crops in hills is very important from the point of view of nutritional and economic security of the people of the regions. Incorporation of pulses in the cropping systems also contributes to environmental security to the large organic farming areas of hills. Judicious blending of available technology with the new one would be more promising in the development of high yielding varieties with satisfactory level of resistance or tolerance to biotic and abiotic stresses coupled with appropriate duration through the innovation tools of plant breeding. In order to improve the utilization of pulses in human consumption, optimum processing conditions need be worked out for various pulses-based products. In such ways, the NWH can be made self-sufficient in pulses production.

REFERENCES

Agricultural Statistics at a Glance. (2019). Government of India Ministry of Agriculture Department of Agriculture and Cooperation Directorate of Economics and Statistics. Published in India by Oxford University Press YMCA Library Building, 1 Jai Singh Road, New Delhi 110 001, India.

Blum, A. (1988). Plant Breeding for Stress Environments. CRC Press, Florida.

Chong, K., Wynne, V., Elkan, G.H. and Schneewies, T.G. (1987). Effects of soil acidity and aluminium content on Rhizobium inoculation, growth and nitrogen fixation of groundnut and other legumes. *Tropical Agriculture* 64, 97-104.

Choudhary, M., Panday, S.C., Meena, V.S., Singh, S., Yadav, R.P., Pattanayak, A., Mahanta, D., Bisht, J.K., and Stanley, J. (2020). Long-term tillage and irrigation management practices: Strategies to enhance crop and water productivity under rice-wheat rotation of Indian mid-Himalayan Region. *Agricultural Water Management* 232, 106067.

Choudhary, A.K. (2013). Technological and extension yield gaps in pulses in Mandi district of Himachal Pradesh. *Indian Journal of Soil Conservation* 41(1), 88–97.

Choudhary, M., Patel, B.A., Meena, V.S., Yadav, R.P. and Ghasal, P.C. (2017). Seed bio-priming of greengram with Rhizobium and levels of nitrogen and sulphur fertilization under sustainable agriculture. *Legume Research* LR-3837,1-6.

Choudhary, M.A. and Pandey, R.K. (1986). Extending limits to legume crop establishment in rainfed lowland rice. In: *Food legumes improvement for Asian farming systems: proceedings of an international Workshop held in KhonKaen, Thailand, 1 -5 September 1986.* (eds) Wallis, E.S., and Byth, D.E. ACIAR Proceedings no. 18. Canberra, Australia: Australian Centre for International Agricultural Research.235-239pages.

DES, (2020). Directorate of Economics and Statistics, Department of Agriculture, Cooperation & Farmers Welfare. https://eands.dacnet.nic.in/

DoC, (2020). Department of Commerce, Ministry of Commerce and Industry. https://commerce.gov.in/.

FAI, (2018). Fertilizer Association, of India http: //www.faidelhi.org/statisticaldatabase.

Gan, Y, Hamel. C, O'Donovan, J.T., Cutforth, H., Zentner, R.P., Campbell, C.A., Niu, Y. and Poppy, L. (2015). Diversifying crop rotations with pulses enhances system productivity. *Scientific Reports* 5, 14625.

Gowda, C.L.L., Srinivasan, S., Gaur, P.M. and Saxena, K.B. (2013). Enhancing the Productivity and Production of Pulses in India. In: Climate Change and Sustainable Food Security, (eds) P.K. Shetty, S. Ayyappan and M.S. Swaminathan. National Institute of Advanced Studies, Bangalore and Indian Council of Agricultural Research, New Delhi.

Kumar, P., Peshin, R., Nain, M.S. and Manhas, J.S. (2010). Constraints in pulses cultivation as perceived by the farmers. *Rajasthan Journal of Extension Education* 17&18, 33-36.

Kumar, S. and Bourai, V.A. (2012). Economic analysis of pulses production their benefits and constraints" (A case study of sample villages of Assan valley of Uttarakhand, India). *Journal of Humanities and Social Science* 1(4), 41-53.

Mahajan, V. and Shukla, S.K. (2006). Approaches to Improvement in Pulses and Oilseeds in Sustainable Production of Agricultural Watersheds in North Western Himalaya. In: *Recent progress in Hybrid Wheat research in India.* (eds) H.S. Gupta, A.K. Srivastva, J.C. Bhatt. Vivekananda Parvatiya Krishi Anusandhan Sansthan, Almora, India. 152-162 pages.

Mann, A., Kumar, A., Sanwal, S.K. and Sharma, P.C. (2020). Sustainable Production of Pulses under Saline Lands in India. In: *Legume Crops:Prospects, Production and Uses*. (eds) M. Hasanuzzaman. IntechOpen. DOI: http://dx.doi.org/10.5772/intechopen.91870.

Mudryj, A.N., Yu, N. and Aukema, H.M. (2014). Nutritional and health benefits of pulses. *Applied Physiology, Nutrition and Metabolism* 39(11), 1197-1204.

Mukherjee, A., Joshi, K., Joshi, P., Shubha, Roy, M.L., Jethi, R. and Chandra, N. (2018). Status of major pulses crop and its importance in nutritional security in North-Western Himalaya. *Climate Risks Management: Sustainable Pulse Production*. (eds) A.K. Srivastava and Yogranja. Astral International Pvt. Ltd., New Delhi. 169-180 pages.

Pal, R.S., Bhartiya, A., Kant, L., Aditya, J.P., Mishra, K.K. and Pattanayak, A. (2020). Common and lesser-known pulses from North-Western Himalaya: A comparison study for quality traits. *Legume Research* LR-3997, 1-8. DOI: 10.18805/LR-3997

Panday, S.C., Choudhary, M., Singh, S., Meena, V.S., Mahanta, D., Yadav, R.P., Pattanayak, A. and Bisht, J.K. (2018). Increasing farmer's income and water use efficiency as affected by long-term fertilization under a rainfed and supplementary irrigation in a soybean-wheat cropping system of Indian mid-Himalaya. *Field Crops Research* 219, 214-221.

Rajkhowa, D.J., Das, A., Ngachan, S.V., Sikka, A.K. and Lyngdoh, M. (2015). Integrated Soil and Water Resource Management for Livelihood and Environmental Security. ICAR Research Complex for NEH Region, Umiam – 793 103, Meghalaya, India.

Paul, J., Suri, V.K., Sandal, S.K. and Choudhary, A.K. (2011). Evaluation of targeted yield precision model for soybean and toria crops on farmers' fields under sub-humid sub-tropical NW Himalayas. *Communication in Soil Science and Plant Analysis* 42(20), 2452-2460.

Reddy, A.A. (2009). Pulses production technology: status and way forward. Economic & Political Weekly, 26 December, vol. xliv no 52. pp 73-80.

Sharma, D. and Jodha, N.S. (1982). Pulses Production in Semi-Arid Regions of India: Constraints and Opportunities. Economic and Political Weekly, vol. 17, no. 52, 1982, pp. A135–A148. JSTOR, www.jstor.org/stable/4371699. Accessed 15 Dec. 2020.

Singh, P., Shahi, B. and Singh, K.M. (2017). Trends of pulses production: a study on current scenario and strategies in India with special reference to Bihar. *Economic Affairs* 62(3), 389-398.

Sinha, P., Kumari, M., Kumar, S. and Ray, R.K. (2019). Constraints in Pulse Cultivation Perceived by the Farmers of Tal Land in Patna District of Bihar, India. *International Journal* of *Current Microbiology and Applied Sciences* 8(08), 2991-2997.

Srivastva, A.K., Bisht, J.K., Kumar, M., Singh, K.P., Pandey, S.C. and Gupta, H.S. (2009). Productivity Enhancement of Mountain Farming Systems in NW Himalaya. In Abstracts: 4th World Congress on Conservation Agriculture. 4-7 February 2009, New Delhi, India. Pp : 267.

Subbarao, G.V., Kumar Rao, J.V.D.K., Kumar, J., Debb, U.K., Ahmed, I., Harris D. (2001). Spatial Distribution and Quantification of Rice-Fallows in South Asia- Potential for Legumes. International Crops Research Institute for Semi-Arid Tropics (ICRISTAT), National Remote Sensing Agency (NRSA) and Department for International Development (DFID). pp 74-80. ISBN 92-9066-436-3.

Islam, M.T., Ullah, M.M., Amin, M.G.M. and Hossain, S. (2017). Rainwater harvesting potential for farming system development in a hilly watershed of Bangladesh. *Applied Water Science* **7:** 2523–2532.

Vishwakarma, R.K, Jha, S.N., Dixit, A.K., Kaur, A., Rai, A. and Ahmed, T. (2019). Assessment of harvest and post-harvest losses of major pulses in India. *Agricultural Economics Research Review* 32(2), 247-258.

White, J.W., Ochoam, R., Ibarrap, F. and Singh, S.P. (1994). Inheritance of seed yield, maturity and seed weight for common been (*Phaseolus vulgaris*) under semi –arid rainfed conditions. *Journal of Agricultural Science* 122, 265-273.

Williams, J.H. (1992). Concepts for the application of crop physiological models to crop breeding. In: Groundnut-A Global perspective, (eds) S.N. Nigam. ICRISAT, Patencheru, India.

Yadav, R.P., Bisht, J.K., Mondal, T., Meena, V.S. and Choudhary, M. (2018). Peach based horti-pastoral system of forage production on marginal land in Indian Himalaya. *Indian Journal of Agroforestry* 20(2), 63-67.

Yadav, R.P., Gupta, B., Bhutia, P.L., Bisht, J.K., Pattanayak, A., Meena, V.S., Choudhary, M. and Tiwari, P. (2019). Biomass and carbon budgeting of sustainable agroforestry systems as ecosystem service in Indian Himalayas. *International Journal of Sustainable Development & World Ecology* 26(5), 460-470.

Yadav, A.S., Kumar, S., Kumar, N. and Ram, H. (2019b). Pulses Production and Productivity: Status, Potential and Way Forward for Enhancing Farmers Income. *International Journal* of *Current Microbiology and Applied Sciences* 8(04), 2315-2322.

CHAPTER 15

Mitigation strategies for abiotic stresses in arid legumes

L.K. Rolaniya[1*], R.L. Jat[1], Narendra Kumar[2], Monika Punia[1] and S.K. Meena[1]

1. *ICAR- Indian Institute of Pulses Research, Regional Centre, Bikaner-334004*

2. *ICAR- Indian Institute of Pulses Research, Kanpur- 208024*

*Corresponding author email: lalitrolaniya@gmail.com

Abstract

The productivity of arid legumes is often threatened by abiotic stresses due to unusually high temperature and drought especially during podding stage. Majority of the pulses growing regions are vulnerable to abiotic stress as maximum threshold temperature for tolerance of pulses has already been reached beyond 35°C. Since drought, heat and salt are invariably recurrent phenomenon in pulse-growing regions, gene exploration from required germplasm for stress tolerance would remain the possible future hope for evolving newer varieties tolerant to multiple-abiotic stresses. Therefore, strategies have been made to develop climate-resilient varieties based upon screening of large number of germplasms. Early flowering, short duration, faster biomass accumulation, deep root-system, high water-use efficiency and high root proliferation before onset of terminal drought and heat have been found to be the desired strategies to escape the abiotic stresses. Improved agronomic strategies such as restructuring plant types, improved varieties, changing cropping pattern, weed management, efficient nutrient and water management, conservation agriculture, seed bank for alternate legume crops, watershed management, micro-irrigation etc. are some of the innovative options to mitigate the abiotic stresses in arid legume.

Keywords: Abiotic stress, Conservation agriculture, Climate-resilient varieties, Cropping pattern, Intercropping, Tillage

1. INTRODUCTION

There is now unequivocal evidence that global climate is changing and it poses the gravest challenge to agriculture and allied activities that are vulnerable to climatic hazards. Due to climate change and associated environmental stresses, agriculture in general and pulses (Shetty *et al.*, 2013) in particular, are likely to be confronted with new forms of risks and uncertainties which have serious implications for global food security. By 2080, the concentration of CO_2 in the atmosphere is accepted to increase to 735 ppm from the present level of 380 ppm and the temperature may rise by 3-4°C which may result in 10-40% reduction in crop production in South Asia (IPCC). Though impact of climate change is global, its adverse effects are likely to be more severe under Indian agro-ecological conditions as nearly 60% of the net area sown in the country is under dryland farming and most of the standard climate models predict severe impact of climate change to this sector (Pathak *et al*., 2012). The net result of this will be decreased productivity and stability of field crops of dryland farming, particularly pulses which are more sensitive to weather fluctuations and changes. Negative effects of climate change on dryland crops will pose a serious threat of food and livelihood insecurity.

The arid zone of India covers about 12% of the country's geographical area. A total of 32 M ha is hot arid and is spread over parts of Rajasthan (61%), Gujarat (20%), Andhra Pradesh and Karnataka (10%), and Punjab and Haryana (9%). The hot arid zone lying west of Rajasthan occupies the largest area (20 M ha) of the arid ecosystem of the country, and is popularly known as 'Thar Desert'. The hot arid region is characterized by scarce natural resources and an inhospitable climate. Rainfall varies from less than 100 mm to 400 mm, and is highly erratic and unpredictable. The region experiences extremes of temperature (-2 to 48°C), long day sunshine, high wind speed (35-40 km/h) and thus very high evaporation. Pearlmillet, cluster bean, mungbean, mothbean, seasame, and mustard are major crops grown under rainfed conditions or with limited irrigation (Behra and France, 2016).

Abiotic stresses are one of the major constraints to crop production and food security worldwide. The situation has aggravated due to the drastic and rapid changes in global climate. Heat and drought are undoubtedly the two most important stresses having huge impact on growth and productivity of the crops. It is very important to understand the physiological, biochemical, and ecological interventions related to these stresses for better management. A wide range of plant responses to these stresses could be generalized into morphological, physiological, and biochemical responses. Crop growth and yields are negatively affected by sub-optimal water supply and abnormal temperatures due to physical damages, physiological disruptions, and biochemical changes. Both these stresses have multi-lateral impacts and therefore, complex in mechanistic action. A better understanding of plant responses to these stresses has pragmatic implication for remedies and management.

1.1 Current scenarios of arid legumes

Generally, the annual food legumes grown in environments encountered with poor soil moisture, high temperature, low agronomic inputs and least aftercare are referred to as arid legumes. However, for convenience, arid legumes in India represent four minor legumes (Fig 1) viz., cluster bean or guar (*Cyamopsis tetragonoloba* (L.) Taubert), cowpea (*Vigna unguiculata* (L.) Walp), horsegram (*Macrotyloma uniflorum* (L.)) and mothbean (*Vigna aconitifolia* (L.) Marechal) exhibiting drought hardy and deep-rooted nature of great adaptive significance. These legumes exhibit wide geographic distribution in India from western dry states of Gujarat and Rajasthan to dry-wet states of Orissa and West Bengal in far-east and from Jammu & Kashmir in northern frontier regions to deep south in Karnataka and Kerala.

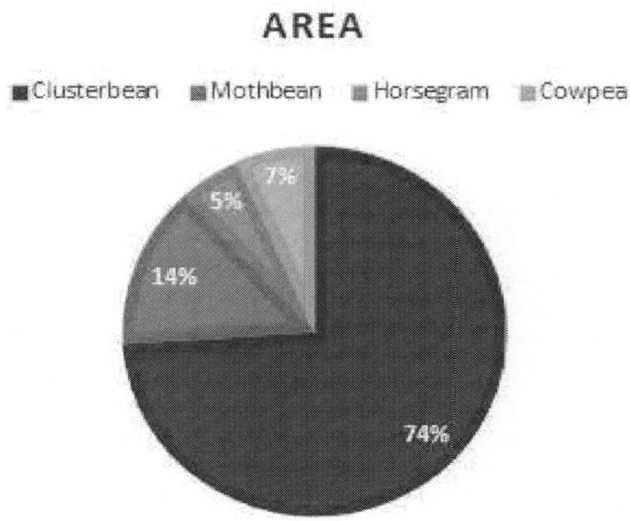

Fig. 1. Percentage share of area under arid legume in India

These crops are adjudged for sustaining growth and production, particularly, in drier-eco-systems confronted with harsh and hostile growing environments of unpredictable intensity and interval. The droughts and unhospitable conditions of arid environments are the ways of life in habitats of these crops. Century long need-based adaptation of these hardy legumes towards the fluctuating growing conditions have stabilized them as the most important and cheapest source of livelihood of farmers who have been sustaining on resource constraint arid farming for longer past. As a consequent, these are recognized for a variety of uses and products have therefore, occupied specific social and agricultural status in fragile agro-ecosystems. These legumes have particularly inclined towards least input and aftercare for reasonable biomass production. Contribution of legumes towards soil health build-up, diversification of agriculture and the source of organic foods and products have pushed them as front runner food legumes of the developing world. These so-called drought hardy legumes have, thus, assumed special significance in poor productive rainfed habitats.

1.2 Factors of yield reduction in arid legumes

Though irrigated agriculture plays a major role in producing 60% foodgrains from the 48% area, but due to several limitations, it is not possible to convert the whole cultivated area into the irrigated area. Therefore, arid ecosystem will remain important, but the average productivity of arid region is very low, which is less than one t/ha and tend to be relatively low compared to potentiality. Several constraints are responsible for the low productivity of arid legume.

1.2.1 Climatic constraints

Climate is one of the most important determinants of agricultural production. Arid legume grown in harsh and variable climatic condition, which is a major determinant of crop yields.

Low and erratic rainfall, high temperature, high solar incidence, high wind velocity, low humidity, weather aberrations, drought, cyclones and burning winds with high potential evapotranspiration rate make the unfavorable condition for the crop cultivation. Severe moisture stress can cause partial or complete crop failure. The growth of crops and the food production of the country are strongly influenced by the total rainfall. Rainfall aberrations during south-west monsoon continue to be a major factor contributing to instability in kharif crops production. The rainfall distribution is becoming more skewed with less number of rainy days, with high intensity causing more soil erosion. Besides rainfall variations, the distribution of rainfall is also changing in key arid agricultural locations and occurrence of the rainfall determines the sowing of the crops under dry condition. Delay in the onset of monsoons severely affects the crop yield due to delay in the sowing of the crops. Further, loss of soil moisture and its duration of availability in the soil, crop growth and duration of the crop is directly governed by the temperature. It highly influences germination, growth and development, physical and chemical processes within the plants. Solar radiation governs all physical process taking place in the soil, plant and environment and controls distribution of temperature. Therefore, climate is the most important factor that significantly influences the growth and yield of arid legume crops.

1.2.2 Edaphic constraints

Soils of the arid region are pre-dominantly poor and marginal lands, which are inherently poor in soil organic matter. (Singh *et al.*, 2011) reported 2.13 Pg carbon (C) stock in the 0-100 cm depth of which 1.23 Pg as soil organic C and 0.90 Pg as soil inorganic C in arid and semi-arid regions of Rajasthan. All these soils have a wide diversity in texture, structure and organic matter status, and soil depth. Variations found in the different soils induce differences in infiltration rate, water holding capacity, bulk density, aggregation, aeration, permeability and moisture, and nutrient release capacity. Moreover, arid and dry areas have uneven topography and poor quality of groundwater, which contains high soil pH, EC, and SAR. The abundance of sand makes the soil prone to erosion, as a result of low organic matter content. Many soil factors like soil moisture, soil temperature, soil air, nutrients, and soil reactions significantly governed by the soil type those directly or indirectly affect the plant growth and yield.

1.2.3 Technological and socioeconomic constraints

Adoption of modern agricultural technology by farmers is an important factor for improving the agricultural productivity in any kind of production systems. Limitations of adoption of improved production technologies and low adoption of technologies in the rainfed regions cause low crop production. The non-availability of the quality seeds of improved and latest released varieties is most important. Since, arid legumes are grown basically without plant protection and agronomic inputs, quality seed becomes the most crucial, critical and vital input for raising the productivity of these crops. Availability of the quality seed of the appropriate variety at the right time and at affordable price in the remotest corner of the country will help the farmers in increasing the productivity of these crops.

1.3 Reduction in yield due to abiotic stress

Various abiotic stresses, such as temperature, drought and salt, affect the growth of legumes at different developmental stages (Suzuki *et al.*, 2014). Abiotic stresses are the primary cause of crop losses worldwide, reducing the yield of most plants by >50% (Rodríguez *et al.*, 2006). Abiotic stresses result in a series of morphological, physiological, biochemical and molecular alterations, which negatively influence plant growth, productivity and yield (Bita and Gerats, 2013). Plants experience multiple effects of these stresses including physiological functions such as photosynthesis, respiration, nitrogen fixation, reproduction, and oxidative metabolism (Farooq *et al.*, 2008). The yield losses occurred in pulses exposed to abiotic stresses at different growth stages are given in Table 1.

Table 1. Yield losses (%) in important legumes under various abiotic stress.

Pulse Crops	Growth Stage	Yield loss (%)	Reference
Chickpea	Reproductive phase	45-69	Nayyar *et al.*, 2006
	Ripening stage	49-54	Samarah *et al.*, 2009
	Anthesis stage	27-40	Mafakheri *et al.*, 2010
	Ripening stage	50	Varshney *et al.*, 2014
Cowpea	Reproductive phase	60	Ogbonnaya *et al.*, 2003
	Reproductive phase	34-66	Ahmed *et al.*, 2010
	Pod filling stage	29	Kyei-boahen *et al.*, 2017
Mungbean	Reproductive phase	26	Al *et al.*, 2012
	Flowering stage	31–57	Ahmad *et al.*, 2015
Mothbean	Vegetative	52.4	Garg *et al.*, 2004
	Flowering	59.3	Garg *et al.*, 2004
Horsegram	Vegetative	52	Maruthi *et al.*, 2018

2. TYPES OF ABIOTIC STRESS

Climate change, food shortage, water scarcity, and population growth are some of the threatening challenges being faced in today's world. Abiotic stresses, such as high temperature (heat stress), deficient water (drought stress), and high salinity are hostile to plant growth and development, leading to great crop yield penalty worldwide.

2.1 Heat stress

Temperature is one of the major factors affecting the yield and quality of legumes (Christophe *et al.*, 2011). Heat stress often is defined as high temperature that causes irreversible damage to plant function or development after a certain period of exposure. Plants can be damaged in different ways by either high day or high night temperatures, and by either high air or high soil temperatures. Also, crop species and cultivars differ in their sensitivity to high temperatures. High temperature may negatively affect photosynthesis, respiration, water relations and membrane stability, and also modulate levels of hormones and primary and secondary metabolites. Furthermore, throughout plant ontogeny, enhanced expression of a variety of heat-shock proteins, other stress-related proteins, and production of ROS constitute major plant responses to heat stress (Hasanuzzaman *et al.*, 2013).

Heat-stressed plants show shorter vegetative and pod-filling periods, poor crop stand and consequently reduced yield. High-temperature stress affects reproductive development, as reported in legumes such as chickpea (Kaushal *et al.*, 2013), mungbean (Bindumadhava *et al.*, 2017) and cowpea (Ahmed *et al.*, 1992). High temperature negatively affects flower initiation, pollen viability (germination and tube growth), stigma receptivity, ovule viability, ovule size, fertilization, seed/fruit set, seed composition, grain filling, and seed quality (Barnabás *et al.*, 2008). The effects of heat stress at reproductive stage are summerised in Fig 2.

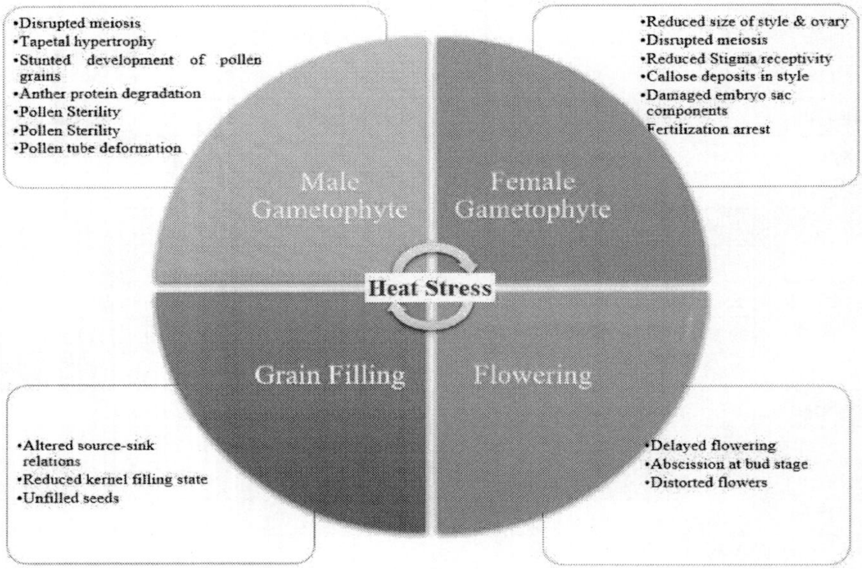

Fig. 2. Various effects of heat stress on reproductive development stages in legumes

2.2 Salinity stress

The adverse effect of excess minerals such as Na+ and/or Cl− on plant is called salt stress (Munns, 2005). Salt stress is one of the most serious limiting factors for crop growth and production. Grain legumes are grown in tropical and subtropical regions, while others are grown in temperate regions and semi-arid areas. Soybean (*Glycine max*), mungbean (*Vigna radiata*), cowpea (*Vigna unguiculata*), Mothbean (*Vigna aconitifolia*) and horsegram (*Macrotyloma uniflorum*) are important crops in many tropical countries, but in some regions of the semi-arid tropics, soil salinity can limit grain legume productivity. In general, salinity stress has a negative effect on growth, nodulation, quality and quantity of seed, and final yield in the grain legumes (Fig 3). As such, salinity stress affects reproductive attributes, such as flower number, pod development, seed number, and seed size (Flowers *et al.*, 2010).

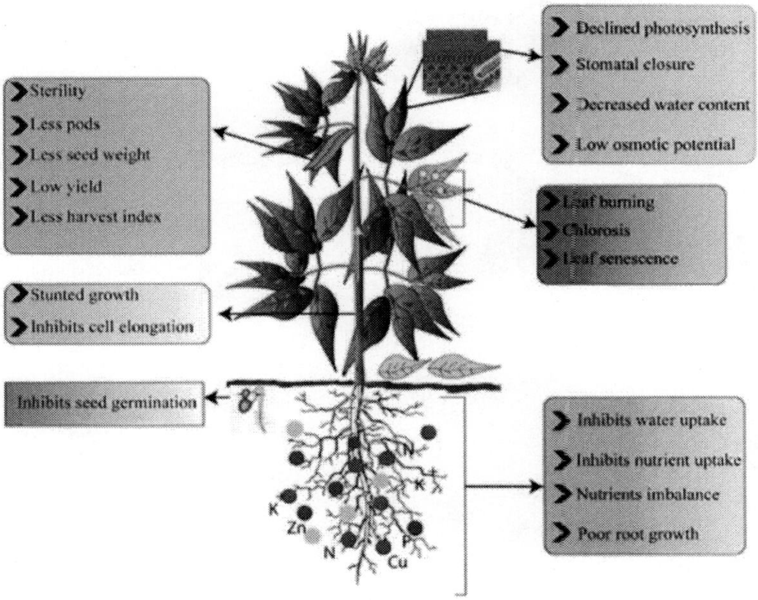

Fig. 3. Schematic representation of plant response to salt stress

In addition, salinity inhibits plant growth due to low external water potentials, ion toxicity, and ion imbalance. Excess ions result in an "osmotic effect" on plant water relations, and specific ions can have individual or synergistic effects on ionic relations/toxicity and nutritional imbalances. Ion toxicity and cation imbalance are the most destructive for grain legumes, as these species are sensitive to salinity levels with only modest levels of osmotic potential (Khan *et al.*, 2016). In clusterbean salt, stress negatively affects all the major processes such as germination, growth, photosynthetic pigments and photosynthesis, water relation, nutrient imbalance, oxidative stress, and yield. Increased salinity caused harmful effects on germination and seedling growth in terms of germination per cent, speed of germination, root length, shoot length, dry matter production and vigour index (Prakash *et al.*, 2021).

2.3 Drought stress

Legumes vary in their responses/sensitivity at the onset of drought, but in all cases, final yield is significantly decreased. This is linked with reduced germination and declined photosynthetic activity, decreased assimilate translocation and carbon fixation (Mondal *et al.*, 2011), repressed flowering time and an effect on reproductive organs, pollen grain sterility, fewer pods and lower grain set and declined sink activity (Andersen *et al.*, 2015). Drought affects several aspects of legume growth and development, including germination, shoot and root development, photosynthesis, and the reproductive stage. Due to global climate change, drought has become one of the most uncontrolled and an unpredicted factor which is continuously limiting crop production and has adverse effects on legume crops (Fig 4).

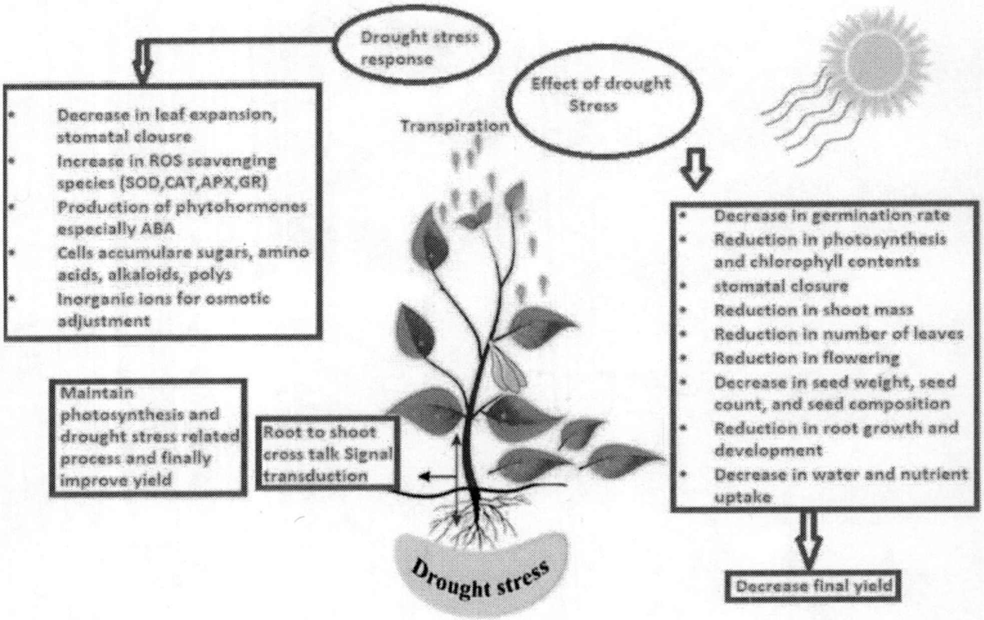

Fig. 4. Adverse effects of drought on legume crops

Studies showed that severe drought conditions disturb plant morphology, physiology, and growing period, whereas moisture contents play an essential role in enzyme activation during germination which could help to elucidate the sensitivity of plants to drought at the germination stage. Germination and reproductive stages are highly sensitive to water deficit. Under drought stress, the germination rate was significantly reduced in soybean (Heatherly, 1993). Awari and Mate (2015) noted a decrease in the germination rate in chickpea under a water deficit. In mothbean most of the flowers wilted and dropped off after opening due to the drought stress, as a result less fruit set takes place and ultimately reduces the yield (Malambane *et al.*, 2014).

3. MITIGATION OPTIONS FOR ABIOTIC STRESSES

The key aspect of sustainable development in agriculture involves resource conservation-based strategies, cropping system-based strategies and exploitation of genetic resources. Soil degradation should be prevented by practices and techniques, such as no-till sowing of crops, drip irrigation, crop rotation and leaving land fallow. Suitable farming systems that have potential to increase food production and promote soil conservation for each agroecological zones should be identified. Holistic land management and soil health restoration are also one of the key aspects of suitable development towards managing food security. In addition, maintenance of soil life and soil quality through practices such as organic fertilizer supplementation and judicious use of chemical fertilizers and pesticides should be a priority. Judicious management of water on a water harvesting and conservation basis should be undertaken to make maximum use of available water. Biotechnological improvement of crops and evolution of crop varieties suitable for abiotic stresses also form the key for sustainable development. Tapping of genetic resources to identify varieties and new crops that can adapt and cope with stresses risk also forms a key approach in sustainable development to tackle the problem of food security. A comprehensive understanding of abiotic stress, especially the mechanism and tolerance aspects for adaptation strategies, across the full range of warming scenarios and regions is essential for preparing for climate change. Therefore, a judicious mixture of basic and applied research outlooks has been presented on developments in management of abiotic stresses in arid legumes.

3.1 Plant breeding and biotechnological approaches

Identifying stress-tolerant traits is critical to develop stress-tolerant cultivars through plant breeding. Adaptive mechanisms of plants responding to environmental stress include escape, resistance by avoidance, and resistance by tolerance. The adaptive trait responses to various abiotic stresses are listed in Table 2. The most effective strategy is to escape stress. Early flowering and/or early maturing cultivars may escape the heat and/or drought conditions. Management and breeding approaches, such as shifting planting date and genetic manipulation of crop phenology, have been used to facilitate plants to escape from heat stress

Table 2. Potential traits/characters for screening legumes for drought resistance

Legume crops	Trait	Reference
Chickpea	Prolific root system, rooting depth, root length	Ramamoorthy *et al.*, 2017
	Shoot biomass, leaf area index, canopy temperature decrease	Ramamoorthy *et al.*, 2016
	Smaller leaf area	Serraj *et al.*, 2003
	Grain size, early maturity and short stature	Singh *et al.*, 1995
Cowpea	Short duration and erect plant type	Hall, 2012
Mungbean	Dry matter partitioning	Hossain *et al.*, 2010
Mothbean	Semi erect plant habit and early maturity	Vir and Singh, 2015

Selection for one abiotic stress tolerance in the field is very challenging due to interactions among the different stresses. Thus, the only strategy to identify the traits to be applied in field for breeding tolerant genotypes is by performing experiments under controlled environment conditions. Regardless of the screening method, a key objective for plant breeders is to develop an effective set of stress combination markers that can be used to improve legume crop species. Controlled environmental conditions allow the dissection of each one of different stress effect and the identification of principal targets affecting plant tolerance.

Breeding for stress tolerance requires efficient screening procedures, identification of key traits in diverse donor or tolerant lines and understanding their inheritance and molecular genetics. Statistical package applied to plant breeding will facilitate the identification of markers in a multi-trait multi-environment way (Malosetti *et al.*, 2004). Several quantitative trait locus (QTL) studies relating to various abiotic stress tolerances have already been reported showing it is possible to improve and accelerate the breeding process in plant species without sequenced genomes (Chandra *et al.*, 2004). In order to transfer these traits, classical breeding requires the establishment of rapid and cost-effective screening procedures and implementing these using breeding approaches such as association mapping or genomic selection procedures. For the complete sequencing of the different important legumes, genome opens the possibility of fine mapping of the QTLs. In this perspective, gene identification for combined stress tolerance in legumes using genetic map information and genome data is an achievable goal (Heffner *et al.*, 2009; Hirayama and Shinozaki, 2010). Phenotypic and physiological characterization along with RNA sequencing analysis of plants subjected to drought, heat and salt stress or their combination would confirm that the simultaneous imposition of different types of stress presents unique but varied aspects that includes alteration of respiration rate, decreased photosynthesis, stomatal closure, high leaf temperature and redox homeostasis. Thus, deep phenotyping methodologies, genome-based selection and massive RNA sequencing technologies emerge as a promising avenue for the development of multiple abiotic stress-tolerant crops.

3.2 Physiological approaches

Recently, exogenous applications of protectants in the form of osmoprotectants (proline, glycine betaine, brassiosteroids, SA, phytohormones (ABA), signaling molecules, trace elements (selenium, etc.) and nutrients (phosphorus, potassium etc.) have been projected and found effective (Akhtar *et al.*, 2015) in mitigating heat stress-induced damage in plants. These molecules show promise in protecting plants from adverse effects of temperature stresses (Awasthi *et al.*, 2015) and impart defense by managing the ROS through upregulation of antioxidant capacity. In a few controlled experiments, application of proline and ASC conferred protection to heat-stressed plants of chickpea (Kaushal *et al.*, 2011) and mungbean (Kumar *et al.*, 2011), respectively. The positive effects of these molecules may be associated with reduction in damage to leaf and root tissues due to oxidative stress and activation of various antioxidants.

Proline and ASC improved the growth of heat-stressed plants significantly, thus indicating the potential of these thermo-protectants. The bean seeds pre-treated with glycine-betaine led to plants with lower membrane damage, better photosynthetic rate, improved leaf water potential and greater shoot dry biomass compared to untreated seeds. In tomato, exogenous application of 4 mM spermidine improved heat resistance through better chlorophyll fluorescence properties, under heat stress and higher level of Ca^{2+} is required to mitigate adverse effects of the stress.

The exogenous foliar application of 30 μM MeJA at the reproductive stage ameliorated drought tolerance of cowpea and led to increased growth and seed yield under drought stress conditions. The beneficial role of MeJA in increasing drought tolerance of cowpea was related to improving water status and chlorophyll value, enhancing osmolytes (proline and soluble sugars) accumulation and reducing membrane lipid peroxidation Sadeghipour (2018).

The sulphydryl compounds sprayed plants had 24-70, 13-40 and 13-20% higher activities of antioxidant enzymes (glutathione reductase, GR; gluthathione-s-transferase, GST), photosynthetic parameters (net photosynthetic rate, PN; stomatal conductance, gs; transpiration rate, E) and seed yield respectively compared to unsprayed plants. The application of sulphydryl compounds mitigate moisture deficit induced negative effects on growth, and photosynthetic parameters which lead to better seed yield of mothbean (Nathawat *et al.*, 2021).

3.3 Agronomic interventions

Agronomic approaches such as adjustment in time of sowing, plant geometry, and fertilizer management can help to improve drought tolerance (Siddique *et al.*, 2012). The main objectives of these management approaches are to increase rainfall use efficiency, to improve total crop water use via transpiration and to lessen the loss of water by evapotranspiration and weeds. Soil structure may be improved by the application of gypsum, which helps flocculate soil particles to enhance water infiltration and root growth. Other practices such as mulching, zero tillage, and deep ploughing in the rainy season are critical agronomic practices to alleviate the adverse effects of abiotic stress in legumes (Johnsona *et al.*, 2018).

3.3.1 Sowing time

Timely sowing of crops plays a vital role in maintaining adequate plant stand, growth and development of the crop. In arid regions, sowing time considerably influences the thermal time requirement of the crop as it modifies the photoperiod, temperature and radiation regimes besides, photo-climate of the crop. However, sowing time in rainfed conditions depends on onset of monsoon. The normal onset time of monsoon in arid zone of Rajasthan is last week of June to first week of July. Delayed sowing reduces the seed yield of greengram, mothbean (Bhati, 1989), cowpea and clusterbean (Sharma *et al.*, 1984). Therefore, sowing time is an important variable that significantly influences the yield of the crops.

3.3.2 Selection of crops

Probability of occurrence of different amounts of rainfall and corresponding length of growing season could be taken as a criterion for selecting suitable crops and cropping systems in an agroclimatic region for sustain the agriculture production and livelihood (Fig 5). Crops and cultivars differ not only for total water requirements but also in their soil moisture extraction pattern. For instance, mothbean (*Vigna aconitifolia*), horsegram (*Macrotyloma uniflorum*) are the most drought tolerant crop for kharif season. Clusterbean (*Cymaopsis tetragonoloba*), is also drought resistant and utilizes moisture quite effectively due to its deep tap root system. Suitable short duration, high yielding cultivars of pulses adapted to different rainfed areas/regions were identified and recommended. Purity of seed assumes greater significance during years of low rainfall.

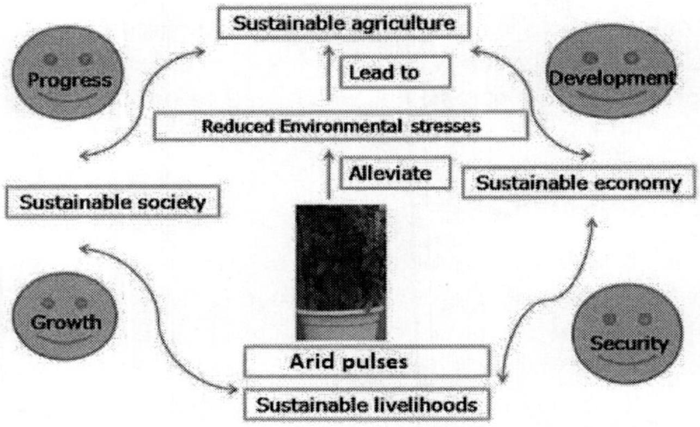

Fig. 5. The 'sustainable agriculture' as sustained by the natural drought or other abiotic stress tolerant plants like arid pulses.

3.3.3 Selection of suitable variety

The genetic manipulation to encounter the drought stress like, drought escape, drought avoidance and drought tolerance is necessary to develop drought tolerant or resistant crop plant (Maqbool *et al.*, 2017). The development of drought tolerant variety is necessary to fulfill the food demands of ever-growing world population which possible through continuous evaluation of genotypes for stress tolerance in drought prone regions and selection of genotypes on the basis of yield performance. Choice of a suitable variety plays paramount role for obtaining efficient and successful crop production in arid zone (Table 3). Peak rainfall is received during July and August and generally recedes by first fortnight of September. Thus, the rainy season in the region remains only for 50-60 days (Rao and Singh, 1998) showing the limitation of cropping period. Hence, it is pre-requisite that variety should mature within this limited period of moisture availability.

Table 3. Abiotic stress tolerant varieties of arid legumes

S.N.	Crop	Variety	Specific traits
1.	Cowpea	RC-101	Extra-early in maturity drought tolerant, determinate type, non-viny, white seeded, suited to low rainfall regions, high yielder
		DCS 47-1	Suited for cultivation under light and loamy soil, sowing time during late kharif, resistant to rust
		Phule Rakhumai PCP 0306-1	Early maturing, suitable for optimum sowing under rainfed and irrigated condition, moderately resistant to *Cercospora* leaf spot
2.	Clusterbean	RGC-936	Early, branched type, dwarf wide adoption, drought hardy, high yielding, medium sized grains with light pink colour, most popular variety of Rajasthan
		RGC-197	Branched type, suitable for arid region
		RGC-1003	Short duration, Branched type, suitable for arid and semi arid regions

3.	Horsegram	CO-1	Suited for rainfed areas, trailing type, brown seeded,
		Pratap Kulthi-2 (AK 53)	Extra-early maturing, lowest tannin content
		Phule Sakas (SHG-0628-4)	Resistant to YMV and escape drought due to earliness.
4.	Mothbean	RMMO-40	First early variety, erect growth, suited to low rainfall, widely grown in low rainfall zones, short stature, less biomass
		CAZRI Moth-3	Erect, early drought tolerant, escapes YMV diseases, heavy podding behavior
		RMO-257	Early maturing, moderately tolerant to YMV and bacterial blight

3.3.4 Seed rate and spacing

Seed rate plays a vital role in maintaining the adequate plant population, which is a pre-requisite for obtaining better yields. Seed rate depends on several factors, viz. Plant architecture, method of sowing, soil moisture, seed size and purpose for which crop is grown. Use of seed in adequate amount would be helpful in maintaining the optimum plant stand and yield. Row spacing of 45 cm in early-sown clusterbean (Rana *et al.*, 1991), 30-40 cm in cowpea (Anitha *et al.*, 2004), and 45 cm in mothbean (Yadav and Beniwal, 2006) are recommended in arid region.

3.3.5 Conservation agriculture

FAO defined goals of conservation agriculture (CA) as follows: "CA aims to conserve, improve, and make more efficient use of natural resources through integrated management of available soil, water, and biological resources combined with external inputs. It contributes to the environmental conservation as well as to the enhanced and sustained agricultural production. Therefore, it can also be referred to as resource efficient or resource effective agriculture".

Three principles of CA:

- Minimal soil disturbance by direct planting through the soil cover without seedbed preparation;
- Maintenance of a permanent vegetative soil cover or mulch to protect the soil surface;
- Diversified crop rotations in the case of annual crops or plant associations in case of perennial crops.

Conservation agriculture incorporates basis to minimize runoff and greater water infiltration, mulching to reduce water loss by soil evaporation, timely planting on the first rains, application of manure to provide better soil structure, top dressing with fertilizer depending on rainfall, rotation with legumes, and timely weed control, can increase yields (upto 100%) and profitability over conventional farming (Mazvimavi *et al.*, 2008). Legumes are ideal crops for two components of CA viz., soil cover and crop diversification, either as a growing crop or as residue. The benefits of food legumes in cereal-based cropping systems have been discussed in detail by Siddique *et al.* (2008) and are summarized as follows. The major advantage of legumes is the addition of fixed nitrogen (N) to the soil and non-depletion of existing soil N reserves because of their ability to meet their own N needs through fixation. Therefore, zero/minimum tillage and maintenance of soil cover in the form of crop residues or cover crops are important elements of CA. At the same time varied crop rotations involving legumes, are important to manage pest and disease problems and improve soil quality through biological nitrogen fixation and addition of organic matter (Baudron *et al.*, 2009).

3.3.5.1 *Tillage management*

Proper soil tillage is a prerequisite for good crop stand, growth and yield. Excessive tillage, particularly in light textured soils, disintegrates clods and exposes soil to wind and water erosion. Reduced tillage implies economy intime, labour and energy besides reduced soil moisture losses, maintenance of soil structure and increased cropping intensity. Sandy soils are vulnerable to erosion due to excessive tillage, whereas reduced tillage results in compaction due to low organic matter content of sandy soils, which seriously affects crop growth and yield. In zero tillage, seed is put in the soil without any soil disturbance through any kind of tillage activity or only with minimal soil disturbance, with time soil life takes over the functions of traditional soil tillage like loosening the soil and mixing the soil components. In addition, increased soil biological activity creates a stable soil structure through accumulation of organic matter. As against this, mechanical tillage disturbs this process.

3.3.5.2 *Residue management*

Crop residues are important source of nutrients and also improve the physical and biological properties of soils. Decomposition rate and release of N from residues depends on soil, climatic conditions and the C: N ratio of plant residues. Crop residues decomposition rate is slow in arid climates because of limited water availability. Consequently, the entire N requirement of a crop can seldom be met from decomposition of crop residues during the growing season. Therefore, an integrated nutrient supply from inorganic and organic N sources may be a more compatible approach under arid conditions. Amgain (2018) found that clusterbean after chickpea and wheat with Leucaena twigs mulching could be high yielding and economically feasible under semi-arid dry land condition.

3.3.5.3 *Cropping system*

To sustain the arid agriculture, efficient cropping system can contribute to a great extent. An effective crop rotation not only helps to increase the crop productivity and soil fertility, but also improves the water use efficiency by reducing weeds, providing conducive micro-climate for plant growth and development, reduction in soil thermal regime and improving physical properties of the soil. The study carried out by Singh (2006) indicated that greengram - mustard cropping sequence resulted in 18.2% higher weed control efficiency as compared to fallow-mustard sequence under western Rajasthan conditions. Pearlmillet - clusterbean rotation gave 11% higher pearlmillet yield compared to continuous pearlmillet cultivation (Singh *et al.*, 1985).

3.3.5.4 *Intercropping*

Since the rainfall is quite low and erratic, farmers of the region hesitate to invest money in agriculture due to risk of crop failure. For minimizing the risk of climatic vagaries, farmers traditionally practice mixed cropping. Though mixed cropping has many demerits and limitations, yet it is practised in arid region as it provides higher equivalent yields and benefit: cost ratio than sole cropping (Singh, 1995). Growing of two or more crops simultaneously in rows instead of mixing the seed of all crops can give better yield without affecting the goal of mixed cropping. Intercropping improves soil fertility and has yield advantage in comparison of cereal crops, besides providing insurance against complete crop failure and high sustainable yield compared to the sole cropping. Intercropping provides higher moisture use efficiency and grain yield under moisture stress conditions as it minimizes the water losses at early vegetative stage over mixed cropping. Intercropping of mothbean with pearlmillet in 3:1 row ratio gave higher seed yield of mothbean than 2:1 and 2:2 row ratios without much affecting pearlmillet grain yield. This combination also provided highest monetary returns. Double rows of dew gram, clusterbean and greengram planted in the inter-spaces of paired rows of pearlmillet (30/70 cm) yielded 381, 381 and 458 kg/ha, additional seed yield respectively without affecting yield of pearlmillet (Kushwah *et al*., 2017).

3.3.6 Water management

Irregular or insufficient rainfall can be a serious limitation to agricultural production causing low yields and even crop failure. This is particularly true in drylands, where productivity levels are generally low. In most cases, a great deal can be done to improve the efficiency of rainwater use. In arid and semiarid regions, uncertain erratic and scanty rains coupled with meager irrigation facilities leads to low and unstable yields. Low and erratic rainfall, high evaporation rate and limited water holding capacity of surface soils are the main constraints in agricultural production

The most important way to increase the grain yield of food legumes per unit area under stress environment should be consist of proper technological backup with infrastructure, timely availability of quality inputs along with policy support. Efficient water management is one of the critical inputs as in general perception is that legumes need no supplementary water, whereas research finding revealed that need based watering at critical stages are capable to improve production by 15-25% depending up how much stress has already been faced by the standing crop till now.

3.3.6.1 *Moisture and water conservation*

Water is the most precious input for arid agriculture. The availability of surface water resources is very limited due to hostile climatic conditions and poor water production efficiency of the sandy terrain in the region. The underground water is very deep, limited and mostly brackish in nature. Irrigation with saline water restricts the choice of the crops and varieties. The only major source of good quality water in arid region is rainfall. Therefore, its proper conservation, distribution and use need special consideration for realizing sustainable crop production in the region. Efficient rainwater management acts as an insurance for the crop during the rainfall deficit periods. Management techniques that increase infiltration and soil water storage, and decrease water losses by runoff, evaporation and evapotranspiration by weeds would lead to increase the amount of water retained in the soil for subsequent use by crops. The rainfall with intensity of 56-65 mm/h for 18-20 minutes in arid zone produces significant runoff. The excess rainwater is lost as runoff causing soil erosion. Low moisture retention capacity of sandy soils further deteriorates the situation. Therefore, rainwater harvesting will not only conserve the soil, its fertility and vegetation; but also, could be utilized as supplementary irrigation that will be advantageous in enhancing total water supply available to crop plants during low rainfall period

Rainwater harvesting enhances the supply of water to crop plants. Different techniques for increasing the supply of water to crop plants used in arid and semi-arid regions are: (a) Inter row water harvesting (b) Inter paired row water harvesting (c) Inter plot- water harvesting (d) Pit or micro-catchment water harvesting

A unique traditional method of runoff farming known as khadin cultivation has been practiced in the driest tract of hot arid region of Jaislmer since ages. The khadin cultivation is based on rainwater harvesting and its conservation to grow agricultural crops and consists water harvesting from the shallow rocky surfaces into low lying farm land during monsoon period and subsequently growing crops when water recedes (Kolarkar *et al.*, 1980). Runoff harvesting through macro catchment can be used with reduced tillage which increase the yield and water productivity of cowpea grown under arid climate (Adekalu *et al.*, 2009). Applying harvested runoff water through supplemental irrigation provides the twin benefits of alleviating the prevailing slack periods and improving the yields of smallholder farming systems.

3.3.6.2 *Irrigation management*

The success of rainfed crop production in arid regions depends on efficient utilization of rainwater. It has been reported that evaporation is most important unproductive water loss [evaporation accounts for 20-43% of ET in different crops in hot-arid region (Singh and Singh, 1993)]. Therefore, practices which cause a reduction in evaporation like elimination of weeds, use of mulches, planting geometry and population, use of organic manure/ fertilizers, biofertilizers and wind breaks and intercropping practices should be used for maximizing productive use of collected water. Gupta and Gupta (1983) reported significant increase in yield of greengram, mothbean and clusterbean with the application of grass mulch @ 6 t/ha. Mulches have been found suitable for winter crop raised on rainy season conserved moisture. Besides reduction in evaporation loss, mulching helps in suppression of weeds (Gupta and Gupta, 1985), and decrease surface runoff, increasing infiltration and moderating the thermal regimes. Higher planting density is advocated as a strategy to minimize the evaporation in relation to transportation in moisture adequate environment. In moisture constrained-environment, the higher planting densities may be disadvantageous as it may exhaust the soil moisture quickly and exposed plant to moisture stress.

Therefore, in such condition deciding optimal plant density and planting geometry is cautionary affair. Paired row planting was found more productive and water use efficient than uniform row planting in number of crops under arid conditions. Conducive microclimate for plant growth and development, efficient use of water, better weed suppression, reduction in evaporation and better utilization of moisture at reproductive stage (Singh, 1988) are factors responsible for higher productivity under paired row planting. Adequate nutrition helps to improve the availability and utilization of water by crops. It has been demonstrated that proper nutrient application encourages the extension of root system, which increase absorption of water and nutrients and thus lead to attain higher yield and water productivity. Areas where water is limiting factor the application of higher dose of inorganic N fertilizer should be restricted. Under these conditions the application of FYM and bio-fertilizer are more appropriate. The collection of rainwater in khadin and farm ponds, and using the collected water for supplemental irrigation to the crops during long dry spells showed considerable yield improvement in rainfed arid region. The harvested water can be used through sprinkler and drip irrigation in crops as a lifesaving irrigation at critical stages of growth which enhance the yield.

3.3.7 Nutrient management

Managing micro- and macro-nutrient application is essential for developing drought-tolerant plants as better nutrition can efficiently mitigate the harmful effects of water deficiency (Hansel *et al.*, 2017). Management of nutrient fertilizer has a positive influence on rainwater use efficiency and hence on final yield. Both phosphorus and nitrogen fertilizers can improve crop water use efficiency and result in a decreased evaporation rate. Adequate K application increases photosynthesis efficiency in legume crops by maintaining higher tissue water potential during drought (Sangakkara *et al.*, 1999). Nitrogen supply tends to enhance plant protein concentration, yield and yield-related traits in chickpea (Palta *et al.*, 2005). Application of recommended dose of nutrients through suitable sources can have a significant effect on the moisture utilization and crop yield as well. A judicious use of nutrients significantly increased the seed yield and WUE in chickpea grown in Khadin. Singh (2011) found significant increase in the seed yield of mothbean due to integration of FYM, chemical fertilizer and biofertilizer under arid regions of Rajasthan.

Microbial inoculants with the high cost of fertilizers and low fertilizer use efficiency, biofertilizers have become an alternate source of nutrient supply in arid agriculture. Biofertilizers contain micro-organisms that play vital role in soil and thus influence the availability of major nutrients like N, P, K and S. Rhizobium inoculation alone in greengram obviates the need for fertilizer N and P under natural conditions in arid ecosystem. Increase in seed yield of clusterbean and mothbean upon inoculation with efficient strains of Rhizobium varied from 8-15%, depending on the intensity and distribution of rainfall. These legumes fix about 20-40 kg N/ha and a residual effect equivalent to about 15-20 kg N/ha.

Application of Se can promote the growth of ageing seedlings and delay leaf senescence (Grams *et al.*, 2007). Silicon (Si) application significantly improved the relative water content (RWC) in plants by improving the levels of glycine betaine and proline. Silicon nutrition enhanced nodule activity for active fixation of nitrogen in cowpea (Mali and Aery 2008). In chickpea, addition of Si in combination with K significantly improved shoot dry matter during drought stress (Kurdali *et al.*, 2016). Zinc (Zn) application significantly increased yield attributes in drought-stressed chickpea (Shaban *et al.*, 2012). Thalooth *et al.* (2006) noted a significant improvement in growth, yield and related traits in mungbean with foliage application of zinc sulphate. Iron (Fe) and Zn addition can improve RWC and also have a positive impact on micronutrients and protein content of grains (Yadavi *et al.*, 2014).

3.3.8 Weed management

Weeds reduce the productivity of crops by competing for essential resources, by allelopathic effects, inducing attack of diseases, pests and by reducing the efficiency of farm operations. Hence, weed control improves crop productivity and imparts more stability to crop production. Timely weeding is not possible during rainy season due to various reasons. In years of low rainfall, weed control assumes greater importance due to scarcity of moisture. Traditional methods of weeding are laborious and time consuming. Hence, improved tillage implements should be used where moisture at sowing is adequate, chemical weed control along with one hoeing may provide the most desired benefits. Among the various biotic stresses, weeds are the major one which causes severe yield loss in pulses. Due to initial slow growth of pulses, weeds emerge first and gain competitive advantage over the crop and exhibit smothering effect on crop. Moreover, major area of pulses (84%) is under rainfed condition and grown in combination with non-legume crop. As a result, pulses are subjected to various types of biotic and abiotic stresses. Weeds besides causing direct loss in yield (Table 4) also hinder farm activities and serve as alternate host to many pests. Weed management in pulses is essential to bring the weeds below the threshold level to maximize the seed yield and quality.

In general, critical period of weed competition for short duration pulses is up to 30 days and for long duration pulse crops it is up to 60 days. The major three types of weeds viz., grasses, broad leaved weeds and sedges were found in association with pulses. Intensity of weed infestation varies with agroecological conditions and crop management practices followed.

Table 4. Yield reduction in arid legumes due to weeds

Crop	Yield reduction (%)	Reference
Mothbean	30-40	Bathar *et al.*, 2005
Chickpea	77.8	Patel *et al.*, 2006
Horsegram	30	Rao *et al.*, 2014
Cowpea	12.7-60	Li *et al.*, 2004

Studies revealed that pre-emergence application of herbicide (imazethapyr/pendimethalin or diclosulam) followed by manual weeding or post-emergence application of herbicides (quizalofop-p-ethyl or imazethapyr) or mulching followed by post-emergence application of herbicides (quizalofop-p-ethyl or imazethapyr) were found to be more effective for the broad-spectrum weed control in cowpea (Sinchana and Raj, 2020). In another study, manual weeding given at 30 days after sowing was most effective for realizing higher seed yield of mothbean in arid region, but monetary benefit was significantly lower as compared to herbicides application. Ratnam *et al.* (2011) reported that pre-emergence application of oxyfluorfen 100 g/ha followed by one hand weeding at 30 days after sowing gave maximum seed yield of chickpea, which was at par with two hand weeding given at 15 and 30 days after sowing. Results of various studies indicated that a system approach is necessary to maintain the weed population below the economic threshold level thereby reducing the yield loss. Integrated weed management (IWM) which has been proved to be more effective than any single method in alleviating the buildup of weeds in arid crops.

4. CONCLUSION

Arid legumes are largely drought-tolerant crop species, and are well adapted to rainfed situation, requiring little conserved soil moisture to sustain and produce reasonably good yield. However, inadequate rainfall under water-limited arid areas is often posing threat to pulses which leads to substantial loss of grain yield. The drought tolerance characters of these legumes are well characterized and genotypic variability of the drought-tolerant morpho-physiological traits has been established. Yield stability of pulses requires less genotype × environment interaction and wider adaptability across the environment. The physiological traits such as water-use efficiency is considered to be the most important to contribute grain yield and biomass under drought condition. In addition to physiological and breeding approaches to enhance yield, efficient agronomic practices need to be evolved to minimize the irrigation water and conserve soil moisture to increase the water productivity. To improve pulse productivity in the present scenario, gene mining for tolerance to abiotic stresses, restructuring plant types for climatically vulnerable regions, changing cropping pattern, efficient nutrient and water management, seed bank for alternate legume crops, watershed management, micro-irrigation facilities are some of the better options to address climate change-related issues. Therefore, improved agronomic practices hold tremendous potential to combat adverse impact of climate change on pulses production. By the adoption of recommended management practices, agriculture contributes not only to soil and water conservation, but also for enhancing the amount of soil organic carbon in soil.

5. WAY FORWARD

The following area must be focus to improve arid pulses production:

(i) Degraded and barren lands which are in low productivity that must bring under pulses cultivation. In this way, the soil fertility and productivity are maintained by reducing the soil erosion and also take part in the contribution of pulses production.

(ii) Drought resistance pulse crops can be grown in area where the stresses are too common. There is need of today to give more attention over the production of arid legume like clusterbean and mothbean.

(iii) Instead of taking sole cereal-based cropping, inclusion of pulses under cereal based cropping systems may be the viable option to produce extra yield.

(iv) An extensive and comprehensive research priorities should be focus on underutilized pulse crops viz., *Vigna umbellate* (Rice bean), *Vigna aconitifolia* (Mothbean), *Vigna angularis* (Adzuki bean), *Mcrotyloma uniflorum* (horsegram), so that these crops become an integral part of cropping system. It is also needed to create awareness about underutilized pulse crops among farmers so that farmer can take advantages by producing and fetching good price.

(v) The lot of research work has been concluded every year, but their reliability and applicability are not properly seen on the ground. Therefore, frame working of suitable policy and their implementation is prime concern toward increasing pulses production in arid region.

REFERENCES

Adekalu, K.O., Balogun, J.A., Aluko, O.B., Okunade, D.A., Gowing, J.W. and Faborode, M.O. (2009). Runoff water harvesting for dry spell mitigation for cowpea in the savannah belt of Nigeria. *Agricultural water management* 96, 1502-1508.

Ahmad, A., Selim, M.M., Alderfasi, A.A. and Afzal, M. (2015). Effect of drought stress on mungbean (*Vigna radiata* L.) under arid climatic conditions of Saudi Arabia. *Ecosystems and Sustainable Development* 192, 185-193.

Ahmed, F.E., Hall, A.E. and DeMason, D.A. (1992). Heat injury during floral development in cowpea (*Vigna unguiculata*). *American Journal of Botany* 79, 784-791.

Ahmed, F.E. and Suliman, A.S.H. (2010). Effect of water stress applied at different stages of growth on seed yield and water-use efficiency of Cowpea. *Agriculture and Biology Journal of North America* 1, 534-540.

Akhtar, S., Hazra, P. and Naik, A. (2015). Harnessing heat stress in vegetable crops towards mitigating impacts of climate change. *Climate Dynamics in Horticultural Science* 1, 173-200.

Al, R., Dahanayaka, N., Ugs, A., Wdrj, R. and Utd, R. (2012). Effect of water stress on growth and yield of mung bean (*Vigna radiata* L.) *Tropical Agricultural Research and Extension* 14, 2-6.

Amgain, L. (2018). Effect of Preceding Winter Crops and Residue Management on Growth Performance, Weed, Nutrient Uptake and Economics of Clusterbean (*Cyamopsis Tetragonoloba* L. Taub.] Under Zero-Till Semi-Arid Condition. *Journal of the Institute of Agriculture and Animal Science* 35, 67-78.

Andersen, M.N., Asch, F., Wu, Y., Jensen, C.R., Næsted, H., Mogensen, V.O. and Koch, K.E. (2015). Soluble invertase expression is an early target of drought stress during the critical, abortion-sensitive phase of young ovary development in maize. *Plant Physiology* 130, 591-604.

Anitha, S., Srinivasan, E. and Singh, Raj. (2004). Cowpea Agronomy. In: *Cowpea in India*. pp. 136–167. D. Kumar and N.B.Singh (Eds.). *Scientific Publishers*, Jodhpur.

Awari, V.R. and Mate, S.N. (2015). Effect of drought stress on early seedling growth of chickpea (*Cicer arietinum* L.) genotypes. *Journal of Life Sciences* 2, 356-361.

Awasthi, R., Bhandari, K. and Nayyar, H. (2015). Temperature stress and redox homeostasis in agricultural crops. *Frontiers in Environmental Science* 3, 11.

Barnabás, B., Jäger, K. and Fehér, A. (2008). The effect of drought and heat stress on reproductive processes in cereals. *Plant, Cell & Environment* 31, 11-38.

Bathar, V.M., Patel, M.M., Patel, I.C. and Tikka, S.B.S. (2005). Weed management in mothbean [*Vigna aconitifolia* (Jacq.) Marechal], pp. 177-178. (In) *Arid Legumes for Sustainable Agriculture and Trade*, Vol. 1. Henry, A. and Kumar, D. (Eds). Scientific Publishers, Jodhpur, Rajasthan.

Baudron, F., Corbeels, M., Monicat, F. and Giller, K.E. (2009). Cotton expansion and biodiversity loss in African savannahs, opportunities and challenges for conservation agriculture: A review paper based on two case studies. *Biodiversity and Conservation* 18, 2625-2644.

Behera, U.K., and France J. (2016). Integrated farming systems and the livelihood security of small and marginal farmers in India and other developing countries. *Advances in Agronomy* 138, 235-282.

Bhati, D.S. (1989). Effect of sowing date on mothbean. *Farming System* 5, 52-54.

Bindumadhava, H., Nair, R.M., Nayyar, H., Riley, J.J. and Easdown, W. (2017). Mungbean production under a changing climate-insights from growth physiology. *Mysore journal of agricultural sciences* 51, 21-26.

Bita, C.E. and Gerats, T. (2013). Plant tolerance to high temperature in a changing environment: scientific fundamentals and production of heat tolerance crops. *Frontiers in Plant Science* 4, 273.

Chandra, S., Buhariwalla, H.K., Kashiwagi, J., Harikrishna, S., Rupa, Sridevi K., Krishnamurthy, L., Serraj, R. and Crouch, J.H. (2004). Identifying QTL-linked markers in markers-deficient crops. In: Fisher T, Turner N, Angust J., eds. *Proceeding of the fourth international crop science congress.* The Regional Institute Ltd., Gosford, Australia.

Christophe, S., Jean-Christophe, A., Annabelle, L., Alain, O., Marion P. and Anne-Sophie, V. (2011). Plant N fluxes and modulation by nitrogen, heat and water stresses: a review based on comparison of legumes and non-legume plants," In *Abiotic Stress in Plants–Mechanisms and Adaptations*, eds Shanker A., Venkateswarlu B. (Rijeka: Intech Open Access Publisher), 79-118 pages.

Farooq, M., Basra, S.M.A., Wahid, A., Cheema, Z.A., Cheema, M.A. and Khaliq, A. (2008). Physiological role of exogenously applied glycine betaine to improve drought tolerance in fine grain aromatic rice (*Oryza sativa* L.). *Journal of Agronomy and Crop Science* 194, 325-333.

Flowers, T.J., Gaur, P.M., Gowda, C.L.L., Krishnamurthy, L., Samineni, S., Siddique, K.H.M., Turner, N.C., Vadez, V., Varshney, R. K. and Colmer, T.D. (2010). Salt sensitivity in chickpea. *Plant, Cell & Environment* 33, 490-509.

Garg, B.K., Burman, U. and Kathju, S. (2004). Effect of water stress on mothbean [*Vigna aconitifolia* (Jacq.) Marechal] genotypes. *Indian Journal of Plant Physiology* 9, 29-35.

Grams, T.E.E., Koziolek, C., Lautner, S. and Matyssek, R. (2007). Distinct roles of electric and hydraulic signals on the reaction of leaf gas exchange upon re-irrigation in *Zea mays* L. *Plant, Cell & Environment* 30, 79-84.

Gupta, J.P. and Gupta, G.N. (1983). Effect of grass mulching on growth and yield of legumes. *Agricultural Water Management* 6, 375-383.

Gupta, J.P. and Gupta, G.N. (1985). Effect of mulches on hydrothermal environment of soil and crop production in arid western Rajasthan. *Annals of Arid Zone* 24, 131-142.

Hall, A.E. (2012). Phenotyping cowpeas for adaptation to drought. *Frontiers in Physiology*, 3,155.

Hansel, F.D., Amado, T.J.C., Diaz, D.A.R., Rosso, L.H.M., Nicoloso, F.T. and Schorr, M. (2017). Phosphorus fertilizer placement and tillage affect soybean root growth and drought tolerance. *Agronomy Journal* 109, 2936-2944.

Hasanuzzaman, M., Nahar, K., Alam, M.M., Roychowdhury, R. and Fujita, M. (2013). Physiological, biochemical, and molecular mechanisms of heat stress tolerance in plants. *International Journal of Molecular Sciences* 14, 9643-9684.

Heatherly, L.G. (1993). Drought stress and irrigation effects on germination of harvested soybean seed. *Crop Science* 33, 777-781.

Heffner, E.L., Sorrells, M.E. and Jannink, J.L. (2009). Genomic selection for crop improvement. *Crop Science* 49, 1-12.

Hirayama, T. and Shinozaki, K. (2010). Research on plant abiotic stress responses in the post-genome era: past, present and future. *Plant Journal* 61, 1041-1052.

Hossain, M., Hamid, A. and Khaliq, M. (2010). Evaluation of mungbean (*Vigna Radiata* L.) genotypes on the basis of photosynthesis and dry matter accumulation. *Journal of Agriculture & Rural Development* 7, 1-8.

Johnsona, Y.K., Fredrick, O.A., Josiah, M.K. and Sijali, I.V. (2018). Effects of tillage practices on water use efficiency and yield of different effects of tillage practices on water use efficiency and yield of different drought tolerant common bean varieties in Machakos county, Eastern Kenya. *American Scientific Research Journal for Engineering, Technology* 40, 217-234.

Kaushal, N., Awasthi, R., Gupta, K., Gaur, P., Siddique, K.H. and Nayyar, H. (2013). Heat-stress-induced reproductive failures in chickpea (*Cicer arietinum*) are associated with impaired sucrose metabolism in leaves and anthers. *Functional Plant Biology* 40, 1334-1349.

Kaushal, N., Gupta, K., Bhandhari, K., Kumar, S., Thakur, P. and Nayyar, H. (2011). Proline induces heat tolerance in chickpea (*Cicer arietinum* L.) plants by protecting vital enzymes of carbon and antioxidative metabolism. *Physiology and Molecular Biology of Plants* 17, 203-213.

Khan, H., Siddique, K. and Colmer, T. (2016). Salt sensitivity in chickpea is determined by sodium toxicity. *Planta* 244, 623-637.

Kolarkar, A.S., Murthy, K.N.K. and Singh, N. (1980). Water harvesting and runoff farming in arid Rajasthan. *Indian Journal of Soil Conservation* 8, 113-118.

Kumar, S., Kaur, R., Kaur, N., Bhandhari, K., Kaushal, N. and Gupta, K. (2011). Heat-stress induced inhibition in growth and chlorosis in mungbean *(Phaseolus aureus* Roxb.) is partly mitigated by ascorbic acid application and is related to reduction in oxidative stress. *Acta Physiologiae Plantarum* 33, 2091-2101.

Kurdali, F., Al-chammaa, M. and Mouasess, A. (2016). Growth and nitrogen fixation in silicon and/or potassium fed chickpeas grown under drought and well-watered conditions. *Journal of Stress Physiology & Biochemistry* 9, 385-406.

Kushwah, A.S., Rawat, G.S., Gupta, S., Patil, D. and Prajapati, N. (2017). Production and profitability assessment of clusterbean [*Cyamopsis tetragonoloba* (L.) Taub.] based intercropping systems under different row arrangement. *Legume Research: An International Journal* 40, 916-919.

Kyei-boahen, S., Savala C.E.N., Chikoye D., Abaidoo, R. and Kyei-boahen, S. (2017). Growth and yield responses of cowpea to inoculation and phosphorus fertilization in different environments. *Frontiers in Plant Science* 8, 646.

Li *et al*. 2004. Damage loss and technology of weed in cowpea field. *Weed Science* 2, 25-26.

Mafakheri, A., Siosemardeh, A., Bahramnejad, B., Struik, P.C. and Sohrabi Y. (2010). Effect of drought stress on yield, proline and chlorophyll contents in three chickpea cultivars. *Australian Journal of Crop Science* 4, 580-585.

Malambane, G. and Bhatt, K.V. (2014). Agro-morphological performance of mothbean (*Vigna aconitifolia*) lines in severe drought stressed and rain-fed condition. *Direct Research Journal of Agriculture and Food Science (DRJAFS)* 2, 107-112.

Mali, M. and Aery, N.C. (2008). Silicon effects on nodule growth, dry-matter production, and mineral nutrition of cowpea (*Vigna unguiculata* L.). *Journal of Plant Nutrition and Soil Science* 171, 835-840.

Malosetti, M., Voltas, J., Romagosa, I., Ullrich, S.E. and Eeuwijk, F.A. (2004). Mixed models including environmental covariables for studying QTL by environment interaction. *Euphytica* 137, 139-145.

Maqbool, M.A., Aslam, M. and Ali, H. (2017). Breeding for improved drought tolerance in chickpea (*Cicer arietinum* L.). *Plant Breeding* 136, 300-318.

Maruthi, V., Reddy, P.R., Reddy, K.S., Reddy, B.M.K. and Saroja, D.G.M. (2018). Agronomic management of CRIDA-18R-a new variety of horse gram (*Macrotyloma uniflorum*) for South India matching monsoon patterns of rainfall. *Legume Research:An International Journal* 41, 102-107.

Mazvimavi, K., Twomlow, S., Belder, P. and Hove, L. (2008). An assessment of the sustainable uptake of conservation farming in Zimbabwe. *Global Theme on Agroecosystems Report No. 19.* International Crops Research Institute for the Semi-Arid Tropics, Bulawayo, Zimbabwe, pp 51.

Mondal, M.M.A., Fakir, S.A., Juraimi, A.S., Hakim, M.A., Islam, M.M. and Shamsuddoha, A.T.M. (2011). Effects of flowering behavior and pod maturity synchrony on yield of mungbean [*Vigna radiata* (L.) Wilczek] *Australian Journal of Crop Science* 5, 945-953.

Munns, R. (2005). Genes and salt tolerance: bringing them together. *New Phytologist* 167, 645-663.

Nathawat, N.S., Rathore, V.S., Soni, M.L., Singh, J.P. and Yadava, N.D. (2021). Exogenous application of sulphydryl compounds enhances growth, photosynthetic efficiency and yield of mothbean (*Vigna aconitifolia* L.) under water limiting environment. *Legume Research: An International Journal* 44, 67-73.

Nayyar, H., Kaur, S., Singh, S. and Upadhyaya, H.D. (2006). Differential sensitivity of Desi (small-seeded) and Kabuli (large-seeded) chickpea genotypes to water stress during seed filling: effects on accumulation of seed reserves and yield. *Journal of the Science of Food and Agriculture* 2082, 2076-2082.

Ogbonnaya, C.I., Sarr, B., Brou, C., Diouf, O., Diop, N.N. and Roy-Macauley, H. (2003) Selection of cowpea genotypes in hydroponics, pots, and field for drought tolerance. *Crop Science* 43, 1114-1120.

Palta, J.A., Kumari, S., Spring, C. and Turner, N.C. (2005). Foliar nitrogen applications increase the seed yield and protein content in chickpea (*Cicer arietinum* L.) subject to terminal drought. *Australian Journal of Agricultural Research* 56, 105-112.

Patel, B.D., Patel, V.J., Patel, J.B. and Patel, R.B. (2006). Effect of fertilizers and weed management practices on weed control in chickpea (*Cicer arietinum* L.) under middle Gujarat conditions. *Indian Journal of Crop Science* 1(1–2), 180-183.

Pathak, H., Aggarwal, P.K. and Singh, S.D. (2012). Climate change impact, adaptation and mitigation in agriculture: methodology for assessment and applications. Indian Agricultural Research Institute, New Delhi, pp 302.

Prakash, M., Pallavamallan, S., Sathiyanarayanan, G. and Rameshkumar, S. (2021). Effect of seed pelleting with botanicals on germination and seedling growth of cluster bean under induced saline condition. *Legume Research: An International Journal* 44, 88-93.

Ramamoorthy, P., Lakshmanan, K., Upadhyaya, H.D., Vadez, V. and Varshney, R.K. (2017). Root traits confer grain yield advantages under terminal drought in chickpea (*Cicer arietinum* L.). *Field Crops Research* 201, 146-161.

Ramamoorthy, P., Lakshmanan, K., Upadhyaya, H. D., Vadez, V. and Varshney, R. K. (2016). Shoot traits and their relevance in terminal drought tolerance of chickpea (*Cicer arietinum* L.). *Field Crops Research* 197, 10-27.

Rana, D.S., Yadav, B.D., Dhokia, R.S. and Midha, L.K. (1991). Effectof row spacing and inter row spacing on seed yield of clusterbean under late sown condition. *Guar Research Annals* 7, 54-56.

Rao, A.N., Wani, S.P. and Ladha, J.K. (2014). Weed management research in India - an analysis of the past and outlook for future. pp. 1 -26. In: Souvenir (1989-2014). DWR Publication No. 18. Directorate of Weed Research, Jabalpur, India.

Rao, A.S. and Singh, R.S. (1998). Climatic features and crop production. In: *Fifty Years of Arid Zone Research in India*, pp. 17-38. Faroda, A. S. and Singh, M. (Eds.). CAZRI, Jodhpur.

Ratnam, M., Rao, A.S. and Reddy, T.Y. (2011). Integrated Weed Management in Chickpea (*Cicer arietinum* L). *Indian Journal of Weed Science* 43, 70-72.

Rodríguez, M., Canales, E., Borroto, C.J., Carmona, E., Lopez, J. and Pujol, M. (2006). Identification of genes induced upon water-deficit stress in a drought-tolerant rice cultivar. *Journal of Plant Physiology* 163, 577-584.

Sadeghipour, O. (2018). Drought tolerance of cowpea enhanced by exogenous application of methyl jasmonate. *International Journal of Modern Agriculture* 7, 51-57.

Samarah, N.H., Haddad, N. and Alqudah, A.M. (2009). Yield potential evaluation in chickpea genotypes under late terminal drought in relation to the length of reproductive stage. *Italian Journal of Agronomy* 3, 111-117.

Sangakkara, U.R., Frehner, M. and Nosberger, J. (1999). Effect of soil moisture and potassium fertilizer on shoot water potential, photosynthesis and partitioning of carbon in mungbean and cowpea. *Journal of Agronomy and Crop Science* 185, 201-207.

Serraj, R., Chauhan, Y.S., Bidinger, F.R., Seetharama, N., Nigam, S.N. and Saxena N.P. (2003). Water Productivity in Agriculture: Limits and Opportunities for Improvement. CAB International; Wallingford, UK. Management of drought in ICRISAT cereal and legume mandate crops. 127-144 pages.

Shaban, M., Lak, M., Hamidvand, Y., Nabaty, E. and Khodaei, F. (2012). Response of chickpea (*Cicer arietinum* L.) cultivars to integrated application of Zinc nutrient with water stress. *International Journal of Agriculture and Crop Sciences* 4, 1074-1082.

Sharma, B.D., Taneja, K.D., Kairon, M.S. and Jain, Veena. (1984). Effect of dates of sowing and row spacing on yield and qual-ity of clusterbean (*Cyamopsis tetragonoloba* L. Taub). *Indian Journal of Agronomy* 29, 557-558.

Shetty, P.K., Ayyappan, S. and Swaminathan, M.S. (2013). Climate change and sustainable food security. National Institute of Advanced Studies, Bangalore and Indian Council of Agricultural Research, New Delhi. ISBN: 978-81-87663-76-8.

Siddique, K.H.M., Johansen, C. and Alghamdi, S.S. (2012). Innovations in agronomy for food legumes. A review. *Agronomy for Sustainable Development* 32, 45-64.

Siddique, K.H.M., Johansen., Kumar, C., Rao, J.V.D.K. and Ali, M. (2008). Legumes in sustainable cropping systems. In: Kharkwal MC (ed) Food legumes for nutritional security and sustainable agriculture Proceedings of the Fourth International Food Legumes Research Conference (IFLRC-IV), October 18–22, 2005, New Delhi, India, vol 1. Indian Society of Genetics and Plant Breeding, New Delhi. 787-819 pages.

Sinchana, J.K. and Raj, S.K. (2020). A review on integrated approach for the management of weeds in Cowpea (*Vigna unguiculata*). *Journal of Applied and Natural Science* 12, 504-510.

Singh, K.B., Bejiga, G., Saxena, M.C. and Singh, M. (1995). Transferability of chickpea selection indices from normal to drought-prone growing conditions in a Mediterranean environment. *Journal of Agronomy and Crop Science* 175, 57-63.

Singh, R. (1995). Mixed cropping in pearlmillet (*Pennisetum glaucum*) in arid zone of Rajasthan. *Indian Journal of Agricultural Sciences* 65, 276-278.

Singh, R. (2011). Production potential, moisture use efficiency and monetary benefit as influenced by intercropping pattern of mothbean (*Vigna aconitifolia*) and pearlmillet (*Pennisetum glaucum*) in arid region. *Journal of Arid Legumes* 8, 35-38.

Singh, R. (2006). Effect of cropping sequence, seed rate and weed management on weed growth and yield of Indian mustard in western Rajasthan. *Indian Journal of Weed Science* 38, 69-72.

Singh, S. D. (1988). Using pond sediments treated catchments. In: Singh, S.D. (eds), *Water harvesting in Arid Tropics*, CAZRI, Jodhpur. 1-11 pages.

Singh, S. D. and Singh, Y. V. (1993). Yield to water use relationship and use of the concept in managing a limited water supply. In: Singh, S.D. (eds.), *Arid land Irrigation and Ecological Management, Scientific Publisher*, Jodhpur.pp. 103-158 pages.

Singh, S.D., Bhandari, R.C. and Aggarwal, R.K. (1985). Long term effects of phasing phosphate fertilizer on soil fertility and yield of pearlmillet in rotation with grain legumes. *Indian Journal of Agricultural Sciences* 55, 274-278.

Suzuki, N., Rivero, R.M., Shulaev, V., Blumwald, E. and Mittler, R. (2014). Abiotic and biotic stress combinations. *New Phytologist* 203, 32-43.

Thalooth, A. T., Tawfik, M. M. and Mohamed, H. M. A. (2006). Comparative study on the effect of foliar application of zinc, potassium and magnesium on growth, yield and some chemical constituents of mungbean plants grown under water stress conditions. *World Journal of Agricultural Sciences*, 2:37-46.

Varshney, R.K., Thudi, M., Nayak, S.N., Gaur, P.M., Kashiwagi, J., Krishnamurthy, L., Jaganathan, D., Koppolu, J., Bohra, A. and Tripathi, S. (2014). Genetic dissection of drought tolerance in chickpea (*Cicer arietinum* L.). *Theoretical and Applied Genetics* 127, 445-462.

Vir, O. and Singh, A.K. (2015). Mothbean [*Vigna aconitifolia* (Jacq.) Marechal] germplasm: Evaluation for genetic variability and inter characters relationship in hot arid climate of western Rajasthan, India. *Legume Research: An International Journal*, 38:748-752.

Yadav, N.D. and Beniwal, R.K. (2006). Performance of mothbean varieties to spacing and fertilizer P application in arid west-ern Rajasthan under rainfed condition. *Journal of Arid Legumes* 3, 64-65.

Yadavi, A., Movahhedi-dehnavi, M. and Balouchi, H. (2014). Effect of micronutrients foliar application on grain qualitative characteristics and some physiological traits of bean (*Phaseolus vulgaris* L.) under drought stress. *Indian Journal of Fundamental and Applied Life Science* 4, 124-131.

CHAPTER 16

New niches for horizontal and vertical expansion of area under pulses in India

Narendra Kumar*, C.P. Nath, Ashik Dutta and Mrunalini Kancheti
ICAR-Indian Institute of Pulses Research, Kanpur-208024, Uttar Pradesh
**Corresponding author email: nkumar.icar@gmail.com*

Abstract

India is largest producer, consumer and importer of pulses in the world. In last few years, the pulses production in the country reached to 23-25 Mt which is near to the target for achieve self sufficiency in pulses production. But, due to ever increasing human population, it is estimated that country needs about 39 Mt of pulses by 2050. To meet out the projected requirement, there is need of increase in area and productivity of pulses. Estimates also showed that an additional area of 4-5 M ha will be required for pulses production in the future. Considering the shrinking agricultural lands due to urbanization and industrialization, there is an urgent need to identify the new niches for area expansion of pulses. Some of the promising areas, where pulses can be introduced as a new crop or catch crop or in cropping systems as a profitable alternative of existing crops or in fallow lands, are discussed under this chapter.

Keywords: Arid ecosystem, Grass pea, Rabi pigeonpea, Rabi rajmash, Sugarcane intercropping, Summer mungbean

1. INTRODUCTION

Pulses are important group of crops after cereals and oilseeds in the country. These crops are valued for their importance in nutritional security (seeds enriched with high quality protein, minerals and vitamins) and sustainable crop production (BNF, soil amelioration and low input requirements). India is not only one of the largest producer and consumer, but also remained largest importer of pulses in the world. The total world acreage during 2018 under pulses was 96.70 M ha with production of 94.99 Mt at 982 kg/ha yields level. India with 29.16 M ha pulses area is the largest pulse producing country (23.40 Mt) in the world. It contributes 30.8 and 26.76 per cent in area and production, respectively of world pulses.

Large numbers of pulses are being grown throughout year in one or other parts of the country, the major pulses are chickpea, pigeonpea, urdbean, mungbean, lentil, peas, horsegram, mothbean, cowpea and lathyrus. Chickpea continues to be the largest consumed in this complex comprising of 43.3% of the total pulses production from 32.5% pulses area with average productivity of 1073 kg/ha (2018-19). Pigeonpea is the second pulse crop with total production of 3.59 Mt from 4.78 M ha area and productivity of 751 kg/ha. The major pulses producing states in the country are Madhya Pradesh (33.39%), Rajasthan (15.71%), Uttar Pradesh (10.24%), Maharashtra (9.23%) and Karnataka (7.94%) which together share about 76.5% of total pulse production while remaining 23.5% is contributed mainly by Gujarat, Tamil Nadu,Chhattisgarh, West Bengal, Bihar, Jharkhand and Odisha.

Traditionally, pulses were grown on marginal lands of rainfed areas under poor crop management practices, and consequently their yields were quite low (< 500 kg/ha). With the development of new varieties (high yielding, early maturing and resistant/tolerant to major diseases and drought) and matching agro-technology, pulses witnessed a silent revolution in terms of horizontal (area expanded from 20.27 M ha in 1950-55 to 27.35 M ha in 2014-19) and vertical growth (productivity increased from 468 kg/ha in 1950-55 to 766 kg/ha in 2014-19) (Table 1). This happened despite loss of considerable area (5.0 M ha) from most productive cropped zone of north India and expansion in central and southern regions experiencing acute moisture stress and short growing season.

In order to ensure self-sufficiency, the pulse requirement in the country is projected at 39 Mt by the year 2050 which necessitates an annual growth rate of 2.2% (Anonymous, 2015). To meet the projected requirement, the productivity needs to uplift at about 1200 kg/ha and about 4 to 5 M ha additional area has to be brought under pulses besides drastically reducing the postharvest losses. A detailed strategy has been developed by Indian Institute of Pulses Research, Kanpur for increasing area under pulses by 2050 (Table 2). But, there are certain new avenues also available which may be exploited to increase the area under pulses in the country. The details about these new niches are discussed in detailed in this chapter.

Table 1. Changes in area, production and productivity of pulses during 1970-75 to 2014-19

	Year	Total Pulses	Chickpea	Pigeonpea	Mung	Urd	Lentil	Pea
Area	1970-75	22.21	7.50	2.52	2.11	2.09	0.82	0.73
(M ha)	2014-19	27.35	9.26	4.47	4.03	4.45	1.45	0.84
Production	1970-75	10.57	4.59	1.75	0.65	0.64	0.38	0.51
(Mt)	2014-19	21.08	9.06	3.62	1.95	2.66	1.28	0.79
Yield	1970-75	476	611	696	306	308	469	654
(kg/ha)	2014-19	766	971	801	482	596	877	944

Table 2. Potential area projected for expansion under pulses cultivation

S. No.	Potential crop/ cropping systems/niche	Specific areas	Potential area (M ha)	Target area (2025)- M ha	Target area (2050)- M ha
1.	*Intercropping*				
	Mungbean with Sugarcane (irrigated) and with Cotton andmillets (rainfed uplands)	Western, Central and Eastern U.P., Bihar, Maharashtra, A.P. and T.N.	0.70	0.40	0.60
	Pigeonpea with soybean, sorghum, cotton, millets andgroundnut (rainfed upland)	A.P., Malwa Plateau of M.P., Vidarbha of Maharashtra, North Karnataka, T.N.	0.50	0.30	0.50

	Chickpea with barley, mustard,linseed and safflower (rainfed)	South East Rajasthan, Punjab, Haryana, U.P., Bihar, Vidharbha of Maharashtra	0.50	0.20	0.40
	Chickpea/lentil with autumnplanted/ratoon sugarcane	Maharashtra, Uttar Pradesh, Bihar	1.00	0.50	0.70
2.	*Catch crop*: mungbean spring/summer	Western U.P., Central U.P. Haryana, Punjab, Bihar, WB	1.00	0.70	0.80
3.	*Rice fallows*				
	Chickpea	Eastern U.P., Bihar, Jharkhand, Orissa, Chattisgarh, W.B.	0.40	0.30	0.35
	Urdbean/mungbean	A.P., Tamil Nadu, Orissa, Karnataka	0.50	0.30	0.40
	Lentil	Eastern U.P., Bihar, West Bengal, Assam, Jhanrkhand	0.30	0.30	0.30
	Lentil/fieldpea	North-East	0.10	0.10	0.10
4.	Kharif fallow	Urd/mung in UP (Bundelkhand), M.P.	1.20	0.40	0.60
	Total		6.20	3.50	4.75

Source: Anonymous (2015)

2. POSSIBLE NEW NICHES FOR PULSES CULTIVATION

2.1 Rabi rajmash

Rajmash (*Phaseolus vulgaris*) is a main kharif pulse crop in hilly regions of NW Himalaya. But its cultivation became possible with the development of cold tolerant varieties PDR-14 (Udai) during year 1987 in the frost-free irrigated plains of northern India (Ali *et al.*, 2012, 2016). Thereafter, its cultivation practices were developed during late 1980s at ICAR-IIPR, Kanpur and other institutions (Ali, 1988, 1989; Ali and Lal, 1991). Later on some more varieties were developed (Table 1). For realizing the good yield of rajmash in this region, it should be sown in 2nd fortnight of October and maintained the plant population of 2.0-2.5 lakhs/ha. Sadhukhan *et al.* (2008) has reported end of October to early November is optimum time of sowing of rajmash in West Bengal. Rabi rajmash in NEPZ requires 2-3 irrigations, the 30 days stage being most critical. Because of non-nodulating behavior, it requires 120 kg N/ha. The crop has shown good compatibility for intercropping with early potato (Fig. 1). This also enabled crop intensification, the important rotations being maize/sorghum – frenchbean – mungbean (Ali and Kushwaha. 1987).

Table 3. Recommended varieties for rabi cultivation in NEPZ

Variety	Year of release	Area of adaptation	Crop duration (days)	Yield (q ha^{-1})	Kernel colour
Uday (PDR-14)	1985	NEPZ	125-130	30-35	Red spotted
Malviya-137	1991	NEPZ	110-115	25-30	Red
VL Rajmash-63	1982	Uttarkhand hills, can be grown in NEPZ	115-120	25-30	Brown spotted, Resistant to bacterial blight, anthracnose and yellow mosaic virus
Amber (IIPR-96-4)	2002	NEPZ	120-125	20-25	Red spotted, BCMV resistance
Utkarsh (IIPR 98-5)	2005	NEPZ	130-135	20-25	Deeply spotted, Cold tolerance
Arun (IPR 98-3-1)	2001	NEPZ &CZ	115-120	24-25	Attractive dark red medium seeds

Fig. 1. Rajmash + potato intercropping

2.2 Post-rainy /rabi pigeonpea

Due to heavy rains in eastern India and other parts of country, it is difficult to grow pulses crop like pigeonpea in rainy season. Eastern India also faces frequent flood in rainy season which cause heavy loss to normal sown pigeonpea and sometimes crop is completely lost. To avoid this situation, the concept of post-rainy / rabi pigeonpea developed in which long duration of pigeonpea variety like 'Bahar' can be sown during post rainy season. With the development of *Alternaria* blight resistant pigeonpea variety 'Sharad' and 'Pusa 9', the post-rainy pigeonpea became quite successful (Table 4) (Roy Sharma *et al.*, 1980). Consequently, the maize (cobs) -pigeonpea crop rotation received favour of the farmers. It can be also grown after early rice, minor millets, rice seedlings etc. The optimum time of sowing was found to be first fortnight of September at a row distance of 30 cm and plant spacing of 15 cm accommodating about 2 lakh plants per hectare. It requires 40 kg nitrogen and 2 irrigations. Choudhary *et al.* (2020) from Patna, Bihar has also reported pigeonpea yield of more than 3.0 t/ha in varieties 'Pusa 9' and 'IPA 203' when sown during first fortnight of September.

Apart from eastern India (eastern Uttar Pradesh, Bihar and West Bengal), it is now getting popularity as *rabi* pigeonpea in central and southern India. Narayanan and Sheldrake (1979) from a study at Hyderabad reported that by sowing in October or November as a winter crop, the duration of the 'medium' and 'late' cultivars of pigeonpea, was much reduced. The crop sown in October (plant population 3-5 times higher) gave yields comparable to those of the normal season. Due to poor plant growth, the closer spacing found beneficial than wider spacing (Karle and Pawar, 1998; Desai and Intiwala, 1999). Nagamani *et al.* (2015) from Tirupati, Andhra Pradesh has reported that pigeonpea crop sown during second fortnight of September receiving nutrient dose of 30-60-20 kg N, P_2O_5 and K_2O per hectare along with foliar spray of NAA (25 ppm) and DAP (2%) applied twice at 60 and 80 DAS resulted in highest seed yield. However, rabi pigeonpea can be sown between 1st September to 1st October gave highest yield at Lam, Guntur (Venkata Rao *et al.*, 2016).

Table 4. Yield potential of pre-rabi pigeonpea in comparison to other crops

Crop	Mean yield (t/ha)
Wheat	2.41
Barley	2.69
Peas	2.29
Chickpea	1.29
Mustard	1.16
Pigeonpea	3.43

Source: Roy Sharma *et al.*, 1980

2.3 Intercropping with sugarcane

Sugarcane is inherently a slow growing and wide spaced (row to row spacing of 75-90 cm) long duration crop which lends ample scope for short duration, mid-season income generation crops for nutritional and economic security of small and marginal farmers. Pulses are suitable candidates with multiple advantages for intercropping in sugarcane. At present (2018-19), the area under sugarcane is 5.11 M ha with production and productivity of 400.16 Mt and 78.25 t/ha, respectively. Significant area and production of pulses can be increased if 15-20% sugarcane area will be brought under intercropping with pulse crops. Spring planted sugarcane alone accounts for 65-70 % of the total area being adopted in the states of Punjab, Haryana, Uttar Pradesh and Bihar (Singh *et al.*, 2009). Ratoon crop of sugarcane also provides good opportunities for intercropping of spring/summer pulses. The possibilities of intercropping of summer mungbean and urdbean with spring planted sugarcane (Fig 2a) was worked out in year 2000-02 at GBPUA&T, Pantnagar (Rana *et al.*, 2006) and during 2002-04 at ICAR-IISR, Lucknow (Singh *et al.*, 2008). Planting two rows of mungbean/urdbean with spring planted sugarcane spaced at 75 cm was found to be quite profitable. From a study at ICAR-IISR Lucknow, it is established that urdbean and mungbean had synergistic effects on cane yield in spring planted crop and provided 0.4-0.5

t/ha additional yield of pulses grains. Among various pulses, mungbean has been found most profitable (Yadav *et al.*, 1987). It has been observed that growing one row of mungbean gave about half t/ha additional yield of mungbean without affecting the sugarcane yield. But, increase in mungbean rows to 2 -3 makes the systems non profitable. In a study at Pantnagar, cowpea was found more profitable than mungbean and urdbean (Kumar *et al.*, 2006). Similar to spring season, *rabi* pulses like lentil, pea, rajmash and chickpea can be intercropped with autumn planted sugarcane (Fig 2b) (Panwar *et al.*, 1990). In a study at Kushinagar, Uttar Pradesh showed that sowing of two rows of lentil in between sugarcane rows was found remunerative (Singh *et al.*, 2009a).

Fig. 2a. Spring planted sugarcane + mungbean (1:1) intercropping

Fig. 2b. Autumn planted sugarcane + lentil (1:2) intercropping

2.4 Summer mungbean in RWCS in IGP

After Green Revolution, rice-wheat cropping system (RWCS) became dominant cropping system in the country occupy about 10.0 m ha area. The area under RWCS increased many folds in Indo-Gangetic Plains (IGP) of India. The RWCS dominant in many states viz. Punjab, Haryana, Uttar Pradesh, Bihar, Madhya Pradesh, Jharkhand and Chhattisgarh and contributes to 75% of the national food grain production. But, the long-term cultivation of mono-cropping of RWCS after 'Green Revolution' resulted in decline in factor productivity and sustainability. Therefore, need was felt for diversification of RWCS with pulses. Summer mungbean being short duration leguminous crop (55-65 days) can accommodate well in the period available between wheat harvest and rice transplanting. Apart from the yield advantage of mungbean crop, the crop being a legume, provides the additional benefits of system diversification, maintenance of soil health and often used as a source of organic as a green manure crop. Although the first short duration mungbean variety 'T 44' was developed in 1948, but the first variety 'G 65' developed in 1971 was found suitable for summer cultivation in Punjab (Gupta and Pratap, 2016). But the development of 'SML 668' variety of mungbean bought revolution in its cultivation after wheat harvest in RWCS. Thereafter, many varieties have been developed for summer cultivation of mungbean in RWCS. Recently developed extra early variety 'Virat (IPM 205-7)' which matured in 55 days is now becoming popular in IGP and central India under RWCS. In a long-term effect (16 years) study at Kanpur showed the higher system productivity of rice-wheat-mungbean (R-W-M) system over rice-wheat (R-W) (Fig 3).

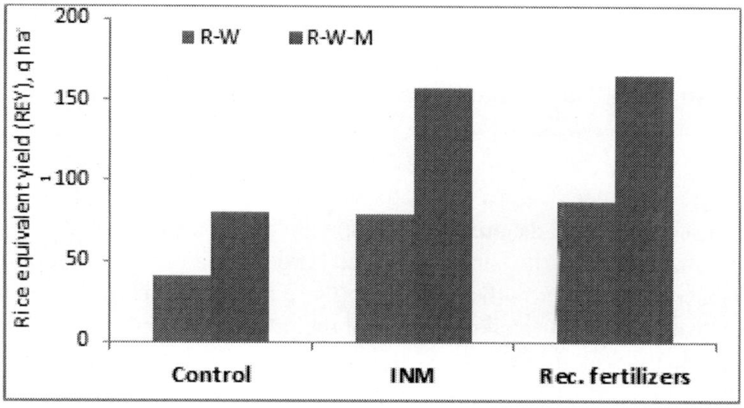

Fig. 3. Long-term effect of inclusion of summer mungbean on system productivity

2.5 Pigeonpea-wheat cropping

Pigeonpea is a long duration crop which takes about 250-270 days to mature. But, the development of early pigeonpea variety during 1975-76 such as Pusa Ageti, UPAS 120, etc led to introduction of pigeonpea in irrigated area of north-west plains under pigeonpea-wheat double cropping system (Ramanujam, 1971). Sandhu *et al.* (1973), Faroda and Singh (1983) and Chandra and Ali (1986) have also reported successful pigeonpea-wheat sequential cropping systems in irrigated areas of northern and central India by advancing sowing date. Its agronomic requirements were comprehensively worked out at IARI New Delhi, PAU Ludhiana, GBPUAT Pantnagar and HAU Hissar under AICPIP as well as in the institutional programmes of IIPR. Later, many other varieties like Pusa 992, PA 291, Manak, ICPL 88039, AL Arhar 1, PA 16 etc have been developed which gave better performance under pigeonpea-wheat system. First fortnight of June was found to be optimum for planting as this enabled timely planting of wheat. A plant population of 1-1.5 lakh/ha was found optimum. Studies at CSUAT, Kanpur showed that a higher productivity from early pigeonpea could be obtained by advancing its planting to mid-April at wider row spacing (75-100 cm) and intercropping of 2 rows mungbean (Panwar and Yadav, 1980). In a long-term study at Kanpur revealed that pigeonpea-wheat cropping system maintained greater total SOC and higher carbon management index over maize-wheat system and thus considered as the ideal system in terms of maintaining of soil health in Inceptisols of the IGP of India (Venkatesh *et al.*, 2013). The early pigeonpea crop is also found suitable for intercropping with short duration pulses like mungbean, urdbean and cowpea.

2.6 Rice fallows

In India, rice is cultivated almost all parts of the country occupying an area of about 44.0 M ha. On an average 11.7 M ha of this area remains fallow during post rainy or winter (*rabi*) season due a varieties of biotic, abiotic and socio-economic constraints prevailing in these areas (Gumma *et al.*, 2016; Kumar *et al.*, 2016). This area is almost equivalent to the net sown area of Punjab, Haryana and Western Uttar Pradesh - the sheet of green revolution in India. The above rice fallow areas is mostly concentrated in eastern and central India (around 80%) covering the states of Assam, Bihar, Chhattisgarh, Jharkhand, Madhya Pradesh, Chhattisgarh, Odisha, West Bengal and North Eastern Hill states (Table 5). The southern states of Tamil Nadu, Karnataka and Andhra Pradesh cover the remaining rice fallow areas (Subbarao *et al*., 2001).

The following of rice cultivation in rice fallow areas for long-run causes the disturbance to the ecology of the system and compactness of the soil. The soil physical constraints following harvest of rice usually present a hostile environment to pulses seed germination and establishment. Drying of the surface soil and hardening and compaction of plough pan layers limit subsequent root growth and the supply of air, water and nutrients to roots. The abiotic factors (water and soil related) also limit the pulses production in rice fallows. Low moisture content in the soil after rice harvest, faster decline in water table with advancement of *rabi* season, and risk of soil moisture stress towards flowering and pod filling stages are some of the water related constraints (Kumar *et al*., 2016). Terminal drought at pod filling stage adversely affects photosynthesis and translocation of food from source to sink. Thus, the stress developed during pod filling stage adversely affects the grain yield of the crop.

Table 5. Estimates of rice area during *kharif* 1999 and rice-fallow area during *rabi* 1999/2000 for major rice-growing states in India

State	Kharif-rice area ('000 ha)	Rabi-fallow ('000 ha)	Rice-fallow area as% of *kharif* rice area	% of total rabi fallow area
Andhra Pradesh	2657	305	11.5	2.6
Assam	2234	539	24.1	4.6
Bihar & Jharkhand	5974	2196	36.8	18.9
Gujarat	469	83	17.7	0.7
Haryana	1109	0	0.0	0.0
Karnataka	984	182	18.5	1.6
Kerala	241	0	0.0	0.0
Madhya Pradesh & Chhattisgarh	5596	4382	78.3	37.6
Maharashtra	1762	629	35.7	5.4
Orissa	3879	1219	31.4	10.5
Punjab	2498	0	0.0	0.0
Rajasthan	214	25	11.7	0.2
Tamil Nadu	1695	20	1.2	0.2
Uttar Pradesh	6255	353	5.6	3.0
West Bengal	4617	1719	37.2	14.8
Total	40,184	11,652	29.0	100

Source: Subbarao *et al*., 2001

Pulses are the ideal crops that can be grown in the areas vacated after rice because of their property to establish with the surface seeding, low water requirement, short duration and more tolerant to soil moisture stress than other crops due to deep tap roots. The incorporation of pulses improves the soil quality in terms of nutrient dynamics, soil organic carbon and biological activities thereby increasing nutritional security and improving sustainability of cropping system. The abiotic stresses under rice fallows can be mitigated by following of resource conservation practices related to soil moisture conservation (Kumar *et al.*, 2018). To exploit these areas, location specific and economicaly viable technologies of pulses production have been developed and upscalled (Singh *et al.*, 2017).

In conclusion, great potential exist for cultivation of pulses under rice fallows (Table 6). The cultivation of pulses in this untrapped area could benefit millions of poor and small farmers solely dependent on agriculture for their livelihood.

Table 6. Potential area in rice fallows for pulses production

Crop	States	Approx. additional area (Lakhs ha)
Chickpea	Eastern Uttar Pradesh, Bihar, Jharkhand, Madhya Pradesh, Chhattisgarh	47.00
Urdbean/mungbean	Andhra Pradesh, Tamil Nadu, Odisha, Andhra Pradesh, Karnataka	23.30
Lathyrus	West Bengal, Jharkhand, Odisha, Chhattisgarh, Bihar, Madhya Pradesh, Maharashtra	3.16
Lentil	Eastern Uttar Pradesh, Bihar, Chhattisgarh, Madhya Pradesh, NE States	25.03

Source: IIPR, 2013

2.7 Grass pea in eastern and central India

The grass pea (*Lathyrus sativus* L.), synonym - Chickling pea, Khesari, Teora and Kasari, has received increased interest since last one decade as a plant that is adapted to arid and semi-arid conditions, and contains high levels of protein. Out of the 187 species and sub-species, only one species *Lathyrus sativus* is widely cultivated as a food crop. Archeological evidence suggests that the domestication of *Lathyrus* dates to the late Neolithic, and precisely to the Bronze Age (Mahler-Slasky and Kislev, 2010; Kumar *et al.*, 2013). Prior to domestication, the crop was presumably present as a weed among other pulse crops. Grass pea is sometimes consumed by boiling whole grains, but is most often processed through a dal mill to obtain split dal. There are many reports that its *dal* or *besan* is used to adulterate pigeonpea *dal* and chickpea *besan* because of their higher market values.

During last four decades, the grass pea area in the country decreased from 1.67 M ha to 0.58 M ha and the production from 0.84 Mt to 0.43 Mt. At present, grass pea is occupying an area of 0.58 M ha with an annual production of 0.43 Mt. It is cultivated primarily in Bihar, Madhya Pradesh, Maharashtra, West Bengal, and Chhattisgarh. Whereas, the majority of this acreage (~70%) is shared by Chhattisgarh and the Vidarbha region of Maharashtra, which is a rice-growing region where supplemental irrigation is available only for rice (Dixit *et al.*, 2016).

The inherent capacity of grass pea to withstand climate extremes viz., temperature, drought, excess soil moisture or even flood has made this crop an '*insurance crop*' as it produces reliable yields when all other crops fail. It has a hardy and penetrating root system suited to a wide range of soil types including very poor soil and heavy clays. This crop has also shown resistant against many pests including storage insect-pests. In addition, the ability to fix atmospheric nitrogen makes the crop resembling as designer crop fit well under adverse conditions.

However, the presence of a neurotoxin [β-N-oxalyl-L-α,β-diaminopropionic acid (ODAP)], prolonged consumption of which leads to the neurological disorder lathyrism in humans and domestic animals, its cultivation and sale has been banned in all states in India except for Chhattisgarh, Maharashtra, and West Bengal under rule 44-A of the Prevention of Food Adulteration Act, 1954 (Dixit *et al.*, 2008). But, with the development of low b-ODAP and high-yielding cultivars such as 'Nirmal', 'Pusa 24', 'Prateek', 'Ratan' and 'Mahateora' has shown the hope to increase its area and production in near future.

The grass pea is mainly cultivated as an '*utera*' crop in areas where rice is grown and water is limited only to give supplemental irrigation to rice (Fig 4). Winter crops are mainly grown on residual soil moisture. This type of situation is prevailing in rice fallow area of eastern and central India, which is about 8.3 M ha, can be effectively utilized for cultivation of grass pea. Under *utera* cultivation, grass pea seed is often broadcasted in standing rice crops 10-15 days before the rice harvest. This allows grass pea to effectively exploit the residual moisture left after the rice harvest. The utilization of part of rice fallow area of eastern and central India for cultivation of grass pea may help to achieve the nutritional and livelihood security of resource poor farmers of this region.

Fig. 4. Grass pea under *utera* cultivation

2.8 Spring / summer rajmash

Rajmash is a crop of *kharif* season in hilly regions of NW Himalaya and NEH. But, the yield of *kharif* season crop is very poor because of several biotic and abiotic constraints. Kharif season rajmash crop is mostly grown under rainfed and thus drought or excessive rains during crop season adversely affect the yield. Many diseases, such as *Rhizoctonia solani*, white rot (*Sclerotinia sclerotiorum*) and anthracnose (*Colletotrichum lindemuthianum*), are commonly observed in rainy season (Banyal and Singh, 2007). Therefore, an initiative taken by ICAR-VPKAS, Almora during 2005 to introduce rajmash in spring/summer season in irrigated areas of NW Himalayas. The results indicated that rajmash can be sown between 15 to 30 March in mid hills of Himalaya (Fig 5). Further, sowing on raised bed gave higher yield. The crop of this season was almost free from all kinds of diseases and less infestation insect-pests.

Fig. 5. Summer season rajmash at Almora

2.9 Pulses in western hot arid ecoregion

The region comes under Western plains and Kutch Peninsula agro-ecological region of the country. This region has hot and arid climate with desert saline soils. It covers 24.7 million ha area which is 7.5% of total geographical area of the country and has <90 days of total growing period. The desert in Hindi literature known as '*Maruasthali*', means the dead land as this region receive very less annual rainfall (<300 mm) and has sparse vegetation cover. In general, eastern part of this region is rocky, while the western part is covered by shifting sand dunes. The soil has 90-95% sands and 5-7% clay contents. The soil has high soluble salts and pH value with higher percentage of calcium carbonate, but very low in organic carbon. The presence of high phosphates and nitrates in soil make the soil fertile for cultivation of crops, if water is available.

Moth bean is predominantly grown in this region followed by cluster bean. Cluster bean is used for both vegetables as well as for grains. However, considering the recent development in early and extra-early varieties of mungbean, urdbean and pigeonpea, it is now right time to introduce these crops in rainy season in this region. Similarly, early maturing heat and drought tolerant varieties of lentil and chickpea can be grown in winter season, wherever facility of giving 1-2 supplemental irrigation is available. At present, early maturing mungbean is becoming popular among farmers of this region. The quality of grains produced in this region was found superior than produced in other parts of the country due to less rainfall and rain free period at the time of harvesting. Thus, large scope exists for horizontal expansion of pulses in this region and to increase pulses production in the country.

2.10 Pigeonpea on rice bunds

About 10% of rice area is available as bunds, which mostly remained unutilized, can be utilized for cultivation of pigeonpea cultivation. The upland conditions of rice bunds are favourable for growth of pigeonpea once the crop is established. The production technology of pigeonpea cultivation on rice bunds is developed by institutions in central India as well as eastern India. Bunds can be prepared before sowing of pigeonpea and seed can be sown by dibbling of 2 seeds/hill. Pigeonpea transplanting can also be followed for maintaining the plant population and planting spacing. Weeds can be controlled by application of herbicides (pendimethalin and imazethapyr). As per an estimate about 20 - 30 thousand hectares area of rice bunds in Chhattisgarh, Madhya Pradesh, Odisha, West Bengal and Jharkhand is available for pigeonpea cultivation for additional yield and income (Singh, 2018).

2.11 Pulses in NER

The north-eastern region (NER) comprising eight states viz., Assam, Arunachal Pradesh, Meghalaya, Manipur, Mizoram, Nagaland, Tripura and Sikkim has a total geographical area of 262180 km^2 which is nearly 8% of the total area with about 4% populations of the country (Das *et al.*, 2016). Out of 4.13 M ha net sown area of the region, roughly 1.3 M ha suffers from serious soil erosion problem. Nearly 0.88 M ha area is under shifting cultivation in NE region. The region receives an average annual rainfall of 2000 mm accounting for around 10% (42.50 M ha m) of the country's total precipitation of 420.0 M ha m (Pattanayak *et al.*, 2011). The soils are rich in organic matter, but the soil of the region is acidic to strongly acidic in reaction. The depth of the soil varies from shallow in *inceptisols* and *antisols* to very deep in *alluvial* soils. The agricultural production system in the region is mostly rainfed, mono-cropped and at subsistence level. Rice is major crop in NER except Sikkim where maize is dominant crop. More than 80% agricultural soils are acidic in reaction and hence the crop diversification with pulses is an important aspect of soil health improvement. It has been estimated that pigeonpea, soybean, cowpea, groundnut, field pea and rice bean can produce 2.0, 1.2, 2.0, 0.5, 0.48 and 1.1 t/ha of leaf fall, respectively, which directly contributes to soil fertility and helps ameliorate soil acidity (Hazarika *et al.*, 2006).

The major pulses grown in the region are mungbean (*Vigna radiata*), urdbean (*Vigna mungo*), pigeonpea (*Cajanus cajan*), rice bean (*Vigna umbellata*) and cowpea (*Vigna unguiculata*) in *kharif* and frenchbean (*Phaseolus vulgaris*) and fieldpea (*Pisum sativum*). In hills, various other beans such as faba bean (*Vicia faba*), adzuki bean (*Vigna angularis*), mothbean (V. *aconitifolia*) and broad bean (*Dolichos lablab*) are also grown and used as pulses. In *jhum* areas of the region, rice bean is still a predominant pulse crop. NER is a seat of diversity for several grain legumes, *viz.* urdbean, cowpea and rice bean (Gupta *et al.*, 1998). This also showed that great ecological potential exist in NER for increasing area under pulses. With the technological interventions and policy support, the area under pulses in NER since 2000 increased by 64%, production by 21% and productivity by 25% (Fig 7). But, the region as a whole has a deficit of 78.79% in pulses, ranging from 6.4% in Nagaland to 91.5% deficit in Tripura (Das *et al.*, 2016). At present (2018-19) in NER, pulses are cultivated in area of 2,76,518 ha with production of 2,44,418 tonnes and productivity of 1062 kg/ha. However, enormous inter-state variation exists in the area and production of pulses in NER. Assam has highest area of 1,50,229 ha followed by Nagaland (37,720 ha), Manipur (31,070 ha), Tripura (25,854 ha), Arunachal Pradesh (13,300 ha), Meghalaya (9,575 ha), Sikkim (5,001 ha) and Mizoram (3,769 ha).

In NER, potential areas exist for horizontal expansion of pulses is *jhum* cultivation (Bujarbaruah, 2004) and rice/maize fallow (Ghosh *et al.*, 2010). Usually, spring (before *kharif* rice sowing) and autumn (after *kharif* rice harvesting) seasons in this region go as fallow. Short duration pulses like mungbean, urdbean and rajmash offer the opportunity for these seasons (Das *et al.*, 2016). Organic pea and lentil cultivation under NT in lowland rice fallow has been reported to be a profitable venturein mid-altitude of Meghalaya (Das *et al.,* 2014). Frenchbean has been identified as one of the most potential crops for organic production owing to its suitability to cultivate twice in a year i.e., during January/ February to April/ May and July/ August to October/November (Patel *et al.*, 2015). Pulses inclusive several mixed cropping or intercropping as well as cropping sequences have been developed by the agricultural institutions of this region can be followed to increase pulses production and to minimize the demand gap in NER. Many pulses varieties have been developed suitable for cultivation in acid soils. Hence, tremendous potential exist for horizontal and vertical expansion of pulses area in NER.

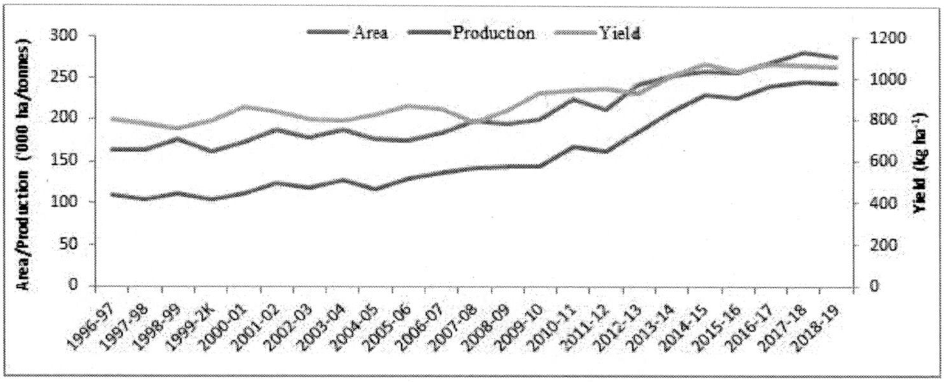

Fig. 7. Trend in area, production and yield of total pulses in North-East Region (1996-2019)

2.12 *Kharif* pulses in Bundelkhand

Bundelkhand comprised of 13 districts, 7 districts of Uttar Pradesh (Banda, Chitrakoot, Hamirpur, Jalaun, Jhansi, Lalitpur and Mahoba) and 6 districts of Madhya Pradesh (Chhatarpur, Damoh, Datia, Panna, Sagar and Tikamgarh) is considered as pulses bowl in the country. The soil of this region is broadly classified into deep and medium black in Southern part (44%), mixed red & black, red & yellow (56%) in middle part and alluvial in north. Heavy black and light red soils are extensively distributed in the region. Agriculture is the predominant occupation and mostly rainfed in Bundelkhand. Double cropping is followed in less than 30% of total cultivated area, except in Jhansi, Lalitpur, Sagar and Tikamgarh districts. The contribution of cereals in total foodgrains production in this region is maximum (about 55%) followed by pulses (32%) and oilseeds (8%). The region produce surplus pulses and therefore, it contribute significantly to the farmer's income.

However, almost all pulse crops are grown under rainfed condition (>95%). In contrast to other parts of the country, rabi sowing (69%) in Bundelkhand is considerably higher than kharif (31%), although the rainfall pattern and the extent of irrigation in Bundelkhand theoretically favours *kharif* (monsoon) cultivation. The area where clay soil is prevailing, monocropping of *rabi* pulses like chickpea, lentil and pea is most common. This area can be utilized for cultivation of *kharif* pulses by following of recommended package and practices. As per an estimate about 1.2 m ha area is available for cultivation under kharif fallow in Bundelkhand region (Tiwari and Shivhare, 2016).

3. CONCLUSION

Despite of increase in production of animal protein, the demand of plant protein is also increasing with time due to food habits and vegetarian population of the country. Pulses are the major source of plant origin quality protein along with minerals and vitamins. Therefore, to meet the demand of burgeoning population and to ensure self sufficiency, the country needs 39 Mt of pulses by 2050 which necessitates an annual growth rate of 2.14%. This can be possible through horizontal and vertical expansion of area under pulses in the country. The new niches discussed above are the avenues where greater scope exit for area expansion and increasing production of pulses in the country that can ensure the nutritional security.

REFERENCES

Ali, M. (1988). Efficacy of herbicides for weed control in winter French bean (*Phaseolus vulgaris*). *Indian Journal of Agricultural Sciences* 58(2), 440-3.

Ali, M. (1989). Response of frenchbean genotypes to population densities during winter season. *Indian Journal of Pulses Research* 2(2), 125-128.

Ali, M. and Kushwaha, B.L. (1987). Cultivation of *rabi* rajmash in plains. *Indian Farming* 37(5), 20-23.

Ali, M., Kumar, N. and Ghosh, P.K. (2012) Milestones on agronomic research in pulses in India. *Indian Journal of Agronomy* 57(3rd IAC Special Issue), 52-57.

Ali, M., Kumar, N. and Praharaj, CS. (2016). Agronomic research in pulses in India: Historical perspective, accomplishments and way forward. *Indian Journal of Agronomy* 61(4th IAC Special Issue), S83-S92.

Ali, M. and Lal, S. (1991). Rabi Rajmash: Potentialities, Limitations and Technology of Production. *Tech. Bull.* (No. 3). Directorate of Pulses Research, Kanpur. pp.17 .

Anonymous. (2015). Vision 2050. Indian Institute of Pulses Research (Indian Council of Agricultural Research), Kanpur208024 (Uttar Pradesh), India. 36 pages.

Banyal O.K. and Singh, A. (2007). Integrated management of rajmash diseases in dry temperate region of north-western Himalayas. Indian Phytopathology 60(3), 317-321.

Bujarbaruah, K.M. (2004). Organic Farming: Opportunities and challenges in North Eastern Region of India. (In) *Nature 2004*, pp. 13–23. Souvenir of International Conference on Organic Food, held during 14–17 February at ICAR Research Complex for NEH

Region, Barapani.

Chandra, S. and Ali, M. (1986). Recent achievements in pulses production. *Technical Bulletin No. 1*. Directorate of Pulses Research, Kanpur, India. pp. 17-18.

Choudhary, A.K., Kumar, R, Saxena, K.B. and Singh, A.K. (2020). Post-Rainy Season Pigeonpea Holds Promise for the Diversification of Cropping Systems. *Journal of AgriSearch* 7(3), 115-120.

Das, A., Babu, S., Yadav, G.S., Ansari, M.A., Singh, R., Baishya, L.K., Rajkhowa, D.J. and Ngachan, S.V. (2016). Status and strategies for pulses production for food and nutritional security in north-eastern region of India. *Indian Journal of Agronomy* 61(Special issue), 43-57.

Das, A., Patel, D.P., Ramkrushna, G.I., Munda, G.C., Ngachan, S.V., Kumar, M., Buragohain, J. and Naropongla. (2014). Crop diversification, crop and energy productivity under raised and sunken beds: results from a seven-year study in a high rainfall organic production system. *Biological Agriculture and Horticulture* 30(2), 73–87.

Desai, N.C. and Intiwala, C.G. (1999). Optimization of sowing time and spacing in post rainy season pigeonpea grown in South Gujarat. *International Chickpea and Pigeonpea Newsletter* 6, 56-57.

Dixit, G.P., Parihar, A.K., Bohra, A. and Singh, N.P. (2016). Achievements and prospects of grass pea (*Lathyrus sativus* L.) improvement for sustainable food production. *The Crop Journal* 4, 407-416.

Dixit, S., Khanna, S.K. and Das, M. (2008). Non-uniform implementation of ban on *Lathyrus* cultivation in Indian states leading to unwarranted exposure to consumers. *Current Science* 94, 570–571.

Faroda, A.S. and Singh, R.C. (1983). Effect of preceding crops on the nitrogen need of succeeding wheat crop. *Indian Journal of Agricultural Sciences* 53, 123-128.

Ghosh, P.K., Das, A., Saha, R., Enboklang, K., Tripathi, A.K., Munda, G.C. and Ngachan, S.V. (2010). Conservation agriculture towards achieving food security in North East India. *Current Science* 99(7), 915–921.

Gumma, M.K., Thenkabail, P.S., Teluguntla, P., Rao, M.N., Mohammed, I.A. and Whitbread, A.M. (2016). Mapping rice fallow cropland areas for short-season grain legumes intensification in South Asia using MODIS 250 m time-series data. *International Journal of Digital Earth* 9(10), 981-1003.

Gupta, H.S., Sanna, B.K. and Sahay, G. (1998). Present status and prospects of pulse production in north eastern hill region. *Indian Journal of Hill Farming* 11(1& 2), 77- 87.

Gupta, S. and Pratap, A. (2016). Mungbean summer cultivation in India (Pocket Guide). *Extension Bulletin No. 1/2016*. AICRP on MULLaRP, Indian Institute of Pulses Research, Kanpur – 208024. 42 pages.

Hazarika, U.K., Munda, G.C., Bujarbaruah, K.M., Das, A., Patel, D.P., Prasad, K., Kumar, R., Panwar, A.S., Tomar, J.M.S., Bordoloi, J.S., Sharma, M. and Gogoi, G. (2006). *Nutrient Management in Organic Farming*, Technical Bulletin No. 30, ICAR Research Complex for NEH Region, Meghalaya.

Karle, A.S. and Pawar, G.G. (1998). Seed yield of pigeonpea genotypes under varying planting geometry during *rabi* season. *Indian Journal of Pulses Research* 11(1), 98-99.

Kumar, N., Hazra, K.K., Singh, S. and Nadarajan, N. (2016). Constraints and prospects of growing pulses in rice fallows of India. *Indian Farming* 66(6), 13-16.

Kumar, N., Yadav, A., Singh, S. and Yadav, S.L. (2018). Growing Pulses in Rice Fallow: Ensuring Nutritional Security in India. In Anup Das *et al* (eds). Conservation Agriculture for Advancing Food Security in Changing Climate. Today & Tomorrow's Printers and Publishers, New Delhi. Vol. 1 (2018), 107-122.

Kumar, S., Gupta, P., Barpete, S., Sarker, A., Amri, A., Mathur, P.N. and Baum, M. (2013). Grass Pea. In: M. Singh, H.D. Upadhyaya, I.S. Bisht (Eds.), Genetic and Genomic Resources of Grain Legume Improvement. Elsevier, Oxford, United Kingdom 2013. pp. 269–292.

Kumar, S., Rana, N.S., Singh, R. and Singh, A. (2006). Production potential of spring sugarcane as influenced by intercropping of dual purpose legumes under Tarai conditions of Uttarakhand. *Indian Journal of Agronomy* 51(4), 271-273.

Mahler-Slasky, Y. and Kislev, M.E. (2010). Lathyrus consumption in late Bronze and Iron Age sites in Israel: an Aegean affinity. *Journal of Archaeological Science* 37, 2477–2485.

Nagamani, C., Sumathi, V. and Prabhakara, R.G. (2015). Performance of RABI Pigeonpea Under Varied Times of Sowing, Nutrient Dose and Foliar Sprays. *Progressive Agriculture* 15(2), 253-258.

Narayanan, A. and Sheldrake, A.R. (1979). Pigeonpea (*Cajanus cajan*) as a winter crop in peninsular India. *Experimental Agriculture* 15, 91-95.

Panwar, B.S., Singh, B.V. and Sharma, J.C. (1990) Feasibility of intercropping in autumn planted sugarcane. *Indian Sugar* 39(10), 755-756.

Panwar, K.S. and Yadav, H.L. (1980). Response of short duration pigeonpea to early planting and phosphorus levels in different cropping systems. pp. 37-44. In: proceedings of the international Workshop on Pigeonpea, Volume1, 15-19 December 1980, International Crops Research Institute for Semi-Arid Tropics (ICRISAT), Patancheru, India.

Patel, D.P., Das, A., Kumar, M., Munda, G.C., Ngachan, S.V., Ramkrushna, G.I., Layek, J., Naropongla, Buragohain, J. and Somireddy, U. (2015). Continuous application of organic amendments enhances soil health, produce quality and system productivity of vegetable-based cropping systems at subtropical eastern Himalayas. *Experimental Agriculture* 51(1), 85–106.

Pattanayak, A., Deka, B.C., Singh, R.K., Tripathi, A.K., Mohapatra, K.P., Das, A., Choudhury, B.U., Khan, M.H., Singh, U. and Ramkrushna, G.I. (2011). Vision 2030. ICAR Research Complex for NEH Region (Indian Council of Agricultural Research), Umiam-793 103, Meghalaya. 69 pages.

Ramanujam, S. (1971). Some salient results of pulse research. *Indian Farming* 21(10), 17-19.

Rana, N.S., Kumar, S., Saini, S.K. and Panwar, G.S. (2006). Production potential and profitability of autumn sugarcane-based intercropping systems as influenced by intercrops and row spacing. *Indian Journal of Agronomy* 53(2), 140-144.

Roy Sharma, R.P., Thakur, H.C. and Sharma, H.M. (1980). Pigeonpea as a *rabi* crop in India. Pp 26-36. In: Proceedings of International Workshop on Pigeonpea, Volume 1, 15-19 December 1980, International Crops Research Institute for Semi-Arid Tropics (ICRISAT), Patancheru, India.

Sadhukhan, R., Bhowmick, M.K. and. Biswas, P.K. (2008). Performance of rajmash (*Phaseolus vulgaris* L.) under varying dates of sowing in West Bengal. *Journal of Crop and Weed* 4(2), 42-44

Sandhu, H.S., Brar, S.S. and Gill, G.S. (1973). Arhar-wheat is good rotation. *Progressive Farmer* 9(10), 8.

Singh, A.K., Lal, M. and Suman, A. (2008). Effect of intercropping in sugarcane (*Saccharum* complex hybrid) on productivity of plant cane - ratoon system. *Indian Journal of Agronomy* 53(2), 140-144.

Singh, A.K., Singh, J., Rao, A.K., Singh, A.K., Singh, S.B., Sharma, M.L. and Singh, R.R. (2009a). Significance of pulse intercropping with sugarcane ratoon. *Indian Journal of Fertilizers* 5(7), 25-27.

Singh, K.K., Ali, M. and Venkatesh, M.S. (2009). Pulses in Cropping Systems. Technical Bulletin, IIPR, Kanpur.

Singh, N.P. (2018). Pulses as a candidate crops for doubling farmers' income. *Indian Farming* 68(01), 36-43.

Singh, S.S., Kumar, N., Praharaj, C.S. and Singh, N.P. (2017). Agro-technologies for enhancing pulses production in rice fallows. ICAR-Indian Institute of Pulses Research, Kanpur - 208 024 (Uttar Pradesh), India. 36 pages.

Subbarao, G.V., Kumar Rao, J.V.D.K., Kumar, J., Debb, U.K., Ahmed, I., Harris D, *et al*. (2001). Spatial Distribution and Quantification of Rice-Fallows in South Asia-Potential for Legumes. International Crops Research Institute for Semi-Arid Tropics (ICRISTAT), National Remote Sensing Agency (NRSA) and Department for International Development (DFID). pp 74-80.

Tiwari, A.K. and Shivhare, A.K. (2016). Pulses in India: Retrospect and Prospects. Publication No.: DPD/Pub.1/Vol. 2/2016. Directorate of Pulses Development, Ministry of Agriculture & Farmers Welfare (GOI), Bhopal, Madhya Pradesh.

Venkata Rao, P., Subbarami Reddy, A., SateeshBabu, J. and Ramana, M.V. (2016). Effect of Sowing Time and Supplemental Irrigation on Yield and Economics of *rabi* Pigeonpea (*Cajanus cajan* (L.) Millsp.). *International Journal of Research in Agricultural Sciences* 3(6), 316-319.

Venkatesh, M.S., Hazra, K.K., Ghosh, P.K., Prahraj, C.S. and Kumar, N. (2013). Long-term effect of pulses and nutrient management on soil carbon sequestration in Indo-Gangetic plains of India. *Canadian Journal of Soil Science* 93(1), 127-136.

Yadav, R.L., Prasad S.R. and Singh, K. (1987). Fertilizer requirement and row arrangements of pulses in sugarcane based intercropping systems. *Indian Journal of Agronomy* 32, 80-84.

CHAPTER 17

Indigenous technical knowledge for enhancing pulses production and livelihood security

Uma Sah*, Hem Saxena, Jitendra Ojha and Shantanu Kumar Dubey[1]

ICAR-Indian Institute of Pulses Research, Kanpur 208024 (Uttar Pradesh)

1. *ICAR-ATARI, Zone III, Kanpur 208001 (Uttar Pradesh)*

**Corresponding author email: umasah@gmail.com*

Abstract

Indigenous knowledge is the pool of knowledge resources accumulated by a community over years of experimentations for surviving the challenges faced. Comprehension of this important knowledge resource give insights into the pathways adopted by indigenous community in solving the problems related to agriculture within the local environment and its integration in the formal research system. This would ensure generating location specific, need oriented and economically affordable technologies rooted in the local environment. Pulses in India are a prominently integrated into the prevailing cropping sequences across different production situations. These crops are primarily cultivated under resource poor conditions and are vulnerable to the climatic fluctuations. It is in this perspective of that the Indigenous Technical Knowledge (ITK) holds high relevance for pulse crops. The chapter comprehensively presents the observations on the ITKs prevalent in different parts of the country, along with the prospects of their integration in the formal research system.

Keywords: Intercropping, Mixed cropping, Nutrient, Soil conservation, Soil moisture

1. INTRODUCTION

Indigenous Technical Knowledge in agriculture refer to the repository of wisdom that farmers have evolved with years of experimentation and trial and error for deriving solutions to the problems faced by them in agricultural production, which form the basis for their survival. It is the sum of experiences and knowledge accumulated by a particular community over period of time, based on which the community or its members take decisions to tackle the problem faced by them. It evolves in the local environment, is creative and experimental. ITKs constantly incorporate the influences to meet the new challenges and are the key elements for the survival of the community. Warren (1991) defined Indigenous knowledge (IK) as the knowledge used by local people to make a living in a particular environment. While Johnson (1992) defined Indigenous knowledge as "a body of knowledge built up by a group of people through generations of living in close contact with nature". Indigenous knowledge has powerful advantages over outside knowledge in terms of cost involved, ease in operations and local availability.

The recent focus on indigenous knowledge base in agriculture is manifestation of the concern for sustainability of agriculture production system and the urgency to search of more sustainable options (Sah, 2019). The recognition of indigenous knowledge for developing appropriate technology gained importance for increasing their adaptability/acceptability and to bring down the dependence on cost intensive green revolution technologies. The very core characteristics of indigenous technical knowledge like their local resource base make the indigenous technologies more relevant.

2. IMPORTANCE OF INDIGENOUS TECHNICAL KNOWLEDGE IN SYSTEM PERSPECTIVE

Understanding of Indigenous knowledge is important for several reasons. Firstly, it gives clue the way in which people have dealt with or experimented with their local environment in the past. It also ensures an enhanced participation of people. Understanding and working with indigenous systems improve the communication between farmers, extension agents and researchers (Tabor, 1993). Secondly, understanding and working with indigenous knowledge and decision-making system can enhance participation, capacity building in a cost effective way (Warren, 1991). It could also contribute to local empowerment and development, increasing self-sufficiency and strengthening self-determination (Thrupp, 1989). Thirdly, understanding ITKs pave the way for generating location specific, need and problem oriented, socially desirable and economically affordable technologies (Sah, 2014). Understanding of farmers' knowledge allows a framework of reference for posing technical and scientific questions in research and also provides the basis for evolving technological options (Scoones, 1989). Use of IK assures that the end user of specific agricultural development projects are involved in developing technologies appropriate to their needs (Warren, 1991). Thus, Indigenous knowledge can play a key role in evolving technologies that are sustainable and can also increase their acceptance of the same among the farming community.

With the rise in concern for sustainability, rethinking and revisiting the traditional crop production practices in search of sustainable options was felt. As a result, the recognition and promotion of indigenous knowledge for developing appropriate technology gained importance. This would also increase their adaptability/acceptability and bring down the dependence on cost intensive green revolution technologies. In all, indigenous practices play a vital role in sustainable agriculture (Karthikeyan *et al.*, 2006).

3. INDIGENOUS TECHNICAL KNOWLEDGE AND PULSES

Pulses are basic to the diet as well as cropping system across India. These crops primarily occupy marginalized rainfed land and are grown as subordinate crops in cropping systems (Nene, 2006). In addition, pulses are usually grown by farmers having little or no access to costly inputs, credits and markets (Ali, 2004). Pulses are grown for "income" and 'not just for home consumption" primarily by small farmers operating under complex, diverse and poor resource conditions. In addition, pulse crops represent a group of crops that are highly vulnerable to the market prices and climatic conditions. It is in this perspective of the constrained conditions in which pulses are grown, Indigenous Technical Knowledge acquire high relevance. Therefore, Indigenous knowledge offer to be an alternative that could be explored for designing of sustainable pulse technologies that easily acceptable for achieving a higher level of yield and production in the country.

3.1 Indigenous knowledge related to soil and moisture conservation

In recent years, the issues of conservation have assumed importance in view of widespread degradation of natural resources, the need to reduce production costs, increase profitability and make agriculture more competitive (Sangar *et al.*, 2004) . Research studies indicate that, in rainfed production situation with unpredictable rainfall and poor productivity, zero/ reduced tillage system without crop residues left on the soil surface fail to generate advantage as much of the rainfall is lost as runoff due to rapid surface sealing nature of soils. It would therefore appear that no tillage alone in the absence of soil cover is unlikely to become a favored practice. Thus, in such semi-arid regions there is need to identify situations where availability of even moderate amount of residues can be combined with reduced tillage to enhance soil quality and efficient use of rainwater (Sangar *et al.*, 2004).

In rainfed areas of Bundelkhand region with heavy soils, *rabi* pulses are grown on conserved soil moisture. The land preparation for the post-rainy season sowing starts with the onset of rains in the months of June- July and is repeated many time after rains during the entire monsoon season. In September, the soil is planked and closed and sowing starts in the month of October. Irrespective of the soil types, the field are ploughed at least 5-6 times under mono-cropped conditions. The rationale given by farmers for use of this procedure for land preparation was conservation of soil moisture, management of weeds and to prevent sealing of soil that would lead to poor water penetration (Sah, 2014).

3.2 Indigenous mixed and intercropping

Mixed and intercropping forms an integral part of traditional farming in India. The objective is to meet the diversified demand of food for the farm family with the given resources as well as risk minimization against crop failure. Intercropping was observed to be essentially a system of small and dry land farmers and the traditional intercropping is highly complex and diverse (Singh, 1981). Intercropping or mixed cropping are the key element of traditional farming systems. Mixed cropping or intercropping stabilized yields, conserve the soil and made it possible to harvest different crops at the same time. Other advantages were reduction in susceptibility of the crops against pests and diseases and a better use of the environment, where the combination of species grown had different light requirements or explored different depths of soil (Spedding, 1979). The system also provides a complete vegetation canopy at different heights and thus broke up heavy rainfall and protected the soil (Mwaura, 2008).

Various crops are mixed and intercropped with pulses in different parts of the country. In Anantpur district of Andhra Pradesh, farmers practice mixed cropping of groundnut pearlmillet, pigeonpea and castor with the rationale that castor serve as trap crop for *Spodoptera litura*, sorghum/ pearmillet serves as guard crop, pigeonpea is a slow growing crop in early stages, offers less competition to groundnut, thus ensuring some returns (Anonymous, 2003a). Intercropping of maize and pigeonpea is practiced in *rabi* season. Maize is a short duration crop and is harvested up till 100 days, and pigeonpea being a long duration crop can efficiently utilize resources (Rao, 2003). Similarly, the tribal farmers of

Nandurbar, Dhule and Ahmednagar dist of Maharastra follow mixed cropping of Nagali, maize, sorghum, and pigeonpea. Pigeonpea is mixed with groundnut and vice versa. Pulses like blackgram and greengram are also mixed with sorghum/maize to cover the risk of crop failure of sole crop (Anonymous, 2003b). While Korku tribal farmers of Amravati district of Maharastra practiced mixed cropping with pigeonpea, upland paddy, sorghum, and sesamum with the rationale that it increases the land use efficiency. The root systems of intercrops are different and they mature at different times and hence a proper utilization of nutrients and soil moisture at different soil depths is made possible. (Anonymous, 1999-2000). Under rainfed conditions, farmers in Dindori district of Madhya Pradesh follow indigenous cropping system with pigeon pea and rice, wherein farmers grow tall varieties of rice and on field bunds pigeon pea (Singh, 2004).

In hilly areas of Himachal Pradesh many mixed cropping systems were reported which include mixed cropping of wheat, mustard, gram, bottle gourd and pea in *rabi* season (Verma, 2003), maize and rajmash by mixing equal amount of seeds and sowing (Ram, 2003), wheat and chickpea (Kumar, 2003) and wheat, mustard, chickpea and lentil (Sharma, 2003).

Similarly in Uttaranchal hills, 'Baraanja' is practiced, which literally means growing twelve kinds of crops together. It involves mixing and sowing twelve crops in a single plot of land involving kidney bean, blackgram, greengram, horsegram, amaranth and others in a jumbled fashion. The rationale mentioned was obtaining optimal and sustained yield and since maturity period of crops are varied, harvested at different times, which help in retaining soil moisture and retain constant supply of food. It enhances fertility of soil leading to increased productivity (Navdanya, 1993).

In Bundelkhand region of Uttar Pradesh, during rainy season, the seeds of the pigeonpea and sorghum crops are mixed in 2:1 ratio and broadcast. The cobs are cut as the sorghum crop matures in the month of November and staggered harvesting of the stems is done for fodder purpose. In addition, sorghum, pigeonpea and greengram were sown as mixed crops whereas no definite ratio was maintained. With a little variation, mixed crop of sorghum, pigeonpea and urdbean was grown wherein the seed were mixed in proportion of 8, 4 and 1 kg, respectively and broadcast in the same field. While during, *rabi* season, in rainfed areas, practice of mixing of seeds and broadcasting was observed in combination of chickpea and linseed or chickpea, wheat, field pea and mustard seeds. This type of mixed cropping was preferred in fields distantly located from the village. Generally, chickpea, linseed and mustard seeds were mixed at a rate of 14 kg, 1 kg and 550 g, respectively. The rationale provided for this preference was a low damage by animals as well as humans in such mixed cropping besides the need to fulfill the varying food demands of farm households. To cover the risk of failure of sole pulse crops and to ensure some returns, farmers' practices mixed/ inter cropping in the country. Intercropping of mustard and chickpea with one line of mustard after every 9 or 18 lines of chickpea was practiced in Bundelkhand region while in Banda district, two rows of linseed was taken every 7 rows of chickpea (Sah, 2014)

3.3 Indigenous method of nutrient management

In Uttar Pradesh state, plants of urdbean are ploughed into the soil after the pods are harvested. This practice is particularly done in the fields where wheat crop follows in rabi season. Farmer perceive that this helps in enhancing productivity of wheat crop (Sah, 2014).

3.4 Sowing in low soil moisture conditions

Soaking of chickpea seed for 7-8 hours prior to sowing is practiced in case of delayed sowing or when the soil moisture is low and when sowing is done with desi plough. The practice is followed with the rational that it would ensure better germination. Under rainfed conditions for sowing of chickpea crop, the field is ploughed in evening and the seed is broadcast. The planking is done in the morning hours. It is perceived that the moisture absorbed by soil and seed from air during night helps in better germination.

4. RESEARCHERS' PERCEPTION ON SCIENTIFIC RATIONALITY OF ITKS

Perception of researchers on the Indigenous knowledge pool is important for its incorporation in the formal research system and generation of technologies that are appropriate and sustainable with faster rate of adoption among farming community. Perception of researchers of national agricultural research system of India was assessed with regard to scientific base of ITKs as well the scientific validity. Researchers (79.41%) agreed that integration of indigenous knowledge in formal agricultural research could lead to easy adoption of technologies among the farming community and perceived that integration of traditional practices/ knowledge with improved agriculture practices can make agriculture more sustainable (82.35%). Researchers (84%) perceived that scientific basis of the indigenous practices needs to be confirmed by on-station experimentations. More than 40% of the researchers disagreed with the statement that traditional or indigenous knowledge is of little relevance to agricultural development while about 23% researchers disagreed that modern agriculture science can only solve the problem of food security and Indigenous knowledge cannot solve the complex problems of pulses production. Therefore, researchers were observed to have favorable perception towards ITKs, however majority of them stressed on confirming the scientific rationality through due experimentations (Table 1).

Table 1. Perception of scientist towards indigenous technical knowledge (N=68)

S. No.	Statement	A	SA	DA	Score
1	Traditional or indigenous knowledge is of little relevance to agricultural development	30 (44.12)	10 (14.70)	28 (41.17)	1.97
2	Integration of indigenous knowledge in formal agricultural research could lead to easily adopted technologies among the farming community.	54 (79.41)	08 (11.76)	06 (08.82)	2.71
3	Modern agriculture science can only solve the problem of food security	20 (29.40)	32 (47.06)	16 (23.52)	1.94
4	Integration of Traditional practices/knowledge with Improved agriculture practices can make the agriculture more sustainable	56 (82.35)	12 (17.65)	0 (00.00)	2.82

5	Indigenous practices are best suited to resource poor farming conditions only	18 (26.47)	28 (41.12)	22 (32.35)	2.06
6	The scientific basis of the indigenous practices needs to be confirmed by on-station experimentations.	57 (83.82)	08 (11.76	03 (04.41)	2.79
7	Indigenous knowledge cannot solve the complex problems of pulses production	22 (32.35)	31 (45.59)	16 (23.52)	1.94
8	Indigenous practices are low cost technologies &they may enhance profitability of pulses cultivation	35 (51.47)	25 (36.76)	08 (11.76)	2.40
9	Farmers use ITKs because they do not have access to improved technologies	20 (29.41)	34 (50.00)	14 (20.59)	1.91
10	Use of ITK in large magnitude and wider geographical area is possible	27 (39.71)	24 (35.29)	17 (25.00)	2.15

*A: Agree, SA: Somewhat Agree, DA: Disagree; Source: Sah *et al.* (2019)

5. RESEARCHERS' PERCEPTION OF ITKS RELATED TO PULSE CULTIVATION

Indigenous practice of soil moisture test by pressing soil in form of ball drawn from 3 inch deep layer was perceived to be having high scientific rationality and recommended it for scientific validation by more than 71% of the sampled respondents (Table 2). Seed soaking before sowing in soils with low moisture level was indigenous practice reported to be performed in situation where soil moisture was not sufficient for sowing of pulses seed. The practice was perceived to have high scientific base by more than 81% of the researchers and equally it was recommended for validation. While ploughing back the urd bean plants after plucking was used as nutrient conservation technique and the practice of irrigation after flowering initiation in mungbean crop was followed by pulse growers of state, these practices were judged to have high scientific base by 75% of the sampled researchers. In addition, more than 81 percent of the researchers recommended these indigenous practices for validation.

Table 2. Researchers' perception of ITKs related to pulse production

Indigenous Technical knowledge related to crop production (n=32)	Scientific base score	Recommended for validation
Soil moisture test before sowing of rabi crop	85	23 (71.87)
Seed soaking before sowing in low soil moisture condition	90	24 (75.00)
Nutrient conservation technique by ploughing back the urd bean plants after plucking	87	26 (81. 25)
Irrigation time in mung bean crop after flowering initiation	83	26 (81. 25)

*Figure in parentheses is in percentage; Source: Sah et al. (2019)

6. CONCLUSION

Indigenous technical knowledge resource is a product of years of experimentation of local community for finding solutions that are basic to their survival. It is a dynamic pool of knowledge that incorporates changes and is thus open to the external influence of technological advancement. It offers solutions that are ecologically and socially sound besides being local resource based. These factors are especially more relevant to pulse crops that are primarily cultivated under marginalized resource conditions in the country. However, promotion of this knowledge pool depends on assessment of the existing indigenous knowledge towards finding solutions. Positive attitude of researchers towards indigenous practices and their recommendation for scientific validation of the most potent ones provides a way forward for integration of this knowledge pool in the structured research efforts for validation and up scaling. This would translate into low cost, viable and sustainable alternative technological interventions for improving pulse productivity on a larger scale.

REFERENCES

Ali, M. (2004). Research achievements and future strategies for increasing pulse production in India- An overview. Proceedings of National Seminar on delineation of thrust areas for pulses development activities under modified ISOPOM scheme held on 27-28 March, 2004. Pp. 28-33.

Anonymous. (1999-2000). Documentation of indigenous technical knowledge in tribal areas of Dharni panchayat samiti, Amravati. Department of Extension education, Dr Punjabrao Deshmukh Krishi Vidyapeeth, Akola. pp. 54.

Anonymous. (2003a). Mixed cropping of groundnut, sorghum, pearl millet, pigeon pea and castor. In: Inventory of Indigenous Technical Knowledge in Agriculture. Document 2, Supplement 1. Compiled by P. Das, S.K. Das, A. Mishra, H.P.S. Arya, K.M. Bujarbarua, R.P. Singh, L.R. Verma, G. Subba Reddy, M. Geetha Rani, H.S. Gupta, Z.D. Kavia and D.P. Ray. Indian Council of Agricultural Research, New Delhi. pp. 21.

Anonymous. (2003b). Use of mixed cropping to cover risk of failure. In: Inventory of Indigenous Technical Knowledge in Agriculture. Document 2, Supplement 1. Compiled by P. Das, S.K. Das, A. Mishra, H.P.S. Arya, K.M. Bujarbarua, R.P. Singh, L.R. Verma, G Subba Reddy, M. Geetha Rani, H.S. Gupta, Z.D. Kavia and D.P. Ray. Indian Council of Agricultural Research, New Delhi. pp 22.

Johnson, M. (1992). Lore: Capturing Traditional Environmental Knowledge. IDRC: Ottawa, Canada.

Karthikeyan, C., Veeraragavathatham, D., Karpagam, D. and Firdouse, S Ayisha. (2006). Indigenous storage practices in pulses. *Indian Journal of Traditional Knowledge* 5(1), 104-107.

Kumar, R. (2003). Mixed cropping of wheat with chickpea. Inventory of Indigenous technical knowledge in Agriculture Document 2. Compiled by P. Das, S.K. Das, H P S Arya,, R.P. Singh 'Ratan', A. Mishra, K.M. Bujarbaruah, , G Subba Reddy, L.R. Verma, M. Geetha Rani, H.S. Gupta, C. Satapathy, and Z.D. Kavia, Indian Council of Agricultural Research, New Delhi. pp 68.

Mwaura, P. (2008). Indigenous Knowledge in Disaster Management in Africa. United Nations Environment Programme, Nairobi, Kenya. Pp 7-8.

Navdanya. (1993). Cultivating diversity: Biodiversity conservation and politics of seed. Report no 1. Research foundation for science. Technology and National Resources Policy, Navdanya, Dehradun, India.

Nene, Y.L. (2006). Indian Pulses through the Millennia. *Asian Agri-History* 10(3), 179–202

Ram, L. (2003). Mixed cropping of maize and rajmash. In: Inventory of Indigenous technical knowledge in Agriculture Document 2. Compiled by P. Das, S.K. Das, H P S Arya,, R.P. Singh 'Ratan', A. Mishra, K.M. Bujarbaruah, , G Subba Reddy, L.R. Verma, M. Geetha Rani, H.S. Gupta, C. Satapathy, and Z.D. Kavia, Indian Council of Agricultural Research, New Delhi. pp 62.

Rao, A.R. (2003). Maize and red gram intercropping system. Inventory of Indigenous technical knowledge in Agriculture Document 2. Compiled by P. Das, S.K. Das, H P S Arya,, R.P. Singh 'Ratan', A. Mishra, K.M. Bujarbaruah, , G Subba Reddy, L.R. Verma, M. Geetha Rani, H.S. Gupta, C. Satapathy, and Z.D. Kavia, Indian Council of Agricultural Research, New Delhi. Pp 65.

Sah, U., Dubey, S.K. and Saxena, H. (2014). Indigenous pulse storage methods in Bundelkhand region of Uttar Pradesh: An exploratory study. *Current Advances in Agricultural Sciences* 6(2), 161-164.

Sah, U., Dubey, S.K., Saxena, H. and Kumar, R. (2019). Scientific Rationality of Indigenous Technical Knowledge related to Pulse Production: Researchers' Perception. *Journal of Community Mobilization and Sustainable Development* 14(1), 20-24.

Sangar, S., Abrol, I.P. and. Gupta R.K. (2004). Conference Report on Conservation Agriculture: Conserving Resources for Enhancing Productivity , held at Centre for Advancement of Sustainable Agriculture , NASC Complex New Delhi , Sept 22-23, 2004.

Scoones, I. (1989). Patch use by cattle in dry land Zimbabwe: Farmer knowledge and Ecological theory. ODI Pastoral Development Network paper no. 28. London: Overseas Development Institute.

Sharma, V.D. (2003). Mixed cropping of wheat, mustard, chickpea and lentil. Inventory of Indigenous technical knowledge in agriculture, Document 2. Compiled by P Das, S.K. Das, HPS Arya, R P Singh Ratan, A Mishra, K.M. Bujarbaruah, G. Subba Raddy, L.R. Verma, M Geetha Rani, H S Gupta, C Satapathy and Z D Kavia. Indian Council of Agricultural Research, New Delhi. Pp 68.

Singh, R.K. (2004). Indigenous cropping system under rainfed agro-ecosystem. Inventory of Indigenous Technical Knowledge in Agriculture. Document 2, Supplement 2. Compiled by P.K. DAs, S.K. Das, A. Mishra, H P S Arya, K.M. Bujarbarua, R.P. Singh 'Ratan', L.R. Verma, G Subba Reddy, M. Geetha Rani, H.S. Gupta, Z.D. Kavia, D.P. Ray, Indian Council of Agricultural Research, New Delhi. pp 20.

Singh, R.P. (1981). Crop Failure and Intercropping in the Semi-Arid Tropics of India , Economics Program Progress Report -21 International Crops Research Institute for the Semi-Arid Tropics ICRISAT Patancheru P.O. Andhra Pradesh 502 324, India.

Spedding, C.R.W. (1979). An introduction to agricultural systems. Applied Science Publishers Ltd.

Tabor, J.A. (1993). Soil surveys and Indigenous soil classification. *Indigenous knowledge and Development Monitor* 1(1), 28.

Thrupp, L.A. (1989). Legitimizing Local Knowledge: From Displacement to Empowerment for Third World People. Agriculture and Human Values. Summer Issue. Pp.13-24.

Verma, L.R. (2003). Mixed cropping in rabi. Inventory of Indigenous Technical Knowledge in Agriculture Document 2, supplement 1. Compiled by P.K. Das, S.K. Das, A. Mishra, H P S Arya, K.M. Bujarbarua, R.P. Singh 'Ratan', L.R. Verma, G Subba Reddy, M. Geetha Rani, H.S. Gupta, Z.D. Kavia, D.P. Ray, Indian Council of Agricultural Research, New Delhi. pp 23.

Warren, D.M. (1991). Using indigenous knowledge in Agricultural development. World Bank discussion papers, No. 127:46.

CHAPTER 18

Post-harvest technology and value addition in pulses

Prasoon Verma

ICAR-Indian Institute of Pulses Research, Kanpur - 208 024, Uttar Pardesh

**Corresponding author email*: pv.kanpur@gmail.com

Abstract

Pulses production in India reached to all time highest level of 25.23 Mt during 2017-18. However, country will need 32 Mt by the end of current decade to fulfill the demand of growing population. As per an estimate, the post-harvest losses in pulses ranges between 25 to 50%, in which storage and milling contributes to the maximum. Pulses are commonly stored in gunny bags in which losses due to insect infestations and fluctuating weather are more. About 75-80% of pulses are consumed as dal (dehusked and splitted cotyledons). The removal of husk from pulses whole grains improves appearance, texture, cooking quality, palatability and digestibility. There are many processes like cleaning, grading, pitting, pre-milling treatment, dehusking, splitting cotyledons and polishing are involved in milling of pulses. In general, dal recovery vary from 65 to 85% depending on pulses, milling process, grains quality, grains size, moisture content, etc. There are many kinds of mini dal mills developed by research institutions in which maximum dal recovery recorded in IIPR mini dal mill. The byproducts obtained in milling can be used for making value added products. Thus, the availability of pulses can be increased by the minimizing post-harvest losses, improving dal milling and utilization of milling byproducts for preparing value added products for human consumption.

Keywords: Dehulling, Dehusking, Milling, Mini dal mill, Pre-treatment, Splitting, Storage

1. INTRODUCTION

Pulses play an important role in meeting protein requirement of the large vegetarian population of India. Certain amino acids cannot be synthesized by human body which need to be supplemented through external sources. Since animals have the ability to synthesize such proteins, often animal source of protein are considered as complete protein. But since time immemorial across the country, Indian dietary pattern utilizes combination of cereals and pulses, which provide all the essential amino acids complementary to each other and in no way the combination proteins, are inferior to animal proteins. Methionine rich pulses are compensated by lysine present in wheat and rice. Probably this is the reason that cereals and pulses are common ingredients in most of the traditional Indian food recipes. Being the cheapest source of protein, pulses are often referred as poor man's meat. Pulses also provide nutritional security to the vegetarian dominant nations. Depending upon regional preferences for different pulses, viz., pigeonpea, chickpea, lentil, greengram, blackgram, field peas and rajmash, are the major pulses grown across the country. Some minor and arid pulses are also consumed in limited locations. Though India is the largest producer of pulses, but despite quantum jump in production of pulses around 23-25 Mt in recent years from 10-15 Mt upto 2015-16, India has to import pulses to meet domestic demand. Since long, per day per capita availability of pulses is much below than recommended level of 80 g by WHO and FAO. Considering intake from other protein sources, ICMR has lowered the requirement of pulses to 47 g but with 35-40 g availability, scenario is still meager to match the nutritional requirements.

Pulses, being rich source of protein, attract insects from the field itself, which get multiplied during storage under hot and humid environmental conditions in commonly adopted gunny bag storage method. In order to improve culinary properties pulses are often consumed after dehusking and splitting in the form of dal. Process of husk removal is referred as pulse milling. These two post-harvest operations i.e., storage and milling, cause maximum 20-30% losses. The gap between demand and availability can be reduced either by increasing production and productivity or by reducing post-harvest losses. Therefore, reduction in post-harvest losses will ensure the greater availability of pulses for human consumption.

2. POST-HARVEST SYSTEM OF PULSES

In the entire value chain of pulses, harvesting is the first stage which contributes to the losses. Harvesting is done at optimal grain maturity, manually or mechanically. Immature grains not only hamper the yield but also result into storage losses. However, harvesting of over mature or dried crop leads to shattering losses. Therefore, crop must be harvested at optimal maturity, at high moisture content 15-18% and further dried to 13-14% before threshing. Threshed grains are cleaned, graded and further dried up to safe storage moisture of 10-12%. Bag storage system, prevalent in India, is vulnerable to ambient environmental conditions. Hot and humid seasons trigger the insect activities and need measures for insect control during storage. From storage godowns, grains are moved to milling units where it is converted into dehusked splits, i.e., dal. Through wholesaler, dealer and retailer network it reaches to the consumers. Since the large capacity mills are located in urban areas, the conversion process from whole grain to dal involves transportation at several stages and each stage causes some losses during loading and unloading. About 60% of population lives in rural areas, therefore, producers of the pulses are consumers as well. However, the entire network of pulses value chain comprises of external players, which control the availability, market prices and makes maximum profit. Storage and processing within production catchment will make cheap pulses available to rural population and minimize transportation cost and losses. Establishment of processing and marketing network will generate employment and income opportunities in the rural areas. For this purpose, small capacity dal mills have been developed by various research organizations. The complex mechanism of production, transportation, storage, milling and marketing network is referred as post-harvest system. An overview of pulses post-harvest chain indicates that only about 60% of total production of pulses is available for human consumption.

Availability of pulses for human consumption

Total pulses produced	**100 units**
Threshing, transport and primary processing	2.00
Storage before milling	7.50
Seed purpose	9.05
Consumed as whole grain	13.57
Pulse available for milling	**67.88 units**
Milling losses	10.18
Husk separated	8.83
Pulse available as *dal*	48.87
Pulse available for human consumption	**13.57 + 48.87 = 62.44**

3. EXTENT OF POST-HARVEST LOSSES

After harvesting and threshing of pulses, qualitative and quantitative deteriorating changes takes place at various post-harvest stages due to poor management, which ultimately lead to economic or monetary losses. However no reliable data is available on post-harvest losses, as it mainly depends upon handling and management. In poor management and unfavourable climatic conditions the losses can be as high as 50%. According to estimates from various research workers, total post-harvest losses in case of pulses ranges between 25 to 50%, in which storage and milling contributes to the maximum (Table 1). In India, still jute bags are used for storage of pulses in conventional godowns, which are prone to insect infestation as grain moisture increases as per ambient conditions, leading to insect and fungal growth. With application of prophylactic and curative chemical control measures, though insect activities can be curbed but regular monitoring, inspection and application of pesticides increases the storage cost. Being rich source of protein, pulses are prone to insect attack. Large scale handling and storage is difficult to manage and monitoring therefore, despite best management practices still the storage losses are reported to the tune of 5-10%. Pre-storage fumigation, cleaning, grading and drying grain of uniform quality at low moisture content and appropriate packaging material, can minimize the possibilities of storage losses.

Similarly, in case of milling despite improved pre-milling treatments and machineries, *dal* recovery is around 70% against the potential *dal* recovery of 85%. Since India has to import pulses to meet domestic demands, it is loss to significant amount of foreign exchange. Abrasive dehusking devices commonly used in pulse processing cause scouring of protein rich outer layer of cotyledons, loss in qualitative terms is much more. In traditional cottage scale milling where attempt is not made for complete husk removal, recovery of unpolished *dal* is more, with less loss in the form of powder. Commonly adopted pre-milling treatments at commercial level are also scaled up versions of traditional household cottage processes. Still there is need to develop more efficient processes and machineries for better and higher *dal* recovery and curb the milling losses. Since India is importing pulses to meet domestic demands, reduction in storage and milling losses will save significant amount of foreign exchange.

Table 1. Estimates losses at different post-harvest stages

Stages	**Losses (%)**
Harvesting	1.0 – 3.0
Handling	1.0 – 7.0
Threshing	0.5 – 5.0
Drying	1.0 – 5.0
Transport	0.5
Primary Processing	1.0
Storage	5.0 – 10.0
Milling	15.0 – 20.0
Total	25.0 – 50.0

4. STORAGE OF PULSES

Storage of agricultural produce is important to ensure round the year availability of commodity at reasonable prices with least deterioration in qualitative parameters. This is true for protein rich pulses as well. If not managed properly, sometimes deterioration is so fast that grain become unfit for human or animal consumption within few months. Sharp rise in pulses prices in 2017, lead to the policy decision of increased minimum support price, assured procurement and maintaining buffer stock to encourage farmers for growing pulses. Availability of quality seeds of improved varieties and policy support resulted into all time high pulses production of 25.23 Mt in year 2017-18. So far, the country used to procure wheat and paddy for national food security. It was first time that storage of pulses was initiated. The prevailing procedures for storage of wheat and paddy were adopted for pulses also. Storing pulses is more difficult in comparison to wheat and paddy as it is rich in proteins and by the time insect activities are noticed, cotyledons are already consumed. Large capacity storage facilities are required so that farmer can store pulses at farm level in order to release grain at appropriate market price instead of distress sale at the time of production peak. Traditional bag storage structures need to be modified for storage of pulses. Large capacity silos and bins are required to replace bag storage system to minimize expenditure being incurred on safe storage of pulses. Preventing field level infestation, through proper drying and fumigation, pulses can safely be stored for longer durations without loss in quality and quantity. During hot and humid seasons, pulse storage is a real challenge. For a safe storage of pulses, apart from initial grain conditions (moisture and infestation level), environmental factors such as temperature, humidity and oxygen level play an important role in activity of insects and moulds. Interaction of these factors hastens or reduces the rate of deterioration. Increase in moisture may also results into microbial growth. It has also been observed that insects lay egg on grain surface and larvae enter into the grain and consume to cotyledons before emerging as an adult ready for mating and multiplying. Dehusking and splitting, ensures removal of egg and eliminates hiding place for larvae, prevents chances of infestation during storage. It can safely be stored in any storage bin where external infestation is protected without use of chemical insecticides.

Despite limitation, bag storage is most prevalent method of pulses storage in India, whereas in developed countries bins and silos are used for bulk storage. Huge initial investment on bulk handling equipment and storage structures is the major reason for non-adoption of the technology which offers better control and monitoring of stored grains. However, bag storage system provides flexibility in stacking, handling and transportation. It can be used within the available space and capacity, even under cover and plinth open storages under sky. It is relatively economical than bulk structures and practiced to suit transport and marketing conditions especially in developing countries. At farm level, low capacity metallic bins and concrete silos are getting popular. In large capacity silos aeration can be done sensing increase in grain temperature and moisture. The stored lot is fumigated as and when oxygen level depletes indicating any kind of insect activity. Such complex structures have high initial and operating costs. In air tight silos, intrinsic gas environment can be manipulated either by decreasing oxygen level or introducing inert gases to inhibit insect activities. Oxygen level also diminishes with respiration of insects, leading to control of insects. Modified and controlled atmosphere silos, earlier used for horticultural crops can also be used for pulses. Concrete silos offer better thermal insulation to stored grain compare to metallic bins. Bulk storage environment control silos are better and well adopted technique, but high initial investment and operational costs, bulk handling and transport system are major bottleneck in adoption of the technology. It is always better to disinfect the storage structures, dry and fumigate the grains prior to storage. Though pulses are harvested at higher moisture content (14%), but it must be dried up to safe limits (10-12%) for long term storage. Low temperature storage has also been found to be effective method of insect control even at high grain moisture but in Indian conditions it is not cost effective for cereals and pulses. However, maintaining grain quality in jute bags at room temperature is real challenge for pulses storage, which requires periodical monitoring to check insect activities and curative measures to control insects, as and when required. Phosphine gas is quite commonly used for bruchid control in pulses, but often used indiscriminately in bag storage godowns. In developed countries, use of chemicals for insect control during storage is very limited. In India, procurement and storage of pulses is being done by NAFED whereas import of pulses is being done through MMTC. The protocols adopted for protection of stored pulses are same as recommended and followed by Food Corporation of India (FCI) for wheat and paddy.

5. MILLING OF PULSES

Pulses are consumed as dehusked and splitted cotyledons i.e., dal, and this process is called milling of pulses. Pulse processing is third largest food processing industry after wheat and rice milling. About 75-80% pulses produced is consumed in the form of dal. The process of husk removal improves appearance, texture, cooking quality, palatability and digestibility of pulses. In dehusked form, cotyledons are less prone to insect infestation. Presence of gum layer in between seed coat and cotyledons makes dehusking difficult. Depending upon nature and quantity of gums, pulses are categorized into easy or difficult to mill kinds of pulses. Usually pigeonpea, greengram and blackgram are difficult to mill whereas chickpea, lentils and fieldpea can be milled easily. Pre-milling treatments are given to loosen the attachment of seed coat from the cotyledons. Cleaning, grading, pre-treatment, husk removal, splitting, polishing are the major steps of pulse processing.

5.1 Cleaning and grading

Raw grains contain various forms of impurities and seeds of different sizes. Cleaning and grading is the first step of pulses processing. It removes dust, dirt foreign matters, and immature, shriveled, infested and damaged grains, and separate over and undersized grains to get the uniform quality of raw material. This process is performed with the help of reciprocating or vibrating screen sets. In commercial mills, destoner and reel graders are also used. Quality of raw material is crucial for the quality of end product, moreover presence of impurities may also damage the machine components while operation.

5.2 Pitting

Pre-milling treatments are being given to whole pulses in order to loosen the seed coat. At cottage scale milling water soaking and at commercial scale oil smearing are common pre-milling treatments. When treatments are applied on whole grains it takes longer duration. Therefore, concept of pitting was introduced for commercial milling, where grain surface is scratch with the help of emery roller. First pass through emery roller scratches the seed coat and eases the absorption of oil or water applied for pre-treatment. Commonly in commercial *dal* mills the abrasive roller made of carborundum/emery-coated surface is used for dehusking. The same roller can be used for further dehusking in subsequent passes. It has been observed that about 30-35% dehusking takes place during pitting.

5.3 Pre-milling treatments

Depending upon attachment of seed coat to cotyledons due to quality and quantity of gummy substances beneath seed coat, different pulses are given different pre-milling treatments. Traditionally overnight soaking in water followed by sun drying for several days is a common practice adopted at cottage scale pulse milling. Soaking and drying results into cup formation is dal which become hard for cooking. Therefore, in commercial scale pulse milling application of oil followed by alternate sun drying and overnight heaping for 2-3 days is adopted throughout the country. In the process, seed coat starts peeling off naturally to produce premium quality of dal. Remaining treated grains are passes repeatedly through roller in subsequent passes to achieve maximum dehusking. These prevalent pre-treatments are often referred as wet (water soaking) and dry (oil application) methods. Both of these methods are time consuming and weather dependent processes. CFTRI, Mysuru has developed thermal treatment method, where grains are heated and tempered for several hours and the cycle is repeated twice before performing dehusking and splitting in abrasive rollers. This method is less labour intensive compare to other pre-treatments, but high initial investment and operational cost are major constraint in adoption of the process.

5.4 Dehusking and splitting

Traditionally at cottage scale after pre-milling treatment, dehsuking and splitting is done manually with the help of horizontal stone chakkies. In commercial level pulse milling, abrasive rollers rotating inside the perforated concave are used for dehusking and splitting. Carborundun and emery are commonly used abrasive materials pasted on the roller surface. For difficult-to mill kind of pulses, first pass is required before pre-milling treatment and additional two more or passes are required to remove husk completely. Sometimes polishing is also preferred to improve consumers appeal. Gap between roller and concave is fixed, therefore, milling losses increased in extra large size grains. Machine parameters must be given due considerations before selecting of milling material. During the process of husk removal, abrasive surface also cause scouring of outer surface of cotyledons which gets mixed with husk in powder form and goes for cattle feed. Broken produced in the process can be utilized for edible purposes. Husk and powder mixture i.e., milling by-product is collected at the bottom of concave. Milled fractions are separated into unhusked whole, dehusked whole, unhusked dal, dehusked dal and broken using appropriate sieves. It is difficult to separate unhusked and dehusked whole, this is main reason for scouring of cotyledons in subsequent passes resulting losses in the form of powder. Grading of finished product is essential to get dal of uniform shape and size and it helps in removal of broken grains which can be utilized in development of other value added products.

5.5 Polishing

Polishing of dehusked splits is quite common to impart uniform appearance to the product. It increases consumer appeal and improves market acceptability. It also removes remaining husk leftover split cotyledons. Coating of edible colours and oil adds luster to the end product. This is additional operation which adds to product cost. It is commonly adopted to hide shortcomings of the product. Therefore, unpolished dal is preferred by quality concerned consumers and fetches better market prices. Different kinds of materials, viz, nylon, oil and water, leather and makhmal are used as polishing surfaces. To give uniform appearance even whole grains are polished with marble dust and non-edible colours which are quite harmful. Therefore, proper washing before cooking is recommended.

6. MILLING OF INDIVIDUAL PULSES

Quality and quantity of gum between seed coat and cotyledons plays an important role in dehusking of grains. Depending upon degree of adherence, pulses are categorized as easy or difficult to mill. Accordingly pre-milling treatments and milling procedures are decided. Due to low gum content lentil, chickpea and field peas are considered as easy to mill pulses. Pigeonpea, greengram and blackgram considered as difficult to mill pulses due to presence of high gum content. Milling losses are more in the pulses where degree of attachment is more and pre-milling treatment is essential to loosen the attachment of husk. Milling process for different pulses is described below:

6.1 Pigeonpea

It is considered as the most difficult pulse to mill. Cleaned and graded grains are passed through emery roller for pitting i.e., scratching of seed coat for ease of oil or water absorption. At cottage scale, water soaking followed by sun drying is adopted as common pre-milling treatment. However in commercial mills, oil is mixed with grains using screw conveyor @ 0.7% and tempered for a day in the bins. Again it is mixed with water @ 4-5% stored overnight followed by sun drying for 2-3 days. Thereafter, treated grains are passed through emery rolls for 2-3 time to get about 90% dehusking. Dal obtained after several passes has edges chipped and rounded off is considered as Grade II dal. Dal without any bruising to the cotyledons is referred as Grade I dal. Dehusked and unhusked grain are separated and mixed with water. After tempering and sun drying, it is splitted using horizontal or vertical chakki or splitting machine. Dal recovery from pigeonpea varies between 68-75%.

6.2 Chickpea

Husk of chickpea is loosely attached with cotyledons; therefore, it falls in category of easy to mill kind of pulses. In commercial mills, dehsuking is done in abrasive rollers after cleaning and grading. Dehusked gota is mixed with water in the ratio of 1:2.5 to 3.0 followed by tempering for 12 hours and splitting in chakkies. Traditionally water soaking and drying is most commonly adopted methods of chickpea milling in stone chakkies. About 78-82% dal is recovered from chickpea which can further be converted into powder or besan. Thus, it has multiples usages and applications. About 98% besan is recovered in grinding of chickpea dal.

6.3 Blackgram

Raw grains are cleaned, graded and pitted in emery rollers before pre-treatment. Pitted grains are mixed with oil @ 0.5% and heaped for 12 hours followed by sun drying for 2-3 days. Again water is mixed in the ratio of 1: 2.5 to 3.0 and tempered overnight. After this pre-treatment, grains are passed through emery rollers for dehusking. For complete dehusking 2-3 passes are required. Dal is separated from milled fraction after each pass. Unhusked and dehusked whole grains are further tempered after water application and subjected to sun drying before splitting in chakkies. Dal obtained from chakki are considered as Grade I dal whereas that obtained after repeated dehusking in emery rollers is considered as Grade II dal. Often polishing with marble dust is done to improve appearance of the product. Blackgram dal is also consumed in the form of split dal with husk, which can be made by passing the whole grain in splitter or chakki without any treatment.

6.4 Greengram

Greengram is also considered as difficult to mill kind of pulses. It is consumed as whole, split with and without husk. Husk of greengram is soft and thin, but due to weak bond between the cotyledons, grains split before dehsuking. In order to loosen the seed coat, oil application is prevalent in milling of greengram. Like pigeonpea and blackgram, pitting, mixing with oil, tempering, and sun drying is done before dehusking and splitting in emery roller mills. Due to thin seed coat, losses in the form of breakage and powdering are more in greengram during abrasive dehusking. Dal with husk can be prepared without dehusking in vertical or horizontal chakkies.

6.5 Lentil

Lentil has thin seed coat lightly attached to cotyledons. It is consumed as whole grain with and without husk. Moisture addition and tempering is required before sun drying as mild pre-milling treatment. About 80% dehusking is achieved in single pass in emery roller. It is mostly preferred as dehusked whole (Malka Masoor), instead of split dal, therefore efforts are made to prevent splitting by reducing residence time inside the roller.

6.6 Field peas

Husk in field peas are loosely attached with cotyledons making it easy to mill kind of pulses. Mostly it is consumed as a whole after cleaning, grading and polishing. However, in some part of the country it is consumed after dehusking and splitting. Cleaned and graded grains are tempered after moisture addition followed by sun drying up to 10-12% moisture before dehusking and splitting in emery roller mills or disk sheller. Due to ease in milling, peas yield 80-82% dal recovery.

7. PULSE MILLING AND ALLIED MACHINERIES

7.1 Stone Chakki

Pulse milling started as household activity using horizontal stone chakkies. Soaking and drying was adopted as pre-milling treatment before milling. Milled product was aspirated manually to remove husk and small broken cotyledons.

7.2 Abrasive roller

Abrasive dehusking surfaces using carborundum or emery are adopted in large scale commercial milling. Husk is removed by scarification with the rough surface of rollers rotating inside perforated screen. Depending upon degree of dehusking required, residence time of grain inside roller can be increased or decreased by opening of outlet. For different pulses, grit of different mesh sizes, ranging from 16 to 36 mesh are used (Table 2). Large scale commercial mills usually design crop specific roller. However, small capacity cottage scale mill grit size of 24-30 mesh accommodated for milling of different pulses within acceptable limits.

Table 2. Recommended emery size for roller *dal* mills for different pulses

Crop	Emery size (No.) under different operations		
	Dehusking/Pitting	Splitting	Polishing
Pigeonpea	16	24	---
Mungbean	24	30	35-40
Urdbean	24	30	35-40
Chickpea	---	24	---

7.3 Abrasive cone

CFTRI, Mysuru has developed cottage scale mini dal mill using vertical cone type emery roller. In horizontal rollers, it is difficult to adjust gap between cylinder and concave. Conical roller, like rice polisher gives this flexibility and gives better dehusking efficiency. The vertical emery cone rotates inside conical perforated casing, which can be lowered or raised depending upon grain size. Gap between the casing and cone can be increased or decreased depending upon grain size.

7.4 Burr mill

Horizontal or vertical burr mills with one stationary and other rotating disks are also used for dehusking and splitting of pulses after pre-milling treatments. Husk and small cotyledon brokens are removed with the help of air aspiration. In IIPR Mini Dal Mill, vertical burr mill with stationary hard rubber disk and rotating steel disk combination is used. Gap between the disks can be adjusted according to grain size. It works well with grains easy to mill and splitting of dehusked whole grains.

7.5 Pulse splitter

To obtain better quality or premium grade dal, dehusking and splitting operations need to be performed separately. In case of good pre-treatment, grains are dehusked and split by movement in elevators or just by dropping grains on hard surface from some height. Dehusked whole grain can be splitted into two halves by impact.

7.6 Rotary grader

Rotary screen graders, commonly used in commercial dal mills, give better performance over reciprocating sieves, where clogging of holes restricts cleaning and grading efficiency. It can be used for cleaning of raw grains and separation of milled fractions.

7.7 Color sorter

Colour sorter is commonly used in rice milling industries to separate white and brown rice after polishing. In pulses milling industry, it is used to separate whole grain with dehusked whole (gota). It is not possible to separate whole dehusked and unhusked grains with perforated sieves. To minimize powdering losses, it is essential to remove dehusked whole after each pass from roller. In case of malka masoor, use of colour sorter is quite common as dehusked whole fetches better price, whereas presence of unhusked grain deteriorates the dal quality. This machine uses high resolution camera and sensors to separate the grains on the basis of colour difference.

7.8 Gota separator

The machine is used to separate dehusked whole grains (gota) with unhusked grains using surface properties of the two in place of colour characteristics. This machine is cheaper in comparison to colour sorter. This unit has been developed by CFTRI, Mysuru

7.9 Treatment mixer

Mixing water or oil to large quantity of grains, horizontal worm conveyor is used. Oil or water is applied to the pitted grains on conveying mechanism at desired rate, drop by drop.

7.10 Polisher

Polisher is used to give uniformity processed dal. Colouring and polishing improves appearance and market acceptability of product. Various kind of surfaces and materials viz., leather belt, nylon or jute ropes, makhmal etc. are used to polish split dal. Marble dust, water, oil and edible colour are common additives of pulses dal polishing. Polishing is an undesirable activity, and these days unpolished dal pay the premium prices in the market.

7.11 Husk aspiration system

In order to prevent air pollution in and around processing area, husk aspiration system is used. During abrasive dehusking, husk and cotyledon powders are generated. Suspended and floating dust particles of husk and cotyledon create unhealthy operational conditions for the workers. An efficient aspiration system removes and collect dust particles and prevents air pollution.

7.12 Elevators

Bucket elevators are used to carry grains to different machines and operational floors.

8. MINI DAL MILLS

Pulses are processed in large scale milling industries situated in urban areas. At the time of harvest, farmers have to sale the produce at prices much below the minimum support prices announced by the government due to lack of storage and processing facilities in production catchments. Low cost raw material is procured from the farmers and reaches to the mill through various intermediaries. Profit margins to intermediaries and processing cost is added to the market price of dal. Transportation cost and losses also add to the dal prices. Therefore, processing of pulses in villages not only reduce the cost but also minimize the losses occurred in the transportations from production sites to mills. Keeping these in views, various research organizations have developed low capacity dal mills for cottage scale pulse milling. Basically, these mills are scale-up version of traditional cottage scale milling methods. Mini dal mills developed by CFTRI, Mysuru; CIAE, Bhopal; PDKV, Akola and ICAR-IIPR, Kanpur are commercially available. All these mills used emery rollers for dehusking and splitting of the grains, except IIPR Mini Dal Mill, where grains are passed through emery roller once for the purpose of pitting. After pre-milling treatment, grains are dehusked and splitted with the help of stationary rubber and rotating steel disk vertical chakki. Gap between the disks can be adjusted according to grain size. Losses in form of powder is less in case of this mill, whereas in other mills grains are passed repeatedly for two to three times over emery roll for higher degree of dehusking leading to more loss in the form of powders. IIPR, CIAE and PDKV dal mills used horizontal emery roller, whereas in CFTRI mini dal mill abrasive surface is coated over vertical cone rotating inside perforated screen. Gap between the abrasive cone and screen can be increased or decreased depending upon pulses and grains size to be milled. Soft dehusking and splitting mechanism in IIPR Mini Dal Mill resulted in higher dal recovery, but with 5-10% husked dal.

IIPR Mini Dal Mill

9. FACTORS AFFECTING THE DEHULLING OF PULSES

Apart from pre-milling treatments and milling machineries, various grains parameters also influence dal recovery in pulses milling. Variety, size, shape, moisture content, grain hardness, duration and method of storage also affect the milling behaviour of pulses.

9.1 Variety

The variations in gum, husk and cotyledon contents observed in different cultivars of same pulse. The quality and quantity of gum in pulses grains influences the dehusking efficiency.

9.2 Size and shape

In emery roller mills, machine parameters, i.e., gap between cylinder and concave, are fixed, hence only very small grains will remain unhusked and the powdering losses will be more in case of large size grains. Therefore, varieties to be milled are selected in such a way which suits to the machine parameters. Similarly, spherical seeds have shown better dehulling ability in comparison to cylindrical, oval or flat grains. Milling of sharp edge grains resulted into higher breakage loss.

9.3 Colour

Grain colour is important to hide the husk over dal. Light husk coloured grains are preferred by the millers as it is difficult to notice on split dal surface. Dark colour grains require higher degree of dehusking leading to more powdering losses.

9.4 Grain moisture

It has been observed that optimum moisture content for pulse milling varies between 8-11%. Low moisture makes husk brittle and easy to remove. Dehusking efficiency decreases with increase in grain moisture, whereas breakage and powdering loss get reduced. Moisture content also influences the grain hardness which is an important parameter to obtain maximum dehusking before splitting of cotyledons. High moisture content weakens the gum. Pulses grains stored in gunny bags led to yield higher dal recovery in comparison to air tight containers or plastic bags.

10. CONCLUSION

In recent past, deficit in supply led to sky rocketing of pulses prices. Some of the measures, like promotion of improved varieties, availability of quality seed, enhanced MSP, and maintenance of buffer stock, implemented during last few years resulted into the all time high pulses production of 23-25.23 Mt during last few years. The future demand projection indicates requirement of 32 Mt of pulses by end of the decade. Therefore, the enhanced production with matching technologies for storage and processing need to develop for achieve self sufficiency in pulses. Storage and milling contributes maximum to post-harvest losses in pulses. Need is being felt to improve long-term storage methods for protein rich pulses in order to control market prices. Milling losses are also high due to tight adherence of seed coat with cotyledon leading to higher loss as brokens and cotyledon powder in case of traditional abrasive dehusking mills. Mini dal mills are now available that needs to be promoted for adoption in villages so that post-harvest losses can be minimized. On production front, challenges can be met with development of insect-pests and disease resistant varieties and better management practices, but the increased production need to be handled with matching efficient storage and processing methods to make more pulse available for human consumption.

11. THRUST AREAS FOR FUTURE RESEARCH

Value addition and processing of pulses had always been third largest food industry after wheat and rice. Increase in minimum support prices and assured procurement for buffer stock of pulses has emerged as an opportunity for income and employment generation in rural areas. Previously, all the produce was sold to mandies or millers through middlemen. Development of infrastructures for storage and processing in production catchments will not add to rural income but also provide processed pulse products at cheaper rates in addition to reduction in losses due to transportation. Being rich source of protein, long term safe storage of pulses is major challenge. Also the present methods of pulse milling, either at cottage or commercial level, are still the scaled-up processes of domestic pulses milling, which have scope for further improvement. There is need to study grain and machine parameters in tandem to achieve higher dal recovery with minimum losses. Based on discussions in the chapter following major thrust areas have been emerged for future research:

(i) **Varietal improvement:** Previously, research programmes were focused for improvement in yield through genetic enhancement and management of biotic and abiotic stresses. Now the time has come for developing pulses varieties with lesser gum and husk content, and higher milling efficiency. High cotyledon to husk ratio will ensure potential for higher dal recovery.

(ii) **Improved storage:** Still pulses are stored in jute bags which are prone to insect infestation during conducive environmental conditions. To maintain the grain quality in bags, pesticides and fumigants need to be applied at regular intervals. This adds to the storage costs. Efforts should be made to minimize the use of chemicals using appropriate method and packaging materials for safe storage.

(iii) **Energy efficient pre-treatments:** Water soaking and oil smearing are common pre-milling treatments adopted at cottage or commercial millings, respectively. Both are time consuming, labour intensive and weather dependent processes. Thermal process requires higher initial investment and operational cost. Thus, there is need to develop pre-milling treatment methods which are energy efficient, less time consuming and weather independent. There is little works done to explore the possibilities to use chemical and biochemical methods in pulses.

(iv) **Dehusking machines:** Abrasive surfaces are commercially used for large scale pulse milling. Different pulses need emery grits of different sizes. With improved pre-milling treatments, there is need to develop new mechanisms for husk removal so that maximum dal can be recovered with least breakage and powdering losses.

(v) **Utilization of milling byproducts:** Pulse milling by-products i.e., husk and cotyledon powders obtained in abrasive roller mills are rich in proteins, phenols, fibers and anti-oxidants. These byproducts can be used for making several food items for human consumption.

REFERENCES

Ali, N. (2003). Country Paper – India. Processing and Utilization of Legumes. Asian Productivity Organization, Tokyo. Pp. 117-145.

Ilyas, S.M. and Wanjari, O.D. (2001). Important issues in post production processing, storage and by-product utilization of pulses and soybean. National seminar on Emerging Trends in Processing, Handling, Storage and by-product Utilization of Pulses and Soybean, G B Pant University of Agric. & Tech., Pantnagar, Jan 18-19.

Karsolia, R.P. and Bal, S. (1980). Dry conditioning of pulses, ISAE Paper No. 80-0420, ISAE, New Delhi.

Krishnamurti, K., Girish, G.K., Ramasivan, T., Bose, S.K., Singh, K. and Tomar, R.P.S. (1972). A new process for removal of husk of red gram using sirka. *Bulletin of Grain Technology* 10 : 181-186.

Kurien, P.P. (1984). Dehulling Technology of Pulses. *Research and Industry* 29, 207-214.

Kurien, P.P. and Parpia, H.B.A. (1968). Pulse milling in India (1) processing and milling of tur arhar (*Cajanus cajan* L). *Journal of Food Science and Technology* 5, 203.

Kurien, P.P., Desikachar, H.R.S. and Parpia, H.B.A. (1972). Processing and utilization of grain legumes in India. Tropical Agriculture Research Series No.6. Symposium on Food Legumes, TARC, Ministry of Agriculture and Forestry, Tokyo, Japan.

Lal, R.R. and Verma, P. (2007). IIPR Mini Dal Mill for Employment Generation, Technical Bulletin, Indian Institute of Pulses Research, Kanpur

Lal, R.R. and Verma, P. (2007). Post-harvest Management of Pulses, Technical Bulletin, Indian Institute of Pulses Research, Kanpur

Mayande, V.M. (1987). Pigeonpea milling technology: biochemical and engineering studies. Thesis, M Tech (Agril. Engg.), G B Pant University of Agriculture & Technology, Pantnagar.

Reddy, V.P.R. (1981). Dehusking of arhar grain (*Cajanus cajan* L). M Tech Thesis. G B Pant University of Agriculture & Technology, Pantnagar.

Saxena R.P. (1985). Milling of Pigeonpea Grain and Associated Aspects. Ph.D. Thesis. G B Pant University of Agriculture and Technology, Pantnagar.

Saxena, R.P., Singh, B.P.N. and Singh, J.K. (1981). Effect of chemical treatment on husk removal of arhar (*Cajanus cajan*) grain. Paper No. 81-PAS-116, New Delhi, India: Indian Agricultural Engineers.

Verma, P., Saxena, R.P., Sarkar, B.C. and More, P.K. (1993). Enzymatic pretreatment of pigeonpea (*Cajanus cajan* L) grain and its interaction with milling. *Journal of Food Science and Technology* 30, 368-370.

Vishwakarma, R.K., Shivhare, U.S., Gupta, R.K., Yadav, D.N., Jaiswal, A. and Prasad, P. (2017). Status of pulse milling processes and technologies: A review. *Critical Reviews in Food Science and Nutrition*, DOI:10.1080/10408398.2016.1274956.